Ideas in Chemistry and Molecular Sciences

Edited by
Bruno Pignataro

Related Titles

Nugent, Thomas C. (ed.)

Chiral Amine Synthesis

Methods, Developments and Applications

2010
ISBN: 978-3-527-32509-2

Boysen, Mike Martin Kwabena (ed.)

Carbohydrates – Tools for Stereoselective Synthesis

2010
ISBN: 978-3-527-32379-1

Pignataro, Bruno (ed.)

Ideas in Chemistry and Molecular Sciences

Where Chemistry Meets Life

2010
ISBN: 978-3-527-32541-2

Pignataro, Bruno (ed.)

Ideas in Chemistry and Molecular Sciences

Advances in Nanotechnology, Materials and Devices

2010
ISBN: 978-3-527-32543-6

Pignataro, Bruno (ed.)

Tomorrow's Chemistry Today

Concepts in Nanoscience, Organic Materials and Environmental Chemistry
2nd edition

2009
ISBN: 978-3-527-32623-5

Carreira, E. M., Kvaerno, L.

Classics in Stereoselective Synthesis

2009
ISBN: 978-3-527-29966-9 (Softcover)
ISBN: 978-3-527-32452-1 (Hardcover)

Christmann, M., Brase, S. (eds.)

Asymmetric Synthesis – The Essentials

2008
Softcover
ISBN: 978-3-527-32093-6

Ideas in Chemistry and Molecular Sciences

Advances in Synthetic Chemistry

Edited by
Bruno Pignataro

WILEY-VCH

WILEY-VCH Verlag GmbH & Co. KGaA

The Editor

Prof. Bruno Pignataro
University of Palermo
Department of Physical Chemistry
Viale delle Scienze
90128 Palermo
Italy

■ All books published by Wiley-VCH are carefully produced. Nevertheless, authors, editors, and publisher do not warrant the information contained in these books, including this book, to be free of errors. Readers are advised to keep in mind that statements, data, illustrations, procedural details or other items may inadvertently be inaccurate.

Library of Congress Card No.: applied for

British Library Cataloguing-in-Publication Data
A catalogue record for this book is available from the British Library.

Bibliographic information published by the Deutsche Nationalbibliothek
The Deutsche Nationalbibliothek lists this publication in the Deutsche Nationalbibliografie; detailed bibliographic data are available on the Internet at <http://dnb.d-nb.de>.

© 2010 WILEY-VCH Verlag GmbH & Co. KGaA, Weinheim

All rights reserved (including those of translation into other languages). No part of this book may be reproduced in any form – by photoprinting, microfilm, or any other means – nor transmitted or translated into a machine language without written permission from the publishers. Registered names, trademarks, etc. used in this book, even when not specifically marked as such, are not to be considered unprotected by law.

Cover Design   Adam Design, Weinheim
Typesetting   Laserwords Private Limited, Chennai, India
Printing and Binding   betz-druck GmbH, Darmstadt

Printed in the Federal Republic of Germany
Printed on acid-free paper

ISBN: 978-3-527-32539-9

Set ISBN: 978-3-527-32875-8

Contents

Preface *XI*
List of Contributors *XVII*

Part I   Innovative Processes in Organic Chemistry   *1*

1   **N-Hydroxy Derivatives: Key Organocatalysts for the Selective Free Radical Aerobic Oxidation of Organic Compounds**   *3*
        Carlo Punta and Cristian Gambarotti
1.1   Introduction   *3*
1.2   General Reactivity of N-Hydroxy Derivatives   *3*
1.3   Aerobic Oxidation Catalyzed by N-Hydroxy Amines   *6*
1.3.1   Aerobic Oxidation of Alcohols to Aldehydes and Ketones   *6*
1.4   Aerobic Oxidation Catalyzed by N-Hydroxy Amides   *9*
1.4.1   Peroxidation of Polyunsaturated Fatty Acids   *9*
1.5   Aerobic Oxidation Catalyzed by N-Hydroxy Imides   *10*
1.5.1   Oxidation of Benzylalcohols to Aldehydes   *10*
1.5.2   Oxidation of Silanes   *12*
1.5.3   Oxidation of N-Alkylamides   *12*
1.5.4   Oxidation of Tertiary Benzylamines to Aldehydes   *13*
1.5.5   Oxidative Functionalization of Alkylaromatics   *14*
1.5.6   Oxidative Acylation of N-Heteroaromatic Bases   *15*
1.5.7   Aerobic Synthesis of *p*-Hydroxybenzoic Acids and Diphenols   *16*
1.5.8   Selective Halogenation of Alkanes   *16*
1.5.9   Aerobic Oxidation of Cycloalkanes to Diacids   *17*
1.5.10 Epoxidation of Olefins   *18*
1.5.11 Oxidation of Alkylaromatics   *19*
1.6   Conclusions   *20*
        Acknowledgments   *20*
        References   *21*

2		**Gold-Catalyzed Intra- and Intermolecular Cycloadditions of Push–Pull Dienynes** *25*
		Manuel A. Fernández-Rodríguez
2.1		Introduction *25*
2.2		Gold-Catalyzed Enyne Cycloisomerizations *26*
2.2.1		General Remarks *26*
2.2.2		Influence of the Electronic Nature of the Alkyne Substituent *26*
2.2.3		Gold-Catalyzed Cycloaromatization of Push–Pull Dienyne Acids: Synthesis of 2,3-Disubstituted Phenols *29*
2.3		Gold-Catalyzed Intermolecular Cycloadditions *34*
2.3.1		Cycloadditions of Enynes, Propargyl Acetylenes, and Alkynyl Cyclopropanes *34*
2.3.2		Gold-Catalyzed Intermolecular Hetero-Dehydro-Diels–Alder Cycloaddition of Push–Pull Dienynes with Nonactivated Nitriles: Regioselective Synthesis of Pyridines *35*
2.4		Conclusions and Future Perspectives *40*
		Acknowledgments *40*
		References *41*
3		***N*-Heterocyclic Carbenes in Copper-Catalyzed Reactions** *43*
		Silvia Díez-González
3.1		Introduction *43*
3.2		Preparation of NHC-Containing Copper Complexes *43*
3.3		Main Applications of [(NHC)Cu] Complexes in Catalysis *46*
3.4		Copper Hydride-Mediated Reactions *49*
3.4.1		Hydrosilylation of Carbonyl Compounds *49*
3.4.2		Mechanistic Considerations *51*
3.4.3		Related Transformations *52*
3.5		[3 + 2] Cycloaddition of Azides and Alkynes *53*
3.5.1		Click Chemistry *53*
3.5.2		Use of Internal Alkynes: Mechanistic Implications *55*
3.6		Concluding Remarks *57*
		Acknowledgments *58*
		References *58*
4		**Supported Organocatalysts as a Powerful Tool in Organic Synthesis** *67*
		Francesco Giacalone, Michelangelo Gruttadauria, and Renato Noto
4.1		Introduction *67*
4.2		L-Proline and its Derivatives on Ionic Liquid-Modified Silica Gels *69*
4.3		Polystyrene-Supported Proline as a Versatile and Recyclable Organocatalyst *73*
4.3.1		Nonasymmetric Reactions *75*
4.3.1.1		α-Selenenylation of Aldehydes *75*
4.3.1.2		Baylis–Hillman Reaction *77*
4.3.2		Asymmetric Reactions *79*

4.3.2.1	Aldol Reaction 79	
4.4	Prolinamide-Supported Polystyrenes as Highly Stereoselective and Recyclable Organocatalysts for the Aldol Reaction 82	
4.5	Outlook and Future Perspectives 84	
	References 85	
5	**The Complex-Induced Proximity Effect in Organolithium Chemistry and Its Importance in the Lithiation of Tertiary Amines** 95	
	Viktoria H. Gessner	
5.1	Introduction 95	
5.2	State of the Art 96	
5.2.1	Structure Formation Patterns of Organolithium Compounds 96	
5.2.2	The Complex-Induced Proximity Effect (CIPE) 97	
5.2.3	Synthesis of α-Lithiated Tertiary Amines 99	
5.3	Latest Developments 101	
5.3.1	Precoordination as Key to Direct Deprotonation of Tertiary Amines 101	
5.3.2	Regioselective α-Lithiation 104	
5.3.3	α-Lithiation versus β-Lithiation 108	
5.4	Conclusions and Outlook 110	
	Acknowledgments 111	
	References 111	
	Part II Predictive Tools in Organic Chemical Reactions 115	
6	**Double Hydrogen Bonding in Asymmetric Organocatalysis: A Mechanistic Perspective** 117	
	Tommaso Marcelli	
6.1	Introduction 117	
6.2	Diols and Amidoalcohols 119	
6.2.1	Single-Point versus Two-Point Activation 119	
6.2.2	The Impact of Acidity on Enantioselectivity 120	
6.3	(Thio)ureas 122	
6.3.1	Mechanistic Duality in Aminothiourea Catalysis 122	
6.3.2	Mono- versus Bidentate Coordination 125	
6.3.3	Catalyst Self-Association 126	
6.3.4	Halide Binding 127	
6.4	Phosphoric Acids 129	
6.4.1	Bifunctionality 130	
6.4.2	Competent Substrates 133	
6.5	Other Catalysts 134	
6.5.1	Guanidiniums 134	
6.5.2	Squaramides 135	
6.5.3	Quinolinium Thioamides 136	

6.6	Conclusions and Outlook	137
	References	138

7   Dynamic Covalent Capture: A Sensitive Tool for Detecting Molecular Interactions 143

Leonard Prins

7.1	Introduction	143
7.2	Hydrogen Bond–Driven Self-Assembly in Aqueous Solution	144
7.3	Measuring Stability and Order in Biological Structures	146
7.3.1	Peptides	146
7.3.2	Bilayer Membranes	149
7.4	Drug Discovery	150
7.5	Catalyst Discovery	152
7.6	The Analysis of Complex Chemical Systems	156
7.6.1	^{1}H-^{13}C HSQC NMR Spectroscopy	156
7.6.2	UV/Vis Spectroscopy	158
7.7	Perspective	160
	Acknowledgments	162
	References	162

Part III   Chemical Reactions, Sustainable Processes, and Environment 165

8   Furfural and Furfural-Based Industrial Chemicals 167

Ana S. Dias, Sérgio Lima, Martyn Pillinger, and Anabela A. Valente

8.1	Carbohydrates for Life	167
8.2	Furfural–Evolution over Nearly Two Centuries	169
8.3	Applications of Furfural	169
8.4	Mechanistic Considerations on the Conversion of Pentosans into Furfural	170
8.5	Production of Furfural	172
8.5.1	Crystalline Microporous Silicates	173
8.5.2	Functionalized Mesoporous Silicas	174
8.5.3	Transition Metal Oxide Nanosheets	179
8.6	Conclusion and Future Perspectives	180
	Acknowledgments	182
	References	182

9   Multiple Bond-Forming Transformations: the Key Concept Toward Eco-Compatible Synthetic Organic Chemistry 187

Yoann Coquerel, Thomas Boddaert, Marc Presset, Damien Mailhol, and Jean Rodriguez

9.1	The Science of Synthesis	187
9.2	Multiple Bond-Forming Transformations (MBFTs)	190
9.2.1	Consecutive Reactions	190

9.2.2	Domino Reactions	*190*
9.2.3	Multicomponent Reactions	*191*
9.2.4	One-Pot Reactions	*191*
9.3	An Account of Our Recent Contributions	*192*
9.3.1	MBFTs Involving a Wolff Rearrangement	*192*
9.3.2	MBFTs Involving Metals	*194*
9.3.3	MBFTs Involving Anions	*196*
9.4	Conclusion	*200*
	References	*200*
10	**Modeling of Indirect Phototransformation Reactions in Surface Waters** *203*	
	Davide Vione, Radharani Das, Francesca Rubertelli, Valte Maurino, and Claudio Minero	
10.1	Introduction	*203*
10.2	Indirect Photolysis Processes in Surface Waters	*205*
10.2.1	Reactions Induced by $^\bullet OH$	*206*
10.2.2	Reactions Induced by $CO_3^{-\bullet}$	*209*
10.2.3	Reactions Induced by $^3CDOM^*$	*211*
10.2.4	Other Reactions	*212*
10.3	Modeling the Photochemistry of Surface Waters	*214*
10.3.1	Modeling the Absorption Spectrum of Lake Water	*214*
10.3.2	Mass versus Concentration Approach in Photochemistry Models	*216*
10.3.3	Radiation Absorption by Photoactive Water Components	*217*
10.3.4	Generation and Reactivity of $^\bullet OH$ upon Irradiation of CDOM, Nitrate, and Nitrite	*218*
10.3.5	Formation and Reactivity of $CO_3^{-\bullet}$ in Surface Waters	*223*
10.3.5.1	Oxidation of $HCO_3^{-\bullet}$ and CO_3^{2-} by $^\bullet OH$	*224*
10.3.5.2	Oxidation of CO_3^{2-} by $^3CDOM^*$	*224*
10.3.5.3	Reactivity of $CO_3^{-\bullet}$ in Surface Waters	*225*
10.3.6	Formation and Reactivity of $^3CDOM^*$ in Surface Waters	*226*
10.4	Conclusions and Perspectives	*230*
	Acknowledgment	*231*
	References	*231*

Part IV   Organic Synthesis and Materials *235*

11	**Bottom-Up Approaches to Nanographenes through Organic Synthesis** *237*	
	Diego Peña	
11.1	Introduction	*237*
11.2	Alternant Polyarenes with 10 Fused Benzene Rings	*239*
11.3	Alternant Polyarenes with 11 Fused Benzene Rings	*244*
11.4	Alternant Polyarenes with 12 Fused Benzene Rings	*248*
11.5	Alternant Polyarenes with 13 Fused Benzene Rings	*249*

11.6	Alternant Polyarenes with 14 Fused Benzene Rings	*252*
11.7	Alternant Polyarenes with More than 15 Fused Benzene Rings	*252*
11.8	Conclusions and Outlook	*256*
	Acknowledgments	*257*
	References	*257*

12 Differentiated Ligands for the Sequential Construction of Crystalline Heterometallic Assemblies *263*
Stéphane A. Baudron

12.1	Introduction	*263*
12.2	Dithiolate Ligands	*268*
12.2.1	Metallatectons	*268*
12.2.2	Infinite Chains	*270*
12.2.3	Discrete Assemblies	*271*
12.3	Dipyrrin Ligands	*273*
12.3.1	Metallatectons	*274*
12.3.2	Heterometallic Architectures Built with the Assistance of an Ag–π Interaction	*274*
12.4	Conclusion and Outlook	*277*
	Acknowledgments	*278*
	References	*278*

13 Water-Soluble Perylene Dyes *283*
Cordula D. Schmidt and Andreas Hirsch

13.1	History and Functionalization of Perylene Dyes	*283*
13.2	Derivatization of Perylene Dyes	*284*
13.3	Properties and Applications of Perylene Dyes	*286*
13.4	Advantages and Applications of Water-Soluble Perylene Derivatives	*286*
13.5	Symmetric Water-Soluble PDIs	*289*
13.6	Unsymmetric PDIs	*292*
13.6.1	Noncovalent Functionalization of Carbon Nanotubes	*293*
13.6.2	PDIs as Reporter Electrolytes for Implantation Technology	*296*
13.7	Chiral Water-Soluble PDIs	*297*
13.8	Conclusion and Outlook	*300*
	Acknowledgments	*301*
	References	*301*

Index *305*

Preface

The idea of publishing books based on contributions given by emerging young chemists arose during the preparations of the first EuCheMs (European Association for Chemical and Molecular Sciences) Conference in Budapest. In this conference I cochaired the competition for the first European Young Chemist Award aimed at showcasing and recognizing the excellent research being carried out by young scientists working in the field of chemical sciences. I then proposed to collect in a book the best contributions from researchers competing for the Award.

This was further encouraged by EuCheMs, SCI (Italian Chemical Society), RSC (Royal Society of Chemistry), GDCh (Gesellschaft Deutscher Chemiker), and Wiley-VCH and brought out in the book "Tomorrow's Chemistry Today" edited by myself and published by Wiley-VCH.

The motivation gained by the organization from the above initiatives was, to me, the trampoline for co-organizing the second edition of the award during the second EuCheMs Conference in Torino. Under the patronage of EuCheMs, SCI, RSC, GDCh, the Consiglio Nazionale dei Chimici (CNC), and the European Young Chemists Network (EYCN), the European Young Chemist Award 2008 was again funded by the Italian Chemical Society.

In Torino, once again, I personally learned a lot and received important inputs from the participants about how this event can serve as a source of new ideas and innovations for the research work of many scientists. This is also related to the fact that the areas of interest for the applicants cover many of the frontier issues of chemistry and molecular sciences (see also *Chem. Eur. J.* 2008, **14**, 11252–11256). But, more importantly, I was left with the increasing feeling that our future needs for new concepts and new technologies should be largely in the hands of the new scientific generation of chemists.

In Torino, we received about 90 applications from scientists (22 to 35 years old) from 30 different countries all around the world (*Chem. Eur. J.* 2008, **14**, 11252–11256).

Most of the applicants were from Spain, Italy, and Germany (about 15 from each of these countries). United Kingdom, Japan, Australia, United States, Brazil, Morocco, Vietnam, as well as Macedonia, Rumania, Slovenia, Russia, Ukraine, and most of the other European countries were also represented. In terms of applicants, 63% were male and about 35% were PhD students; the number of postdoctoral researchers was only a small percentage, and only a couple of them came from industry. Among the oldest participants, mainly born between 1974 and 1975, several were associate professors or researchers at Universities or Research Institutes and others are lecturers, assistant professors, or research assistants.

The scientific standing of the applicants was undoubtedly very high and many of them made important contributions to the various symposia of the 2nd EuCheMs Congress. A few figures help to substantiate this point. The, let me say, "h index" of the competitors was 20, in the sense that more than 20 applicants coauthored more than 20 publications. Some patents were also presented. Five participants had more than 35 publications, and, h indexes, average number of citations per publication, and number of citations, were as high as 16, 35.6, and 549, respectively. Several of the papers achieved further recognition as they were quoted in the reference lists of the young chemists who where featured on the covers of top journals. The publication lists of most applicants proudly noted the appearance of their work in the leading general chemistry journals such as *Science*, *Nature*, *Angewandte Chemie*, *Journal of the American Chemical Society*, or the best niche journals of organic, inorganic, organometallic, physical, analytical, environmental, and medicinal chemistry.

All of this supported the idea of publishing a second book with the contributions of these talented chemists.

However, in order to have more homogeneous publications and in connection to the great number of interesting papers presented during the competition, we decided to publish three volumes.

This volume represents indeed one of the three edited by inviting a selection of young researchers who participated in the European Young Chemist Award 2008. The other two volumes concern the two different areas of nanotechnology/material science and life sciences and are entitled "Ideas in Molecular Sciences: Advances in Nanotechnology, Materials and Devices" and "Ideas in Chemistry and Molecular Sciences: Where Chemistry Meets Life, respectively."

It is important to mention that the contents of the books are a result of the work carried out in several topmost laboratories around the world both by researchers who already lead their own group and by researchers who worked under a supervisor. I would like to take this occasion to acknowledge all the supervisors of the invited young researchers for their implicit or explicit support to this initiative that I hope could also serve to highlight the important results of their research groups.

The prospect of excellence of the authors was evident from the very effusive recommendation letters sent by top scientists supporting the applicants for the Award.

A flavor of these letters is given by the extracts from some of the sentences below:

"Excellent chemist, who has outstanding synthetic skill"; "He was a very creative and dedicated, but at the same time a highly practical and pragmatic researcher"; "The candidate was one of the best associates that I have had in my career"; "... has extraordinary laboratory skills and manages to obtain high quality data as a result of careful experimentation"; "The candidate is thoughtful, thorough, and insightful in chemical judgments"; "I believe that he will someday become a leader in the field"; "The skills of the candidate are such that could contribute to many facets of organic chemistry"; " ... has outstanding quality and intellectual capacity"; "I have no doubt about him to become a major contributor to the chemistry of the first half of the XXIst century"; "The candidate has proven to be a most dedicated, enthusiastic and productive chemist"; "He is a first rate organic chemist"; " ... has a keen intellect"; "He is a leader and outstanding researcher"; "The work of the candidate is characterized by its quality as well as its breadth"; "He has an exceptionally strong publication record and has already won several awards"; "The candidate was an exceptional talented young scientist"; "The pioneering research of the candidate resulted in a number of paper in top journals"; "He performed the research in a professional and highly independent manner and was highly capable of guiding young students during their research projects"; "The exceptional talent is further illustrated in the large number of projects and grants he was involved in"; "It is my firm opinion that he has the capabilities to soon become one of the leaders of the new generation of chemists in Europe"; "One of the most promising young scientists of his generation worldwide. I believe that he has the drive, creativity, and track record of accomplishment that suggest the candidate himself is highly worthy of the European Young Investigator Award"; "Is a very brilliant young researcher"; "With such diversified capability he is able to create and expand the research topic of own interest."

The content of the first part of the volume is largely dedicated to catalysis and catalytic processes including the very important and hot field of organocatalysis. For example, taking into account that the oxidation of organic compounds is one of the most important transformations of chemistry with fundamental involvements in many areas (general synthesis, industrial processes, materials, energy, biology, etc.) and that nowadays, the ecological standards require the development of new catalytic processes characterized not only by oxidant economy, but also by environmental benignity under mild conditions, one contribution (Punta *et al.*) refers to the case of *N*-hydroxyderivatives as organocatalysts for the selective free radical aerobic oxidation of organic compounds. Following the author's suggestion it is hopeful that this synthetic route will soon become a "winning industrial process." In another chapter (Fernández Rodríguez) the gold-catalyzed reactions of push–pull conjugated dienyne are highlighted, while the lithiation of tertiary amines as a powerful tool for the building block processes in organic synthesis (Gessner) is the theme of another stimulating chapter. In particular, this last chapter gives an overview of nitrogen ligands in the coordination chemistry of organolithium compounds with special focus on the direct deprotonation of tertiary amines to α-lithiated species. A combination of structure elucidation and computational studies gives insight into mechanistic features and observed selectivity. The process and selectivities are explained by means of the spatial proximity of reactive groups in intermediate, precoordinated adducts of the amine, and the organolithium compounds according to the complex-induced proximity effect.

A further contribution (Díez-Gonzáles) is dedicated to the field of copper-catalyzed transformations using N-heterocyclic ligands. In particular, recent results concerning Click Chemistry along with related mechanistic studies are highlighted in this chapter.

In the last paper of this area (Giacalone et al.) the following cases are reviewed: (i) L-proline or H–Pro–Pro–Asp–NH$_2$ supported on ionic liquid–modified silica gels as recyclable catalysts for aldol reaction; (ii) polystyrene-supported proline as versatile and recyclable organocatalyst; and (iii) prolinamide-supported polystyrenes as highly stereoselective and recyclable organocatalysts for the aldol reaction.

In connection to the increasing relevance of the field, the second section of this volume is dedicated to the study of chemical interactions and reactions by predictive tools.

A chapter in this section (Marcelli) exploits the computational methods for the study and development of the organocatalysis field. This chapter, in particular, features a gallery of recent examples unraveling important mechanistic aspects of double hydrogen bonding organocatalysis in synthetically relevant transformations. In addition, some recent striking results in the development of novel types of double hydrogen bonding organocatalysts are briefly discussed. I am particularly happy to have a chapter like this in the book since, in agreement with the author, it seems reasonable to assume that, in the coming years, the importance of computational tools in the development of new catalytic systems is destined to grow. The results described in these pages significantly improved our understanding of organocatalytic reactions, often by pointing out important aspects that had been overlooked in the initial catalyst design. Together with the inexhaustible source of inspiration represented by enzymatic processes, the use of these mechanistic insights in catalyst design will likely result in significant progress toward the development of truly biomimetic organocatalytic systems.

Another stimulating chapter (Prins) in this section is dedicated to the molecular recognition area and the emerging strategy of the dynamic covalent capture. This very general concept is potentially applicable for different purposes. An important role of dynamic covalent capture is envisioned in areas where subtle noncovalent interactions are crucial, for example, in protein–protein interactions as well as in catalytic pathways and therefore for catalyst discovery.

Taking into account that (see also above) one of the most important challenges of synthetic chemistry lies in combining efficiency, reduced costs, and environmental impact in the production of relevant molecules for application in different areas such as pharmaceutical, food, agrochemistry, material chemistry, and energy resources, some contributions are particularly dedicated to this area.

The first chapter in this area (Dias et al.) deals just with the theme of environmentally important reactions. In this paper, after an overview of the applications of furfural and the reaction mechanisms of dehydration/hydrolysis of polysaccharides into furfural, some of the most relevant results on the use of solid acid catalysts in the conversion of saccharides (in particular, xylose) into furfural are discussed.

The second chapter (Coquerel et al.) in this area deals with ecocompatible organic synthesis, which, by the way, for the authors means being both economically as well as ecologically compatible. In this last chapter the idea that the next major evolution in synthetic organic chemistry might be the control of multiple bond-forming transformations (MBFTs) is underlined and in particular the nonconcerted MBFTs are presented as reactions whose control would nicely combine with the now required criterion of *ecocompatibility*.

Given the great present concern on environment as mentioned above, a further chapter on modeling of the photochemical reactions in surface fresh water (Vione et al.) is dedicated to the ecology theme.

A modeling approach was presented to describe the indirect photolysis processes that can take place in surface waters, with particular emphasis on the reactions that involve OH, CO_3^-, and $^3CDOM^*$ (triplet states of photoexcited colored dissolved organic matter). The model allows the important assessment of the lifetime of dissolved compounds in water bodies, including hazardous xenobiotics, as far as the indirect photochemical processes are concerned.

The last contributions to the volume can be considered as a bridge between organic synthesis and material chemistry. In particular, the first one (Peña) deals with alternant polycyclic aromatic hydrocarbons (PAHs) that contain more than 10 fused benzene rings and have been prepared through organic synthesis. These bottom-up approaches used to obtain nanographene should now be addressed to prepare larger nanosized polyarenes and nanographenes with potential outstanding electronic properties.

In the second chapter of this section (Baudron), it is shown that the synthetic challenge that represents the elaboration of heterometallic architectures can be faced using a sequential construction strategy. The used ligands can be tuned with respect to the nature and relative arrangement of the coordination poles, having therefore a direct influence on the structure of the architectures. As it is underlined by the author, beyond the structural beauty of these systems, their physical properties (such as the optical and electronic properties) are of prime importance.

In the last chapter (Schmidt), the general properties and applications of perylene dyes are summarized. In this chapter, after an overview of water-soluble perylene dyes and their applications, chiral water-soluble perylene dyes are presented. Particularly interesting is the noncovalent strategy of SWNT functionalization. This strategy has the advantage of implementing multifunctional groups without compromising the main properties of SWNTs as opposed to usual covalent functionalization of this system, in the sense that the sp2 carbon backbone is not altered.

The chapters of this book show, *inter alia*, that, as is written in one of the contributions (Coquerel and coworkers), "the only limitations to the exploration of the molecular world are the creativity of the chemist, and most importantly on a practical point of view, the major limitation is the current knowledge of the science of synthesis. In the current era, however, it is assumed that any three-dimensional molecular architecture, providing it is sufficiently stable, can be prepared by total synthesis (i.e., the laboratory construction of naturally occurring or designed molecules by chemical synthesis from simple starting materials) if the chemist has enough experience, knowledge, time, and money."

I cannot end this preface without acknowledging all the authors and the persons who helped me in the book project together with all the societies (see the book cover) that motivated and sponsored the book. I'm personally grateful to Professors Giovanni Natile, Francesco De Angelis and Luigi Campanella for their motivation and support in this activity.

Palermo, October 2009   *Bruno Pignataro*

List of Contributors

Stéphane A. Baudron
Université de Strasbourg
Institut Le Bel
Laboratoire de Chimie de
Coordination Organique (UMR 7140)
4 rue Blaise Pascal
67000 Strasbourg
France

Thomas Boddaert
Institut des Sciences Moleculaires de Marseille
ISM2 - UMR CNRS 6263
Laboratoire STeReO
Université Paul Cezanne d'Aix-Marseille
Centre Saint Jerome - Service 531
F-13397 Marseille Cedex 20
France

Yoann Coquerel
Institut des Sciences Moleculaires de Marseille
ISM2 - UMR CNRS 6263
Laboratoire STeReO
Université Paul Cezanne d'Aix-Marseille
Centre Saint Jerome - Service 531
F-13397 Marseille Cedex 20
France

Radharani Das
Università degli Studi di Torino
Dipartimento di Chimica Analitica
Via Pietro Giuria 5
10125 Torino
Italy

and

Haldia Institute of Technology
Department of Chemical Engineering
ICARE complex
Haldia 721657
India

Ana S. Dias
University of Aveiro
CICECO
Campus Santiago
3810–193, Aveiro
Portugal

and

Avantium Chemicals BV
Zekeringstraat 29
1014BV, Amsterdam
The Netherlands

Ideas in Chemistry and Molecular Sciences: Advances in Synthetic Chemistry. Edited by Bruno Pignataro
Copyright © 2010 WILEY-VCH Verlag GmbH & Co. KGaA, Weinheim
ISBN: 978-3-527-32539-9

Silvia Díez-González
Department of Chemistry
Imperial College London
Exhibition Road
South Kensington
SW7 2AZ
UK

Manuel A. Fernández-Rodríguez
Universidad de Burgos
Facultad de Ciencias
Pza Misael Bañuelos s/n
09001 Burgos
Spain

Cristian Gambarotti
Politecnico di Milano
Department of Chemistry
Materials and Chemical
Engineering "Giulio Natta"
Via Mancinelli 7
20131 Milan
Italy

Viktoria H. Gessner
Technische Universität
Dortmund
AK Strohmann Anorganische
Chemie
Otto-Hahn-Str. 6
44227 Dortmund
Germany

Francesco Giacalone
Università degli Studi di Palermo
Dipartimento di Chimica
Organica "E. Paternò"
Via delle Scienze
Parco d'Orleans – Padiglione 17
90100 Palermo
Italy

Michelangelo Gruttadauria
Università degli Studi di Palermo
Dipartimento di Chimica
Organica "E. Paternò"
Via delle Scienze
Parco d'Orleans – Padiglione 17
90100 Palermo
Italy

Andreas Hirsch
Friedrich-Alexander-Universität
Erlangen-Nürnberg
Institut für Organische Chemie
Henkestr. 42
91054 Erlangen
Germany

Sérgio Lima
University of Aveiro
Department of Chemistry
CICECO
Campus de Santiago
3810–193, Aveiro
Portugal

Damien Mailhol
Institut des Sciences Moleculaires
de Marseille
ISM2 - UMR CNRS 6263
Laboratoire STeReO
Université Paul Cezanne
d'Aix-Marseille
Centre Saint Jerome - Service 531
F-13397 Marseille Cedex 20
France

Tommaso Marcelli
Politecnico di Milano
Department of Chemistry
Materials and Chemical
Engineering "Giulio Natta"
Via Mancinelli 7
20131 Milan
Italy

Valte Maurino
Università degli Studi di Torino
Dipartimento di Chimica
Analitica
Via Pietro Giuria 5
10125 Torino
Italy

Claudio Minero
Università degli Studi di Torino
Dipartimento di Chimica
Analitica
Via Pietro Giuria 5
10125 Torino
Italy

Renato Noto
Università degli Studi di Palermo
Dipartimento di Chimica
Organica "E. Paternò"
Via delle Scienze
Parco d'Orleans – Padiglione 17
90100 Palermo
Italy

Diego Peña
Universidad de Santiago de
Compostela
Departamento de Química
Orgánica
Facultad de Química
15782 Santiago de Compostela
Spain

Martyn Pillinger
University of Aveiro
Department of Chemistry
CICECO
Campus de Santiago
3810–193, Aveiro
Portugal

Marc Presset
Institut des Sciences Moleculaires
de Marseille
ISM2 - UMR CNRS 6263
Laboratoire STeReO
Université Paul Cezanne
d'Aix-Marseille
Centre Saint Jerome - Service 531
F-13397 Marseille Cedex 20
France

Leonard Prins
University of Padova
Department of Chemical Sciences
Via Marzolo 1
35131 Padova
Italy

Carlo Punta
Politecnico di Milano
Department of Chemistry
Materials and Chemical
Engineering "Giulio Natta"
Via Mancinelli 7
20131 Milan
Italy

Jean Rodriguez
Institut des Sciences Moleculaires
de Marseille
ISM2 - UMR CNRS 6263
Laboratoire STeReO
Université Paul Cezanne
d'Aix-Marseille
Centre Saint Jerome - Service 531
F-13397 Marseille Cedex 20
France

Francesca Rubertelli
Università degli Studi di Torino
Dipartimento di Chimica
Analitica
Via Pietro Giuria 5
10125 Torino
Italy

Cordula D. Schmidt
Friedrich-Alexander-Universität
Erlangen-Nürnberg
Institut für Organische Chemie
Henkestr. 42
91054 Erlangen
Germany

Anabela A. Valente
University of Aveiro
Department of Chemistry
CICECO, Campus de Santiago
3810–193, Aveiro
Portugal

Davide Vione
Università degli Studi di Torino
Dipartimento di Chimica
Analitica
Via Pietro Giuria 5
10125 Torino
Italy

Part I
Innovative Processes in Organic Chemistry

1
N-Hydroxy Derivatives: Key Organocatalysts for the Selective Free Radical Aerobic Oxidation of Organic Compounds

Carlo Punta and Cristian Gambarotti

1.1
Introduction

Oxidative processes of organic compounds represent some of the most important chemical transformations involved in many fundamental areas, including general synthesis, industrial processes, materials, energy, biology, and so on. In particular, eco-friendly standards require oxidants to be able to combine a low environmental impact with an economical convenience. Molecular oxygen and hydrogen peroxide are the ideal oxidants from this point of view [1]. Nevertheless, their use strictly depends upon the employment of catalytic systems, which allow operating with high selectivity under mild and environmentally benign conditions [2–4]. Thus we can say that, in this case also, catalysis represents the key to waste minimization.

N-hydroxy derivatives (NHDs) proved to be of particular importance for this purpose, allowing development of innovative synthetic processes of great relevance, as it has been demonstrated by the several patents involving the use of the two most commonly employed NHD catalysts: 2,2,6,6-tetramethylpiperidine-1-oxyl (TEMPO) and N-hydroxyphthalimide (NHPI) [5].

This chapter will focus on the key role played by NHDs as catalysts in the aerobic oxidation of organic compounds. After a brief overview of the thermochemical and kinetic aspects, that we have contributed to determine and are responsible for the differences in reactivity among the various families of N-hydroxy derivatives, we will discuss our recent results in the field by employing, in turn, N-hydroxy amines, amides, and imides with the aim to develop selective oxidative processes characterized by high conversions and selectivity.

1.2
General Reactivity of N-Hydroxy Derivatives

In spite of their similar structure, TEMPO (**1**) and phthalimide-*N*-oxyl (PINO) radicals (**2**), generated *in situ* from NHPI, show a completely different behavior. In fact, being a persistent radical, TEMPO inhibits radical processes, whereas PINO, with its nonpersistent character (it decays by first order kinetics with a $k = 0.12 \text{ s}^{-1}$) [6], is able to promote free radical chains.

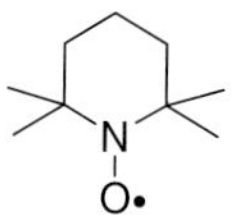

1 (TEMPO) 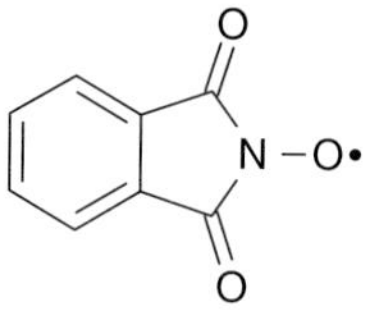 **2 (PINO)**

The consequence of thermochemical studies conducted on a wide range of NHDs (including TEMPO-H (**3**), N-methylbenzo-hydroxamic acid (NMBHA, **4**), and NHPI (**5**)), shows that this opposite behavior has to be ascribed to enthalpic factors.

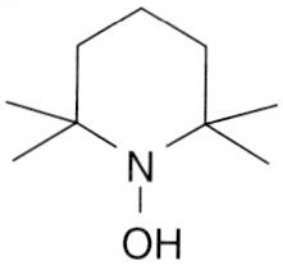

3 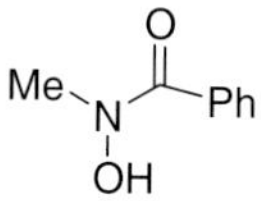 **4** 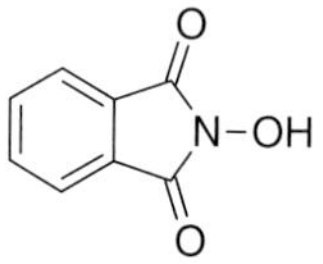 **5**

By means of EPR radical equilibration technique [7], in collaboration with Lucarini and coworkers, it was possible to measure the bond dissociation enthalpy (BDE) of the O–H bonds in hydroxylamine derivatives [8]. The most significant results, reported in Table 1.1, clearly indicated that the carbonyl groups directly bonded to the nitrogen atom strongly increase BDE values.

This effect can be ascribed to the energy difference between the oxygen-centered radicals and the corresponding hydroxyl derivatives. More specifically, the carbonyl group, owing to its electron-withdrawing character, reduces the importance of the mesomeric structure **7** in the resonance equilibrium of the nitroxyl radical Eq. (1.1). As a consequence, the radical is less stabilized and the corresponding O–H BDE increases.

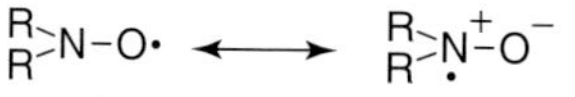

$$R_2N-O\bullet \longleftrightarrow R_2\overset{+}{N}-O^- \qquad (1.1)$$

    6   **7**

On the basis of these results, it is apparent that the general reaction of hydrogen abstraction from a C–H bond by an oxygen-centered nitroxyl radical cannot occur with TEMPO, the process being largely endothermic with any kind of organic substrate.

Table 1.1 BDE Values of the O–H bonds in the hydroxylamines **3**, **4**, and **5**.

Hydroxylamine	BDE (kcal mol^{-1})
3	69.6
4	79.2
5	88.1

Nevertheless, TEMPO has been widely employed as a catalyst for the oxidation of alcohols with a variety of oxidants [9], including aerobic oxidation when used in combination with transition metal salt complexes [10–12]. In these cases, TEMPO plays two key functions: it promotes the oxidation of the alcohols, following an ionic mechanism, but, being a persistent radical, it also inhibits the subsequent free radical oxidation of aldehydes and ketones.

On the contrary, due to the relatively high BDE value of the O–H bond in NHPI, the PINO radical is able to catalyze the aerobic oxidation of a wide range of organic substrates through the formation of a carbon-centered radical by hydrogen abstraction from a C–H bond, according to the general radical chain reported in Scheme 1.1.[13].

Many concomitant aspects make NHPI an intriguing catalyst for selective oxidations. As we have seen, from a thermochemical point of view, the hydrogen transfer reaction from a C–H bond to PINO may be in many cases exothermic or only slightly endothermic. Nevertheless, other factors need to be taken into consideration in order to justify the catalytic role of NHPI. Kinetic experiments, carried out by Lucarini and coworkers using EPR technique [8], have clearly demonstrated that the hydrogen abstraction from a C–H bond (path i) by PINO radical is always faster than by a generic peroxyl radical (t-BuOO$^\bullet$). These results explain why PINO is able to selectively catalyze a classical autoxidation, leading to the formation of a carbon-centered radical which, in turn, reacts fast with oxygen, forming the corresponding peroxyl radical (path ii). The observed behavior cannot be ascribed to enthalpic reasons (the O–H BDEs in NHPI and in *tert*-butyl hydroperoxide are almost identical, ~ 88 kcal mol^{-1}) but, instead, to a polar effect

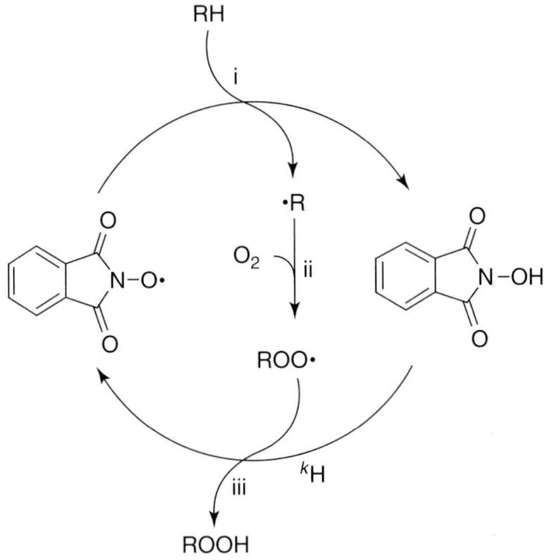

Scheme 1.1

due to a more pronounced electrophilic character of PINO with respect to the peroxyl radical. Such a behavior is common to nitroxyl radicals, but in this case it is considerably enhanced by the presence of the two carbonyl groups in α to the nitrogen atom Eq. (1.2).

$$\text{[phthalimide N-O}\cdot\text{]} \longleftrightarrow \text{[phthalimide N}^+\text{-O}^-\text{ resonance form]} \tag{1.2}$$

Moreover, the same research group determined the rate constant for the hydrogen atom abstraction from NHPI by peroxyl radicals (path iii). The unexpected moderately high value obtained ($k_H = 7.2 \times 10^3$ M^{-1} s^{-1}) [8] allows the complete insight of the catalytic effect of the PINO radical in the aerobic oxidation of organic substrates.

Ishii et al. [14] have reported many examples of oxidations catalyzed by NHPI, based on the *in situ* generation of PINO radical through different methodologies, including the employment of radical initiators [15], transition metal salts (mainly Co(II) and Mn(II)) [13], cerium ammonium nitrate (CAN) [16], acetaldehyde [17], bromine [18], enzymes [19], NO$_2$ [20], and so on.

However, the high potentiality of NHPI, in terms of conversions and selectivity, was evidenced solely upon the mechanistic investigation of the catalytic cycle. In this deep rationalization relies the secret of our success in developing several selective oxidative processes under aerobic conditions, at room temperature and atmospheric pressure.

Finally, N-hydroxy amides, having a halfway O–H BDE value between TEMPO and NHPI, result to be ideal catalysts for free radical aerobic oxidations of organic derivatives bearing weak C–H bonds, whereas NHPI would undergo high exothermic hydrogen atom transfer reactions, negatively affecting the selectivity of the process [21].

1.3
Aerobic Oxidation Catalyzed by N-Hydroxy Amines

1.3.1
Aerobic Oxidation of Alcohols to Aldehydes and Ketones

TEMPO, when used in combination with Mn(II)–Co(II) or Mn(II)–Cu(II) nitrates, is an ideal catalyst for the selective aerobic oxidation of aliphatic and aromatic alcohols to the corresponding aldehydes or ketones Eq. (1.3), under very mild conditions [22, 23].

1.3 Aerobic Oxidation Catalyzed by N-Hydroxy Amines

$$\text{R}_2\text{CH-OH} + 1/2\, O_2 \xrightarrow[\text{Mn(II) and Co(II) or Cu(II)}]{\text{TEMPO, } O_2 \;\; \text{rt}} \text{R}_2\text{C=O} + H_2O \qquad (1.3)$$
$$> 96\%$$

Under the same conditions, in the absence of TEMPO, aldehydes and ketones are readily oxidized to carboxylic acids via free radical chains [24], while the corresponding alcohols are quite inert [25]. This clearly demonstrates that the reaction catalyzed by TEMPO follows a nonradical mechanism, while TEMPO itself, thanks to its persistent character, rapidly traps the forming radicals Eq. (1.4), inhibiting further oxidation of aldehydes and ketones.

$$\text{TEMPO}^\bullet + R^\bullet \longrightarrow \text{TEMPO-OR} \qquad (1.4)$$

Thus, oxidation of alcohols occurs by means of an oxammonium cation Eq. (1.6), formed *in situ* by disproportionation of TEMPO in acidic medium Eq. (1.5), while the metal salts catalyze the reoxidation of the N-hydroxypiperidine (3) to TEMPO Eq. (1.7), so that molecular oxygen results to be the unique consumed oxidant.

$$2\, \text{TEMPO}^\bullet + H^+ \rightleftharpoons \text{TEMPO}^+ + \text{TEMPOH} \qquad (1.5)$$

$$\text{TEMPO}^+ + \text{>CH-OH} \longrightarrow \text{TEMPOH} + \text{>C=O} + H^+ \qquad (1.6)$$

$$\text{TEMPOH} \xrightarrow[\text{Mn(II), Co(II)}]{O_2} \text{TEMPO}^\bullet \qquad (1.7)$$

In spite of the efficiency of this catalytic system in terms of conversion and selectivity, the use of TEMPO has a significant limitation: it is rather expensive, so that recycling of the catalyst is necessary but, at the same time, its recovery from the reaction medium is difficult. Many efforts were devoted to the design of easy-recycling catalysts by anchoring TEMPO to solid supports [26]. However, till now, many drawbacks have been encountered by using TEMPO in heterogeneous systems. In most of these cases [27], NaOCl had to be used as oxidizing agent instead of oxygen, the latter leading to poor conversion in the desired products. Furthermore, in several circumstances, partial degradation of the supported TEMPO catalysts was observed.

Recently, a TEMPO-type catalyst supported on SBA-15 (**6**) (an ordered mesoporous material) was reported by Karimi *et al.* [28].

When **6** is employed in combination with catalytic amounts of $NaNO_2$ and n-Bu_4NBr under an atmosphere of oxygen or air, alcohols are completely and selectively converted to the corresponding aldehydes and ketones Eq. (1.8).

$$\underset{OH}{\overset{H}{\diagdown}} + 1/2\, O_2 \xrightarrow[\substack{AcOH,\, O_2 \text{ or air (1 atm), 50–60 °C} \\ > 99\%}]{\substack{\textbf{6}\ (1–1.5\ \text{mol\%}) \\ NaNO_2\ (10\ \text{mol\%}),\ nBu_4NBr\ (8\ \text{mol\%})}} =\!O + H_2O$$

(1.8)

However, in many cases, homogenous catalysis remains the best solution for the development of selective oxidative processes, due to the usually higher versatility of the catalytic systems in terms of applicability (wider range of substrates) and operative conditions (room temperature). Thus, in order to eliminate the disadvantages in using TEMPO catalysis, we developed, in collaboration with CIBA Speciality Chemicals, a new TEMPO-analogous catalyst [23, 29], characterized by a macrocyclic polypiperidine-N-oxyl radical structure (**7**).

This derivative, which is even more active than TEMPO for the aerobic oxidation of alcohols to the corresponding aldehydes and ketones, has amino groups that confer to **7**, the great advantage of being easily recovered and recycled in the form of its ammonium salt, considering that the catalysis is effective only in acidic medium.

1.4
Aerobic Oxidation Catalyzed by N-Hydroxy Amides

1.4.1
Peroxidation of Polyunsaturated Fatty Acids

Peroxidation of polyunsaturated fatty acids (PUFAs) and esters has attracted increased research attention, due to the mounting evidence that uncontrolled peroxidation is involved in the origin and development of many pathologies such as tumor promotion and the deposition of arterial plaques.

Lipid hydroperoxides are the primary products of free radical chain oxidations and their synthesis is of interest in order to simplify the study and characterization of secondary oxidation products, which seem to be the real promoters of diseases.

In order to provide a diasteroselective synthesis of *trans–cis* hydroperoxides (in place of the undesired *trans–trans* products, deriving from the β-fragmentation of peroxyl radicals), in collaboration with Porter's research group we have introduced a new N–OH derivative, N-methylbenzohydroxamic acid (NMBHA, **4**) [21], for the selective oxidation of PUFA in the presence of a radical initiator (2,2′-azobis(4-methoxy-2,4-dimethylvaleronitrile)) at 37 °C. The O–H BDE value of **4** (79.2 kcal mol^{-1}), is lower when compared with that of NHPI (88.1 kcal mol^{-1}), but higher than that of the C–H bond in the bisallylic position of a fatty acid ($\sim$ 76 kcal mol^{-1}), determining an increase in the value of the rate constant for hydrogen abstraction by peroxyl radical Eq. (1.11), $k_{NMBHA} = 1.2 \times 10^5$ M^{-1} s^{-1}. This suggested that NMBHA might behave as an ideal catalyst for selective lipid peroxidation Eqs. (1.9–1.11) by favoring the hydrogen abstraction from the weaker C–H bond Eq. (1.9) and, being a suitable H donor, by trapping the peroxyl radicals Eq. (1.11) derived from Eq. (1.10).

$$\text{Ph-}\underset{\underset{\text{Me}}{|}}{\overset{\overset{\text{O}}{\|}}{\text{C}}}\text{-N-O}\cdot + \text{R'}\diagdown\diagup\diagdown\diagup\text{R''}$$

$$\downarrow \qquad (1.9)$$

$$\text{Ph-}\underset{\underset{\text{Me}}{|}}{\overset{\overset{\text{O}}{\|}}{\text{C}}}\text{-N-OH} + \text{R'}\diagdown\diagup\diagdown\diagup\text{R''}$$

$$\text{R'}\diagdown\diagup\diagdown\diagup\text{R''} + \text{O}_2 \longrightarrow \text{R'}\diagdown\diagup\diagdown\underset{\text{R''}}{\overset{\text{OO}\cdot}{|}} \qquad (1.10)$$

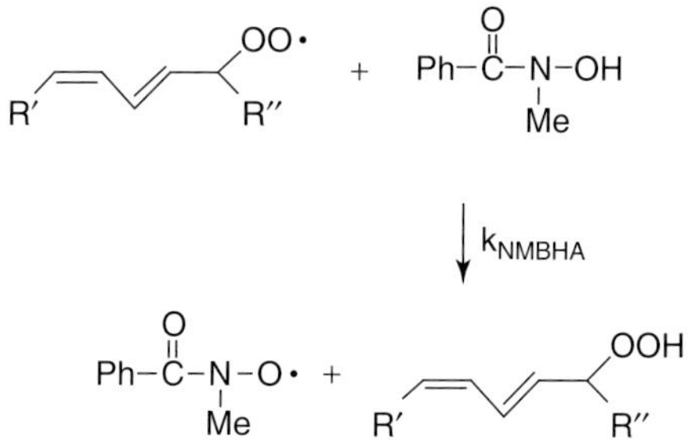

1 N-Hydroxy Derivatives

$$\text{R'}\!-\!\!\!=\!\!\!-\!\!\!\overset{OO\bullet}{\underset{R''}{C}}\quad+\quad \text{Ph}-\overset{O}{\underset{\underset{Me}{|}}{C}}-N-OH$$

$$\Big\downarrow k_{NMBHA}$$

$$\text{Ph}-\overset{O}{\underset{\underset{Me}{|}}{C}}-N-O\bullet\quad+\quad \text{R'}\!-\!\!\!=\!\!\!-\!\!\!\overset{OOH}{\underset{R''}{C}} \qquad (1.11)$$

The same process, conducted in the presence of NHPI instead of NMBHA, did not afford the same interesting products. In fact, in spite of the good conversions observed, the diasteroselectivity of the process, that is, the ratio of *trans–cis* to *trans–trans* oxidation products, was poor. This was because the undesired *trans–trans* hydroperoxides arise from the β-fragmentation of primary peroxyl radicals (Scheme 1.2b), a process for which the rate is competitive with that of the hydrogen transfer from NHPI (Scheme 1.2a).

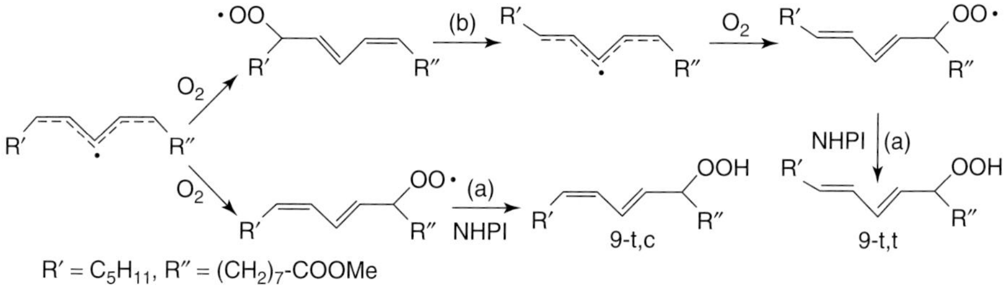

R' = C_5H_{11}, R'' = $(CH_2)_7$-COOMe

Scheme 1.2

1.5
Aerobic Oxidation Catalyzed by N-Hydroxy Imides

1.5.1
Oxidation of Benzylalcohols to Aldehydes

The aerobic oxidation of primary benzylic alcohols, catalyzed by NHPI and Co(II) salts, leads to aromatic aldehydes without appreciable formation of carboxylic acids [30] Eq. (1.12). In contrast, the oxidation of primary aliphatic alcohols leads to carboxylic acids without significant formation of aldehydes, even at low conversions. This selectivity observed in the catalysis with NHPI clearly indicates that benzyl alcohols are much more reactive than the corresponding aldehydes while, in the case of nonbenzylic alcohols, the corresponding aldehydes are much more reactive than the starting alcohols.

1.5 Aerobic Oxidation Catalyzed by N-Hydroxy Imides

$$\text{PhCH}_2\text{OH} \xrightarrow[\text{rt 2 h} \atop 92\%]{\text{NHPI O}_2 \atop \text{Co(II) } m\text{CPBA}} \text{PhCHO} \quad (1.12)$$

Both polar and enthalpic effects present in the NHPI catalysis explain this behavior well. To better understand the reasons of these results, we investigated the effect of aromatic ring substituents on benzyl alcohols in their aerobic oxidation by NHPI catalysis [23]. A good Hammett correlation was obtained (Figure 1.1) with the exception of *p*-nitro and *p*-cyano substituents, which have a negligible effect on the reactivity, while *m*-nitro and *m*-cyano benzyl alcohols were significantly deactivated.

This behavior is due to the captodative effect, which qualitatively suggests that pairs of substituents having opposite polarities both concur to the stabilization of a radical according to the resonance structures showed in Eq. (1.13). While the captodative effect causes a significant decrease in the BDE values for benzylic C–H bonds in *p*-cyano- and *p*-nitrobenzyl alcohols, the favorable enthalpic effect balances the unfavorable polar effect due to the presence of *p*-cyano and *p*-nitro groups.

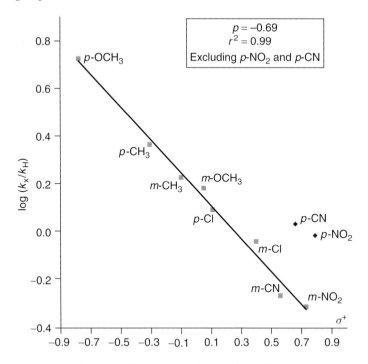

Figure 1.1 Substituent effect in the aerobic oxidation of substituted benzyl alcohols with NHPI catalysis.

$$\text{(structure 1)} \longleftrightarrow \text{(structure 2)} \longleftrightarrow \text{(structure 3)} \longleftrightarrow \text{(structure 4)} \tag{1.13}$$

Lanzalunga et al. also reported the effect of aryl-substituted N-hydroxyphthalimides (X-NHPI) used as catalysts in the aerobic oxidation of primary and secondary benzylic alcohols to the corresponding aldehydes and ketones respectively [31]. It was found that, when X was an electron-withdrawing group, the O–H BDE of X-NHPI, determined by using the EPR radical equilibration technique, increased. Kinetic studies proved that this behavior was reflected in an increasing of the substrate oxidation rate, thus indicating that the hydrogen atom transfer from the alcohol to X-PINO represented the rate-determining step. Besides enthalpic effects, polar effects were also emphasized in the same report for the hydrogen atom transfer process. In particular, a negative ρ value of the Hammett correlation for the oxidation of substituted primary benzylic alcohols and the decrease of the ρ values by increasing the electron-withdrawing properties of the substituents, have been observed.

1.5.2
Oxidation of Silanes

The classic routes for the synthesis of silanols involve the oxidation of silanes by a variety of metal-based oxidants, ozone, and dioxiranes. Most of these methods, however, afford the corresponding siloxanes as undesired side products and use expensive oxidants, which often involve environmental drawbacks. The aerobic oxidation of silanes, catalyzed by NHPI and Co(II) salts, revealed to be particularly effective for the selective synthesis of silanols, without appreciable formation of side products [32] Eq. (1.14).

$$Ph_3SiH + 1/2\, O_2 \xrightarrow[\text{rt 24 h 97\%}]{\text{NHPI } O_2 \text{ Co(II)}} Ph_3SiOH \tag{1.14}$$

1.5.3
Oxidation of N-Alkylamides

The oxidation of N-alkylamides by O_2, catalyzed by NHPI and Co(II) salt Eqs. (1.15–1.17), leads to the corresponding carbonyl derivatives (aldehydes, ketones, carboxylic acids, imides) whose distribution depends upon the nature of the alkyl group and the reaction conditions [33]. Lactams are oxidized to the corresponding imidoderivatives Eq. (1.15). Primary N-benzylamides lead to imides and aromatic aldehydes at room temperature without any appreciable amount of carboxylic acids Eq. (1.16) while, under the same conditions, nonbenzylic derivatives give carboxylic acids and imides with no trace of aldehydes, even at very low conversions Eq. (1.17).

$$\text{2-pyrrolidinone} \xrightarrow[\text{80 °C 5 h}]{\text{NHPI 10\%, O}_2} \text{succinimide (92\%)} \qquad (1.15)$$

$$\text{4-methylbenzyl acetamide} \xrightarrow[\text{mCPBA 5\%}]{\text{NHPI 10\%, O}_2,\ \text{rt 4 h}} \text{imide (67\%)} + \text{4-methylbenzaldehyde (22\%, CHO)} \qquad (1.16)$$

$$\text{CH}_3(\text{CH}_2)_4\text{-CH}_2\text{-NHC(O)CH}_3 \xrightarrow[\text{mCPBA 5\%}]{\text{NHPI 10\%, O}_2,\ \text{rt 4 h}} \text{imide (47\%)} + n\text{-C}_4\text{H}_9\text{COOH 3\%} + n\text{-C}_5\text{H}_{11}\text{COOH 10\%} \qquad (1.17)$$

1.5.4
Oxidation of Tertiary Benzylamines to Aldehydes

Tertiary benzylamines are easily oxidized to the corresponding arylaldehydes under aerobic conditions in the presence of NHPI or N-hydroxysuccinimide (NHSI) and Co(II) salts [34] Eq. (1.18).

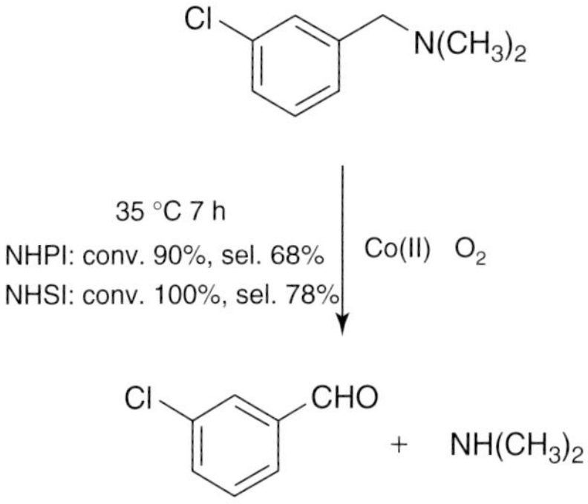

1 N-Hydroxy Derivatives

[Scheme: 3-Cl-C6H4-CH2-N(CH3)2 → 35 °C 7 h, NHPI: conv. 90%, sel. 68%; NHSI: conv. 100%, sel. 78%, Co(II) O2 → 3-Cl-C6H4-CHO + NH(CH3)2] (1.18)

With NHSI, the reaction is slower, but it goes to completion without deactivation of the catalyst. Competitive experiments with NHPI and NHSI in the presence of N,N-dimethyl-m-Cl-benzylamine showed that the former reacts faster than the latter. Moreover, the faster oxidation by NHPI catalysis makes the reaction somewhat less selective compared with the slower NHSI catalysis.

1.5.5
Oxidative Functionalization of Alkylaromatics

Alkylbenzenes are selectively oxidized to the corresponding acetates by nitric aerobic oxidation catalyzed by NHPI and I_2 [35] Eq. (1.19).

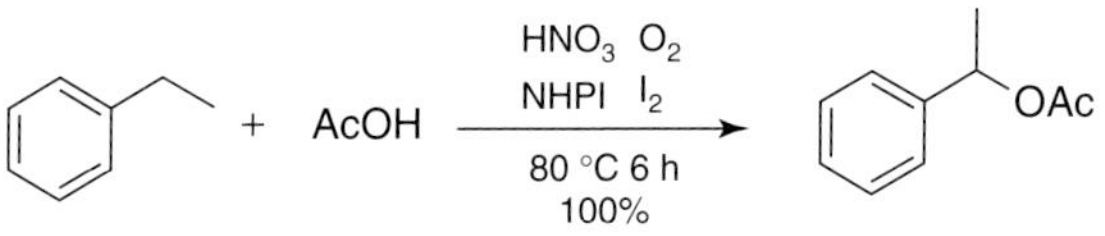

[Scheme: PhCH3 + AcOH → HNO3, O2, NHPI, I_2, 80 °C 6 h, 100% → PhCH2OAc] (1.19)

The winning point of these reactions is the fact that the product is less prone toward further oxidation than the starting hydrocarbon, making it possible to obtain products with high selectivity.

The role of I_2 is to trap the intermediate benzyl radical giving the corresponding aryl iodide Eq. (1.20). Under these reaction conditions the iodide undergoes fast SN2 substitution by the acetic acid, which is used as a solvent, achieving the final acetoxy derivative. The same reaction carried out in cyclohexane gives cyclohexyl acetate and *trans*-iodocyclohexyl acetate: the elimination of HI from the intermediate cyclohexyl iodide leads to cyclohexene which, in the presence of I_2 and AcOH, gives rise to the *trans* adduct Eq. (1.21).

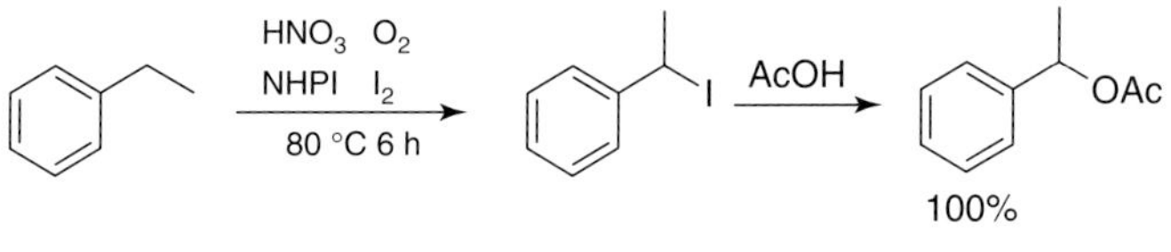

[Scheme: PhCH3 → HNO3, O2, NHPI, I_2, 80 °C 6 h → PhCH2I → AcOH → PhCH2OAc, 100%] (1.20)

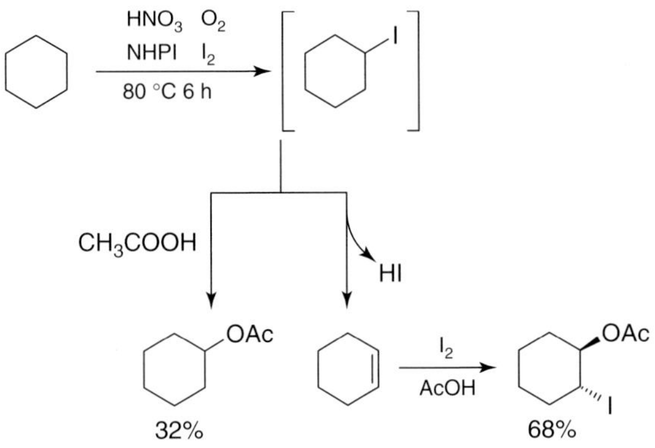

$$\text{Cyclohexane} \xrightarrow[\substack{\text{NHPI } I_2 \\ 80\,°C\,6\,h}]{HNO_3\ O_2} [\text{Cyclohexyl-I}]$$

$$\begin{array}{c} \text{CH}_3\text{COOH} \swarrow \qquad \searrow \text{HI} \\ \text{Cyclohexyl-OAc} \qquad \text{Cyclohexene} \xrightarrow{I_2,\ AcOH} \text{trans-1-OAc-2-I-cyclohexane} \\ 32\% \qquad\qquad\qquad\qquad 68\% \end{array}$$

(1.21)

1.5.6
Oxidative Acylation of N-Heteroaromatic Bases

Protonated N-heteroaromatic bases are efficiently functionalized to the corresponding acyl derivatives by using aldehydes, as acyl radical sources, in the presence of NHPI and Co(II) salts under aerobic conditions [36] Eq. (1.22).

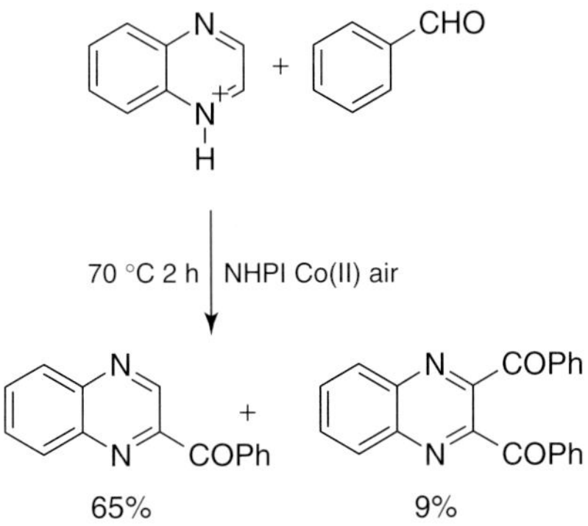

$$\text{Quinoxaline-H}^+ + \text{PhCHO} \xrightarrow[\text{NHPI Co(II) air}]{70\,°C\ 2\,h} \text{2-COPh-quinoxaline (65\%)} + \text{2,3-bis(COPh)-quinoxaline (9\%)}$$

(1.22)

Depending on the operative temperature, the direct aerobic oxidation of the aldehydes may afford carboxylic acids as by-products.

Quinazoline has an anomalous behavior compared with other aromatic bases: no acylation occurs, but 3H-quinazolin-1-one is the sole product Eq. (1.23). This behavior might be ascribed to the oxidation by the peracid, a possible intermediate formed from the aldehyde in the reaction media.

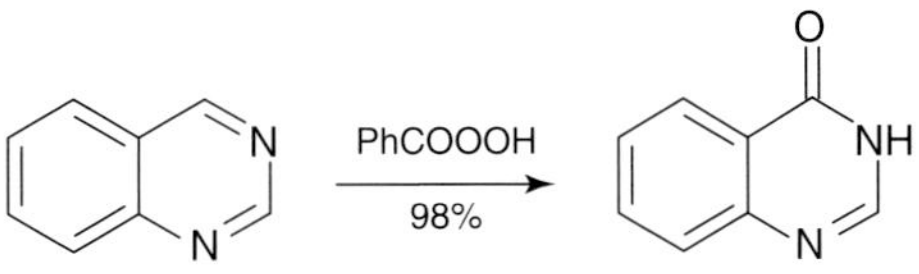

$$\text{Quinazoline} \xrightarrow[98\%]{\text{PhCOOOH}} 3H\text{-quinazolin-1-one}$$

(1.23)

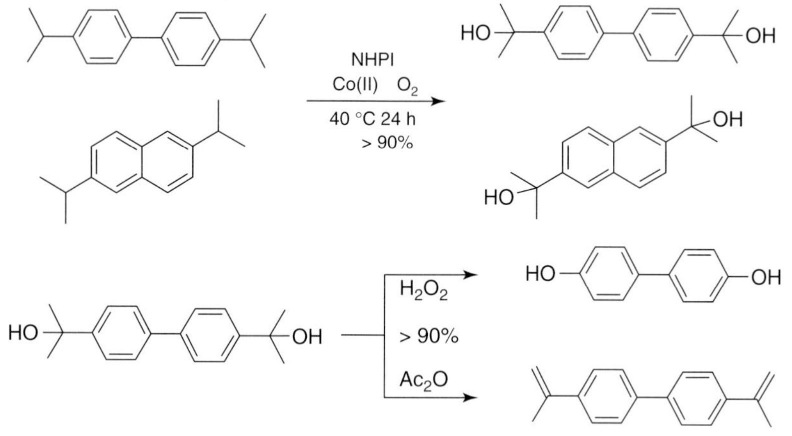

Scheme 1.3

1.5.7
Aerobic Synthesis of p-Hydroxybenzoic Acids and Diphenols

The aerobic oxidation of 4,4-diisopropyldiphenyl and 2,6-diisopropylnaphthalene, catalyzed by NHPI and Co(II) salts, leads to the corresponding tertiary benzyl alcohols with high conversion and selectivity. The latter are efficiently converted either to diphenols (useful in the production of liquid crystals) by reaction with H_2O_2 or to dienes (useful as cross-linking agents) by dehydration [37] (Scheme 1.3).

A screening in different solvents showed that low polar solvents such as chlorobenzene, are particularly convenient for the synthesis of cumyl alcohol. However, the low solubility of NHPI in these solvents does not allow high conversions. On the other hand, high polar solvents increase the solubility of NHPI but, at the same time, reduce the selectivity to benzyl alcohol. Therefore, a compromise has been achieved with acetonitrile, which grants a good solubility of NHPI and allows high yield of dibenzyl alcohol at low temperature.

1.5.8
Selective Halogenation of Alkanes

The strong polar effect due to the two carbonyl groups of NHPI plays a key role in the selective halogenation of alkanes. PINO is generated in the presence of a catalytic amount of HNO_3 when the reaction is carried out under aerobic conditions [38] Eq. (1.24).

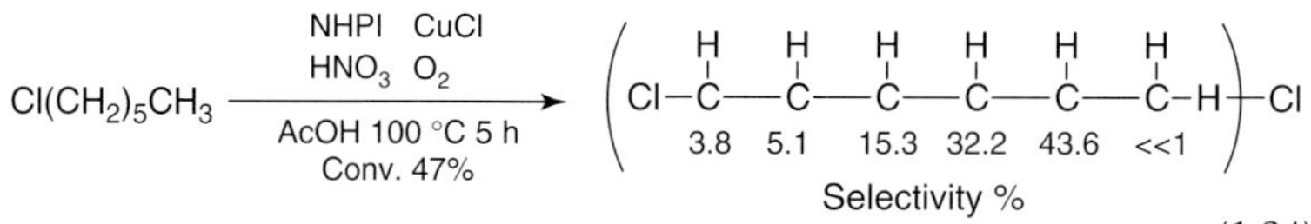

$$(1.24)$$

In the reactions catalyzed by NHPI, the chemoselectivity is much higher than in free radical halogenations by Cl_2: the introduction of an electron-withdrawing group determines a significant deactivation of the substrate, allowing selective monosubstitution even at considerable conversions. The quite different regioselectivity observed for chlorination in the presence of NHPI with respect to the one with Cl_2 in the absence of NHPI, suggests a high polar effect in H abstraction by PINO. Enthalpic effects also considerably affect the selectivity: the methyl group, despite being the less deactivated by a polar substituent, reacts only in traces, because of the higher BDE values of C–H bonds compared with those of $-CH_2-$ groups.

1.5.9
Aerobic Oxidation of Cycloalkanes to Diacids

The aerobic oxidation catalyzed by nitroxyl imides has been applied to the synthesis of aliphatic dicarboxylic acids. Great attention has been devoted to the production of adipic acid, an important intermediate in the synthesis of 6,6-nylon.

Ishii and Daicel Chemical Company patented a method for the direct aerobic oxidation of cyclohexane to adipic acid by using NHPI together with suitable metal salts as cocatalysts Eq. (1.25). The process is currently under evaluation at a pilot scale for further commercial applications [39a]. The best result claimed so far is a 73% conversion of cyclohexane with 73% selectivity for adipic acid.

$$\text{cyclohexane} \xrightarrow[\text{AcOH}]{\text{NHPI } O_2 \text{ Metal salts}} \text{cyclohexanone} + \text{adipic acid (53\%)} \qquad (1.25)$$

These processes are usually carried out in polar solvents such as acetic acid, acetonitrile, or ethyl acetate due to the low solubility of NHPI in nonpolar solvents. Thus, the use of 4-lauryloxycarbonyl-N-hydroxyphthalimide has given the possibility to perform the reaction directly in neat substrate [39b] Eq. (1.26).

$$\text{cyclohexane} \xrightarrow[\substack{O_2 \text{ Co(II) Mn(II)} \\ 100\,°C\ 14\,h}]{\text{CH}_3(\text{CH}_2)_{11}\text{-O-CO-phthalimide-NOH}} \text{cyclohexanol (27\%)} + \text{cyclohexanone (59\%)} + \text{adipic acid (7\%)} \qquad (1.26)$$

Sheldon et al. have reported the use of N-hydroxysaccharin (NHS), as an alternative to NHPI, in the oxidative catalysis of cycloalkanes to dicarboxylic acids. The mechanism is expected to be similar to that of NHPI catalysis [39c] Eq. (1.27).

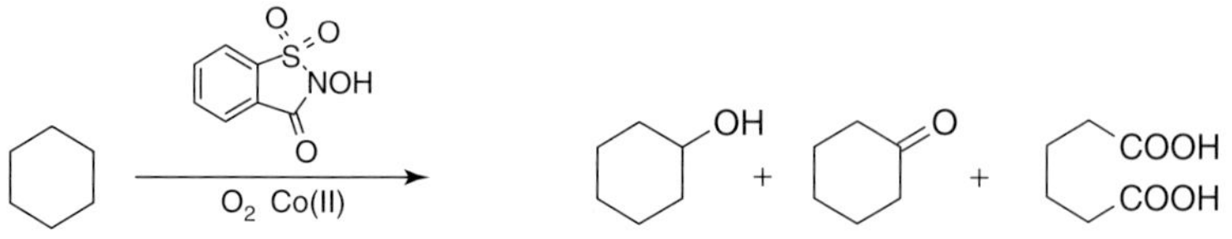

1 N-Hydroxy Derivatives

$$\text{cyclohexane} \xrightarrow[\text{O}_2 \text{ Co(II)}]{\text{NHS catalyst}} \text{cyclohexanol} + \text{cyclohexanone} + \text{adipic acid}$$

100 °C, 6 h
NHPI: conv. 58%
NHS: conv. 64%   Sel. %   5   29   30
                              8   31   16

50 °C, 24 h
NHPI: conv. 0%
NHS: conv. 42%   Sel. %   0   0   0
                              14   47   20

(1.27)

NHS shows greater catalytic activity than NHPI, especially at lower temperatures, because both an enhanced polar effect and an expected higher BDE of the O–H group in NHS hastens the H abstraction from hydrocarbons.

Recently, Xu et al. have reported an efficient metal-free aerobic oxidation of cyclohexane to adipic acid and cyclohexanone using catalytic amounts of NHPI in the presence of o-phenanthroline and Br_2 [39d] Eq. (1.28).

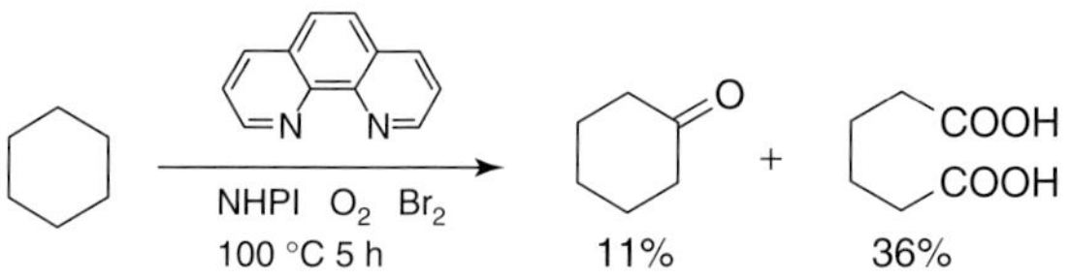

$$\text{cyclohexane} \xrightarrow[\text{100 °C 5 h}]{\text{NHPI O}_2 \text{ Br}_2 \text{/phenanthroline}} \text{cyclohexanone (11\%)} + \text{adipic acid (36\%)}$$

(1.28)

1.5.10
Epoxidation of Olefins

The induced homolysis of NHPI in the presence of peracids or dioxiranes has been employed to promote the aerobic oxidation of olefins to the corresponding epoxides Eq. (1.29) [40].

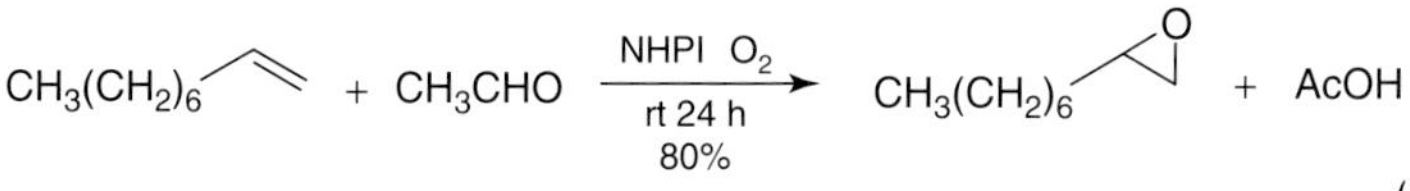

$$CH_3(CH_2)_6\text{-CH=CH}_2 + CH_3CHO \xrightarrow[\text{rt 24 h}]{\text{NHPI O}_2} CH_3(CH_2)_6\text{-epoxide} + AcOH$$
80%

(1.29)

Aldehydes in the presence of oxygen slowly give in situ formation of peracids, which promote the homolysis of NHPI leading to the formation of PINO and H_2O Eq. (1.30).

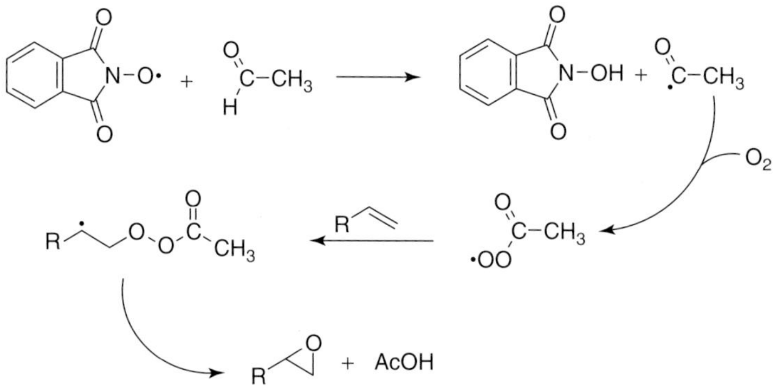

Scheme 1.4

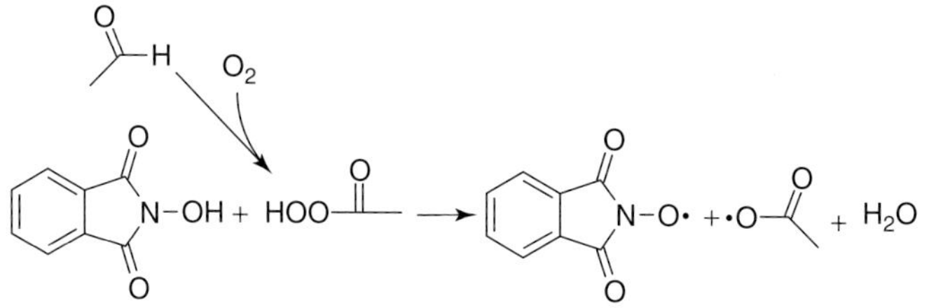

PINO abstracts the formyl-hydrogen from the aldehyde affording the corresponding acyl radical, which is fast trapped by oxygen. The resulting acylperoxyl radical adds to the double bond of the olefin leading to the formation of the epoxide (Scheme 1.4).

1.5.11
Oxidation of Alkylaromatics

The same catalytic system used in the epoxidation of olefins has been successfully applied to the oxidation of alkylaromatic compounds to the corresponding hydroperoxides. It is well known that, in the presence of metal salts, hydroperoxides undergo fast decomposition to the corresponding alcohols. In this case, the catalytic system acts under aerobic conditions in the absence of metal species, allowing hydroperoxides with high selectivity to be obtained.

This system has been applied in the oxidation of cumene to cumyl hydroperoxide (CHP), an important intermediate in the industrial production of phenol [41]. The reaction is carried out at a lower temperature compared with classic autoxidation processes (110–140 °C), affording a similar conversion of the reactant (up to 70% after 24 hours) and a selectivity up to 90% in CHP Eq. (1.31).

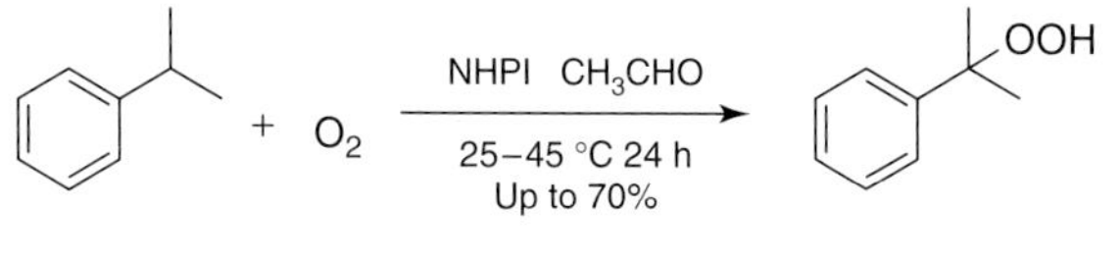

The high efficiency of the process is due to the fast reaction of the intermediate cumyloxyl radical with NHPI to give CHP and PINO. In this way, during the process, the concentration of peroxyl radical is kept very low and the chain termination is strongly disfavored Eq. (1.32).

$$\cdot OO\text{-CMe}_2\text{Ph} \xrightarrow[k = 7.2 \times 10^3 \text{ M}^{-1}\text{s}^{-1} \text{ at } 25\,°\text{C}]{\text{NHPI}} \text{PINO (N-O}\cdot\text{)} + \text{HOO-CMe}_2\text{Ph}$$

(1.32)

The same reaction, carried out with NHSI, instead of NHPI, led to no conversion.

1.6 Conclusions

In the last decade, NHDs were widely studied as organocatalysts for the development of oxidative processes worldwide and we intensely contributed to the design and investigation of the mechanistic aspects of oxidation processes involving NHDs.

However, in spite of the many results previously disclosed within this field, a thorough investigation related to such catalysts and catalytic processes is still mandatory. The potential of NHD catalysts is well documented, but the industrial exploitation of such catalysts is rather absent. This represents the real gap to be filled. Thus, the progress beyond the state of the art, which is expected in the future years, is to transform "interesting synthetic routes" into "winning industrial processes" by acting, in particular, on two fronts: (i) the main limitations to the employment of NHDs as catalysts for industrial processes are due to the fact that they are commonly considered expensive and instable. The reasons for this instability need to be exhaustively investigated and explained once and for all. This should induce to find out new NHDs, which might be employed under milder conditions, in order to be easily recovered and recycled; (ii) the cost and environmental demands, as well as the effect that the metals have on the instability of these catalysts, urge the development of cleaner metal-free routes for the activation of NHDs.

Nevertheless, it is apparent that NHDs will play a key role in the future regarding the development of oxidative industrial processes of strategic relevance.

Acknowledgments

We would like to sincerely thank our past and current coworkers, whose names are within the references, and especially Prof. Francesco Minisci and Prof. Ombretta Porta, who introduced us to the intriguing field of free radical chemistry. Support from MIUR (PRIN 2004 and PRIN 2006) and Polimeri Europa (Eni S.p.a.) is gratefully acknowledged.

References

1. Suresh, A.K., Sharma, M.M., and Sridhar, T.T. (2000) Engineering aspects of industrial liquid-phase air oxidation of hydrocarbons. *Ind. Eng. Chem. Res.*, **39**, 3958–3997.
2. (a) Simandi, L.L. (1991) *Dioxygen Activation and Homogeneous Catalytic Oxidation*, Elsevier, Amsterdam; (b) Parshall, G.W. and Ittel, S.D. (1992) *Homogeneous Catalysis*, 2nd edn, John Wiley & Sons, Inc., New York; (c) Barton, D.H.R., Martell, A.E., and Sawyer, D.T. (1993) *The Activation of Dioxygen and Homogeneous Catalytic Oxidation*, Plenum, New York; (d) Sheldon, R.A. and Kochi, J.K. (1981) *Metal-catalysed Oxidations of Organic Compounds*, Academic, New York.
3. Bäckvall, J.-E. (ed.) (2004) *Modern Oxidation Methods*, Wiley-VCH Verlag GmbH, Weinheim.
4. (a) Centi, G., Trifirò, F., and Cavani, F. (2001) *Selective Oxidation by Heterogeneous Catalysis*, Kluwer Academic/Plenum, New York; (b) Sheldon, R.A. and van Bekkum, H. (2001) *Fine Chemicals through Heterogeneous Catalysis*, Wiley-VCH Verlag GmbH, Weinheim.
5. (a) Zedda, A., Sala, M., and Schneider, A. (2002) Stable free nitroxyl radicals as oxidation catalysts and process for oxidation. WO02/058844; (b) Baucherel, X. and Sheldon, R.A. (2002) Catalytic oxidation process. WO02/100810; (c) Kühnle, A., Jost, C., Sheldon, R.A., Chatel, S., and Arends, I. (2003) Method for producing saturated alcohols, ketones, aldehydes and carboxylic acids. WO03/004447; (d) Kühnle, A., Duda, M., Sheldon, R.A., Sasidharan, M., Arends, I., Schiffer, T., Fries, G., and Kirchhoff, J. (2001) Method for oxidizing hydrocarbons. WO01/74742.
6. Lucarini, M., Ferroni, F., Pedulli, G.F., Gardi, S., Lazzari, D., Schlingloff, G., and Sala, M. (2007) Metal free in situ formation of phthalimide N-oxyl radicals by light-induced homolysis of N-alkoxyphthalimides. *Tetrahedron Lett.*, **48**, 5331–5334.
7. (a) Lucarini, M., Pedulli, G.F., and Cipollone, M. (1994) Bond dissociation enthalpy of α-tocopherol and other phenolic antioxidants. *J. Org. Chem.*, **59**, 5063–5070; (b) Lucarini, M., Pedulli, G.F., Pedrielli, P., Cabiddu, S., and Fattuoni, C. (1996) Bond dissociation energies of O-H bonds in substituted phenols from equilibration studies. *J. Org. Chem.*, **61**, 9259–9263; (c) Lucarini, M., Pedulli, G.F., Valgimigli, L., Amorati, R., and Minisci, F. (2001) Thermochemical and kinetic studies of a bisphenol antioxidant. *J. Org. Chem.*, **66**, 5456–5462; (d) Brigati, G., Lucarini, M., Mugnaini, V., and Pedulli, G.F. (2002) Determination of the substituent effect on the O-H bond dissociation enthalpies of phenolic antioxidants by the EPR radical equilibration technique. *J. Org. Chem.*, **67**, 4828–4832.
8. Amorati, R.R., Lucarini, M.M., Mugnaini, V.V., Pedulli, G.F.G. F., Minisci, F.F., Fontana, F.F., Recupero, F.F., Astolfi, P.P., and Greci, L.L. (2003) Hydroxylamines as oxidation catalysts: thermochemical and kinetic studies. *J. Org. Chem.*, **68**, 1747–1754.
9. de Nooy, A.E.J., Besemer, A.C., and van Bekkum, H. (1996) On the use of stable organic nitroxyl radicals for the oxidation of primary and secondary alcohols. *Synthesis*, **10**, 1153–1174.
10. (a) Sheldon, R.A., Arends, I.W.C.E., Brink, G.J., and Dijksman, A. (2002) Green, catalytic oxidations of alcohols. *Acc. Chem. Res.*, **35**, 774–781; (b) Dijksman, A., Arends, I.W.C.E., and Sheldon, R.A. (1999) Efficient ruthenium–TEMPO-catalysed aerobic oxidation of aliphatic alcohols into aldehydes and ketones. *Chem. Commun.*, 1591, 1592; (c) Dijksman, A., Arends, I.W.C.E., and Sheldon, R.A. (2001) The Ruthenium/TEMPO-catalysed aerobic oxidation of alcohols. *Platinum Metals Rev.*, **45**, 15; (d) Dijksman, A., Marino-Gonzalez, A., Mairata I Payeras, A., Arends, I.W.C.E., and Sheldon, R.A. (2001) Efficient and selective aerobic oxidation of alcohols into aldehydes and

ketones using Ruthenium/TEMPO as the catalytic system. *J. Am. Chem. Soc.*, **123**, 6826–6833.
11. Betzemeier, B.B., Cavazzini, M.M., Quici, S.S., and Knochel, P.P. (2000) Copper-catalysed aerobic oxidation of alcohols under fluorous biphasic conditions. *Tetrahedron Lett.*, **41**, 4343–4346.
12. Brink, G.J., Arends, I.W.C.E., and Sheldon, R.A. (2000) Green, catalytic oxidation of alcohols in water. *Science*, **287**, 1636–1639.
13. (a) Recupero, F. and Punta, C. (2007) Free radical functionalization of organic compounds catalysed by N-hydroxyphthalimide. *Chem. Rev.*, **107**, 3800–3842; (b) Minisci, F., Punta, C., and Recupero, F. (2006) Mechanisms of the aerobic oxidations catalysed by N-hydroxyderivatives. Enthalpic, polar and solvent effects, "molecule-induced homolysis" and synthetic involvements. *J. Mol. Catal. A Chem.*, **251**, 129–149.
14. Ishii, Y., Sakaguchi, S., and Iwahama, T. (2001) Innovation of hydrocarbon oxidation with molecular oxygen and related reactions. *Adv. Synth. Catal.*, **343**, 393–427.
15. (a) Fukuda, O., Sakaguchi, S., and Ishii, Y. (2001) A new strategy for catalytic Baeyer–Villiger oxidation of KA-oil with molecular oxygen using N-hydroxyphthalimide. *Tetrahedron Lett.*, **42**, 3479–3481; (b) Aoki, Y., Hirai, N., Sakaguchi, S., and Ishii, Y. (2005) Aerobic oxidation of 1,3,5-triisopropylbenzene using N-hydroxyphthalimide (NHPI) as key catalyst. *Tetrahedron*, **61**, 10995–10999; (c) Foricher, J., Furbringer, C., and Pfoertner, K. (1986) Catalytic oxidation with N-hydroxydicarboxylic acid imides. EP Patent 0198351; (d) Foricher, J., Furbringer, C., and Pfoertner, K. (1991) Process for the catalytic oxidation of isoprenoids having allylic groups. US Patent 5,030,739.
16. Minisci, F., Recupero, F., Punta, C., Gambarotti, C., Antonietti, F., Fontana, F.F., and Pedulli, G.F. (2002) A novel, selective free-radical carbamoylation of heteroaromatic bases by Ce(IV) oxidation of formamide, catalysed by N-hydroxyphthalimide. *Chem. Commun.*, 2496–2497.
17. Einhorn, C., Einhorn, J., Marcadal, C., and Pierre, J.-L. (1997) Oxidation of organic substrates by molecular oxygen mediated by N-hydroxyphthalimide (NHPI) and acetaldehyde. *Chem. Commun.*, 447–448.
18. Tong, X., Xu, J., and Miao, H. (2005) Highly efficient and metal-free aerobic hydrocarbons oxidation process by an o-phenanthroline-mediated organocatalytic system. *Adv. Synth. Catal.*, **347**, 1953–1957.
19. (a) Baiocco, P., Barreca, A.M., Fabbrini, M., Galli, C., and Gentili, P. (2003) Promoting laccase activity towards non-phenolic substrates: a mechanistic investigation with some laccase–mediator systems. *Org. Biomol. Chem.*, **1**, 191–197; (b) Astolfi, P., Brandi, P., Galli, C., Gentili, P., Gerini, M.F., Greci, L., and Lanzalunga, O. (2005) New mediators for the enzyme laccase: mechanistic features and selectivity in the oxidation of non-phenolic substrates. *New J. Chem.*, **29**, 1308–1317.
20. Sheldon, R.A. and Arends, I.W.C.E. (2004) Organocatalytic oxidations mediated by nitroxyl radicals. *Adv. Synth. Catal.*, **346**, 1051–1071.
21. Punta, C., Rector, C.L., and Porter, N.A. (2005) Peroxidation of polyunsaturated fatty acid methyl esters catalysed by N-methyl benzohydroxamic acid: a new and convenient method for selective synthesis of hydroperoxides and alcohols. *Chem. Res. Toxicol.*, **18**, 349–356.
22. Cecchetto, A., Fontana, F., Minisci, F., and Recupero, F. (2001) Efficient Mn– Cu and Mn– Co– TEMPO-catalysed oxidation of alcohols into aldehydes and ketones by oxygen under mild conditions. *Tetrahedron Lett.*, **42**, 6651–6653.
23. Minisci, F., Recupero, F., Cecchetto, A., Gambarotti, C., Punta, C., Faletti, R., Paganelli, R., and Pedulli, G.F. (2004) Mechanisms of the aerobic oxidation of alcohols to aldehydes and ketones, catalysed under mild conditions by persistent and non-persistent nitroxyl radicals and transition metal salts. polar, enthalpic, and captodative effects. *Eur. J. Org. Chem.*, **1**, 109–119.

24. Minisci, F., Fumagalli, C., and Pirola, R. (2001) Process for the preparation of carboxylic acids. WO01/58845A1.
25. (a) Minisci, F., Recupero, F., Pedulli, G.F., and Lucarini, M. (2003) Transition metal salts catalysis in the aerobic oxidation of organic compounds: Thermochemical and kinetic aspects and new synthetic developments in the presence of N-hydroxy-derivative catalysts. *J. Mol. Catal. A Chem.*, **204-205**, 63–90; (b) Minisci, F., Recupero, F., Fontana, F., Bjørsvik, H.R., and Liguori, L. (2002) Highly selective and efficient conversion of alkyl aryl and alkyl cyclopropyl ketones to aromatic and cyclopropane carboxylic acids by aerobic catalytic oxidation: a free-radical redox chain mechanism. *Synlett*, 610–612.
26. (a) Benaglia, M., Puglisi, A., and Cozzi, F. (2003) Polymer-supported organic catalysts. *Chem. Rev.*, **103**, 3401; (b) Cozzi, F. (2006) Immobilization of organic catalysts: when, why, and how. *Adv. Synth. Catal.*, **348**, 1367.
27. (a) Fey, T., Fischer, H., Bachmann, S., Albert, K., and Bolm, C. (2001) Silica-supported TEMPO catalysts: synthesis and application in the Anelli oxidation of alcohols. *J. Org. Chem.*, **66**, 8154; (b) Bolm, C. and Fey, T. (1999) TEMPO oxidations with a silica-supported catalyst. *Chem. Commun.*, 1795; (c) Verhoef, M.J., Peters, J.A., and van Bekkum, H. (1999) MCM-41 supported TEMPO as an environmentally friendly catalyst in alcohol oxidation. *Stud. Surf. Sci. Catal.*, **125**, 465–472; (d) Ciriminna, R., Blum, J., Avnir, D., and Pagliaro, M. (2000) Sol– gel entrapped TEMPO for the selective oxidation of methyl α-D-glucopyranoside. *Chem. Commun.*, 1441; (e) Dijksman, A., Arends, I.W.C.E., and Sheldon, R.A. (2000) Polymer immobilised TEMPO (PIPO): an efficient catalyst for the chlorinated hydrocarbon solvent-free and bromide-free oxidation of alcohols with hypochlorite. *Chem. Commun.*, 271.
28. Karimi, B., Biglari, A., Clark, J.H., and Budarin, V. (2007) Green, transition-metal-free aerobic oxidation of alcohols using a highly durable supported organocatalyst. *Angew. Chem. Int. Ed.*, **46**, 7210–7213.
29. Minisci, F., Recupero, F., Rodinò, M., Sala, M., and Schneider, A. (2003) A convenient nitroxyl radical catalyst for the selective oxidation of primary and secondary alcohols to aldehydes and ketones by O_2 and H_2O_2 under mild conditions. *Org. Process Res. Dev.*, **7**, 794–798.
30. Minisci, F., Punta, C., Recupero, F., Fontana, F., and Pedulli, G.F. (2002) A new, highly selective synthesis of aromatic aldehydes by aerobic free-radical oxidation of benzylic alcohols, catalysed by N-hydroxyphthalimide under mild conditions. Polar and enthalpic effects. *Chem. Commun.*, 688–689.
31. Annunziatini, C., Gerini, M.F., Lanzalunga, O., and Lucarini, M. (2004) Aerobic oxidation of benzyl alcohols catalysed by aryl substituted N-hydroxyphthalimides. Possible involvement of a charge-transfer complex. *J. Org. Chem.*, **69**, 3431–3438.
32. Minisci, F., Recupero, F., Punta, C., Guidarini, C., Fontana, F., and Pedulli, G.F. (2002) A new, highly selective, free-radical aerobic oxidation of silanes to silanols catalysed by N-hydroxyphthalimide under mild conditions. *Synlett*, **7**, 1173–1175.
33. Minisci, F., Punta, C., Recupero, F., Fontana, F., and Pedulli, G.F. (2002) Aerobic oxidation of N-alkylamides catalysed by N-hydroxyphthalimide under mild conditions. Polar and enthalpic effects. *J. Org. Chem.*, **67**, 2671–2676.
34. Cecchetto, A., Minisci, F., Recupero, F., Fontana, F., and Pedulli, G.F. (2002) A new selective free radical synthesis of aromatic aldehydes by aerobic oxidation of tertiary benzylamines catalysed by N-hydroxyimides and Co(II) under mild conditions. Polar and enthalpic effects. *Tetrahedron Lett.*, **43**, 3605–3607.
35. Minisci, F., Recupero, F., Gambarotti, C., Punta, C., and Paganelli, R. (2003) Selective functionalisation of hydrocarbons by nitric acid and aerobic oxidation catalysed by N-hydroxyphthalimide and iodine under

mild conditions. *Tetrahedron Lett.*, **44**, 6919–6922.

36. Minisci, F., Recupero, F., Cecchetto, A., Punta, C., Gambarotti, C., Fontana, F.F., and Pedulli, G.F. (2003) Polar effects in free-radical reactions. a novel homolytic acylation of heteroaromatic bases by aerobic oxidation of aldehydes, catalysed by N-hydroxyphthalimide and Co salts. *J. Heterocycl. Chem.*, **40**, 235–328.

37. Minisci, F., Recupero, F., Cecchetto, A., Gambarotti, C., Punta, C., Paganelli, R., Pedulli, G.F., and Fontana, F. (2004) Solvent and temperature effects in the free radical aerobic oxidation of alkyl and acyl aromatics catalysed by transition metal salts and N-hydroxyphthalimide: new processes for the synthesis of p-hydroxybenzoic acid, diphenols, and dienes for liquid crystals and cross-linked polymers. *Org. Process Res. Dev.*, **8**, 163–168.

38. Minisci, F., Porta, O., Recupero, F., Gambarotti, C., Paganelli, R., Pedulli, G.F., and Fontana, F. (2004) New free-radical halogenations of alkanes, catalysed by N-hydroxyphthalimide. Polar and enthalpic effects on the chemo- and regio-selectivity. *Tetrahedron Lett.*, **45**, 1607–1609.

39. (a) (2004) Daicel Chemical employs NHPI catalyst method for adipic acid. *Focus Catal.*, **1**, 7; (b) Sawatari, N., Yokota, T., Sakaguchi, S., and Ishii, Y. (2001) Alkane oxidation with air catalysed by lipophilic N-hydroxyphthalimides without any solvent. *J. Org. Chem.*, **66**, 7889–7891; (c) Baucherel, X., Gonsalvi, L., Arends, I.W.C.E., Ellwood, S., and Sheldon, R.A. (2004) Aerobic oxidation of cycloalkanes, alcohols and ethylbenzene catalysed by the novel carbon radical chain promoter NHS (N-hydroxysaccharin). *Adv. Synth. Catal.*, **346**, 286–296; (d) Tong, X., Xu, J., and Miao, H. (2005) Highly efficient and metal-free aerobic hydrocarbons oxidation process by an o-phenanthroline-mediated organocatalytic system. *Adv. Synth. Catal.*, **347**, 1953–1957.

40. (a) Minisci, F., Gambarotti, C., Pierini, M., Porta, O., Punta, C., Recupero, F., Lucarini, M., and Mugnaini, V. (2006) Molecule-induced homolysis of N-hydroxyphthalimide (NHPI) by peracids and dioxirane. A new, simple, selective aerobic radical epoxidation of alkenes. *Tetrahedron Lett.*, **47**, 1421–1424; (b) Punta, C., Moscatelli, D., Porta, O., Minisci, F., Gambarotti, C., and Lucarini, M. (2008) Selective aerobic radical epoxidation of α-olefins catalysed by N-hydroxyphthalimide, in *Mechanisms in Homogeneous and Heterogeneous Epoxidation Catalysis* (ed. S.T.Oyama), Elsevier, pp. 217–229.

41. Minisci, F., Porta, O., Punta, C., Recupero, F., Gambarotti, C., and Pierini, M. (2008) Process for the preparation of phenol by means of new catalytic systems. WO2008037435A1.

2
Gold-Catalyzed Intra- and Intermolecular Cycloadditions of Push–Pull Dienynes

Manuel A. Fernández-Rodríguez

2.1
Introduction

The development of new methodologies that allow the efficient and selective synthesis of complex target molecules from simple and readily available starting materials is a fundamental goal in modern organic chemistry. Transition metal–catalyzed reactions represent an essential pillar among the strategies for this aim. These processes have radically changed the methodologies in organic chemistry and their applications include both the development of new materials and industrial processes and the total synthesis of biologically active compounds.

Reactions mediated by metals such as palladium, nickel, ruthenium, and rhodium have been extensively studied and described in the literature [1]. More recently, platinum and coinage metals have shown a significant relevance as catalysts in the activation of carbon–carbon multiple bonds. In particular, gold derivatives have proven to be the most active catalysts for transformations concerning triple bonds [2]. Gold catalysts act as soft carbophilic Lewis acids promoting the addition of soft nucleophiles to alkynes in carbon–carbon and carbon–heteroatom bond forming reactions. As a result, an impressive array of new gold-catalyzed processes has been described and reviewed [3–5]. The extraordinary activity of these catalytic species can be explained in terms of basic principles in frontier orbitals, relativistic effects [6], and π-acidity [7].

In particular, cycloisomerization reactions have been thoroughly studied and developed in practical methodologies to form a broad range of carbo- and heterocycles. However, gold-catalyzed intermolecular cycloadditions have been scarcely described, although they are synthetically more attractive.

The aim of this chapter is to highlight our recent research in gold-catalyzed reactions of push–pull conjugated dyenines. The ability of these activated substrates to react with nucleophiles has resulted in the development of novel cycloaromatization processes and the discovery of new intermolecular cycloaddition reactions. In addition, a brief overview of relevant aspects related to our research in gold-catalyzed enyne cycloisomerization and intermolecular cycloaddition reactions is presented.

Ideas in Chemistry and Molecular Sciences: Advances in Synthetic Chemistry. Edited by Bruno Pignataro
Copyright © 2010 WILEY-VCH Verlag GmbH & Co. KGaA, Weinheim
ISBN: 978-3-527-32539-9

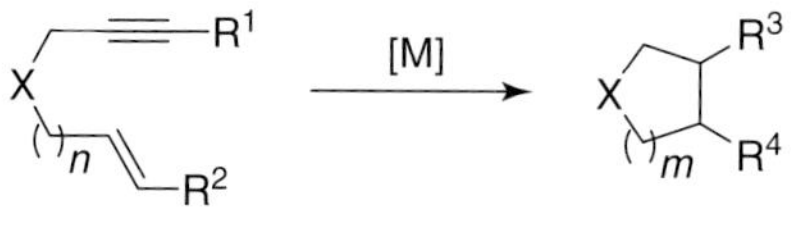

[M] = Pd, Ru, Rh, Ir, Hg, Ti, Cr, Fe, Co, Ni, Ga, In, Cu, Ag, Pt, Au

Figure 2.1 Metal-catalyzed cycloisomerization of 1,n-enynes.

2.2
Gold-Catalyzed Enyne Cycloisomerizations

2.2.1
General Remarks

The transition metal–catalyzed skeletal rearrangement of 1,n-enynes is a fundamental strategy to access functionalized carbo- and heterocyclic compounds. Trost and coworkers first reported an Alder-ene reaction mediated by palladium, and several catalytic systems, including a wide range of transition metals have now been identified for this kind of rearrangements (Figure 2.1) [8]. Among them, gold complexes have shown and extraordinary activity promoting these transformations due to a superior Lewis acidity compared to other metals and, consequently, these rearrangements typically occur under mild conditions and with excellent chemoselectivity [9].

Gold-catalysed enyne cycloisomerizations are initiated by metal coordination to the triple bond to form species **1** (Figure 2.2). These intermediates can evolve by different cyclization pathways depending on the reaction conditions, the structure of the starting 1,n-enyne, as well as on the presence of additional functional groups or nucleophiles. As a result, an extraordinary range of synthetically useful cyclic compounds have been synthesized. Figure 2.2 shows the diversity of cycloadducts that can be obtained from 1,5-enynes. Remarkably, reactions of related 1,6-enynes usually occur by distinct pathways allowing the synthesis of different cyclic structures [9].

2.2.2
Influence of the Electronic Nature of the Alkyne Substituent

The selectivity of gold-catalyzed skeletal rearrangements of enynes is highly dependent on the structure of the starting unsaturated substrates. Therefore, the steric influence of substituents in several parts of the molecule as well as the presence of heteroatoms and nucleophiles in the tether has been studied in detail [8, 9]. On the other hand, enol ethers, silylenol ethers, enamines, as well as furans or indoles have been employed as the nucleophilic counterpart in this type of transformations [10]. However, reactions of substrates containing a heteroatom directly attached to the triple bond of the enyne have been scarcely developed.

In this sense, Kozmin and coworkers first described the cycloisomerization of enynes bearing a triisopropylsilyloxy group in the alkyne moiety. Thus,

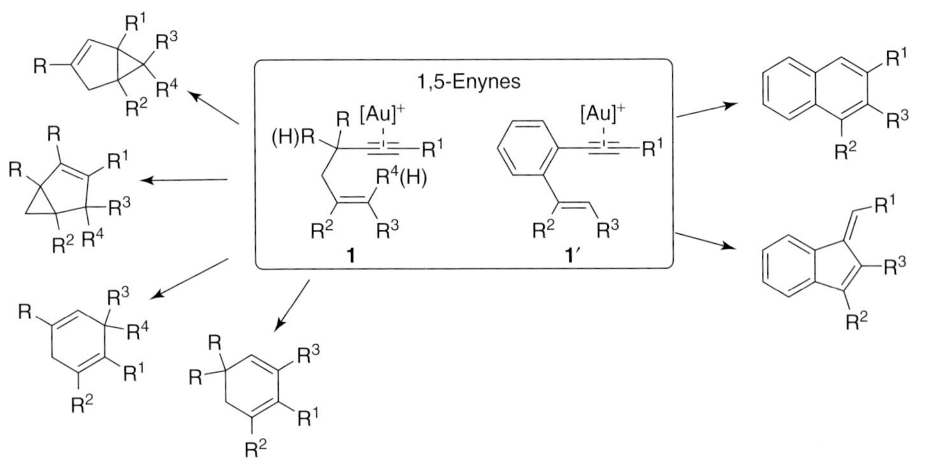

Figure 2.2 Gold-catalyzed cycloisomerization of 1,5-enynes.

1-siloxy-5-en-1-ynes **2** reacted, at room temperature and in the presence of catalytic amounts of AuCl, to furnish 1,4- and/or 1,3-cyclohexadienes **3**, depending on the substitution pattern of the starting enyne, with broad scope and generally in high yields (Scheme 2.1) [11].

The reaction is proposed to occur via a novel reaction mechanism that involves a cascade of 1,2-alkyl shifts (Scheme 2.1). The presence of the electron-donating substituent at the triple bond is critical and presumably responsible for the stabilization of cationic intermediate **B**.

The authors have studied these transformations and mechanism in detail and, consequently, the process could be extended to nonactivated 1,5-enynes **4** having an alkyl-, H-, or aryl-substituted alkyne, provided that the initial substrate contains a 5,5-disubstituted olefin and a quaternary center at C(3) (Scheme 2.1) [12]. However, these reactions required higher temperatures and catalyst loadings of $PtCl_2$ and afforded lower yields of cycloadducts **5** than the corresponding cycloisomerization of the silyloxy derivatives.

On the contrary, reaction of terminal enyne **6** possessing a hydrogen atom at C(3) afforded bicyclo[3.1.0]hexene **7** (Scheme 2.2) as a result of a more conventional rearrangement initiated by a 1,2-hydrogen shift (Scheme 2.2, via intermediate **D**) [13].

On the other hand, gold-catalyzed cycloisomerization of ene-ynamide **8** has also been developed [14]. This process leads to the formation of cyclobutanone derivative **9** as a major product, albeit in moderate yield (Scheme 2.3). However, the analogous reaction of the nonactivated 1,6-enyne **10**, which is an isomer of **8** that presents the sulfonamide group in the tether but not directly linked to the triple bond, furnished a six-membered cycloadduct **11** as a result of an *endo*-skeletal rearrangement (Scheme 2.3) [15]. The different ring opening pathways of the initially formed cyclopropyl intermediates **E** or **G'** could be explained by the stabilization effect of

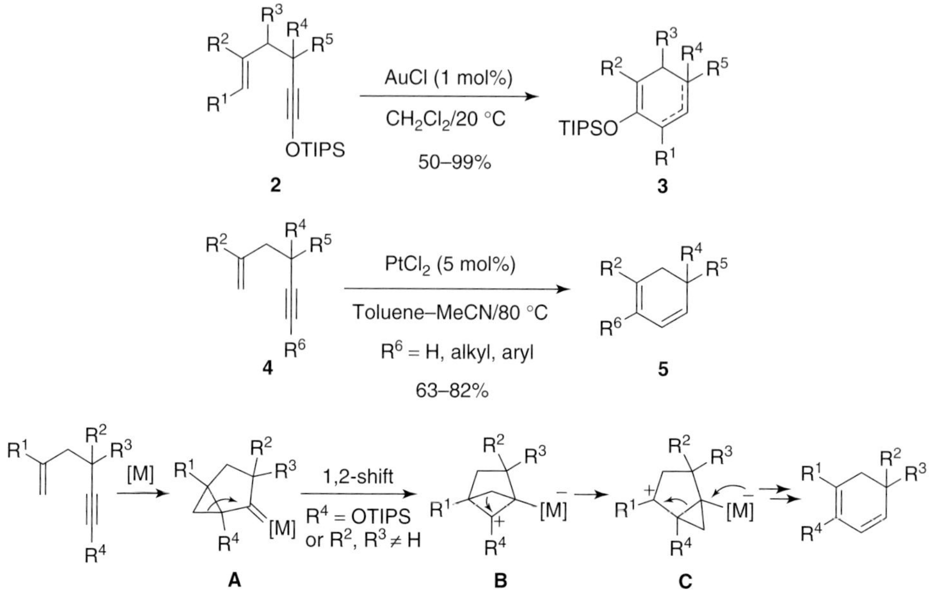

Scheme 2.1 Gold- and platinum-catalyzed cycloisomerization of 1,5-enynes **2** and **4** to cyclohexadienes **3** and **5**.

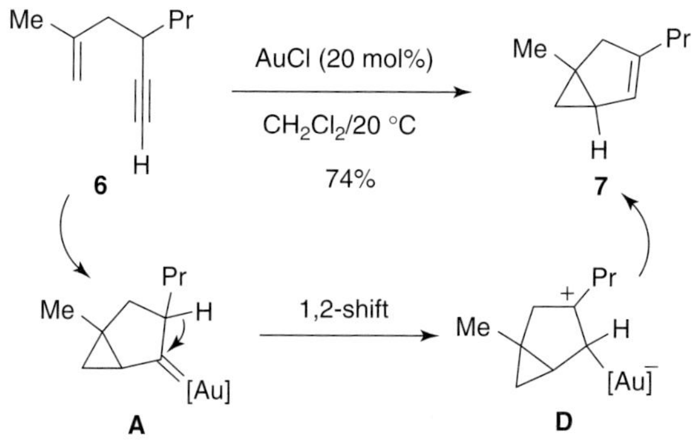

Scheme 2.2 Gold-catalyzed isomerization of terminal 1,5-enyne **6**.

the heteroatom in the cationic intermediate **F**, which is only possible in the first case (Scheme 2.3).

More recently, the influence of an electron-donating group linked to the triple bond was also pointed out by Hashmi et al. in the synthesis of dihydroindole, dihydrobenzofuran, chroman, and tetrahydroquinoline derivatives (Scheme 2.4) [16]. Remarkably, the stabilizing effect of the heteroatom allows these reactions to occur in shorter reaction times and with higher selectivities than the corresponding reactions of enynes bearing nonhetereoatom-substituted alkynes.

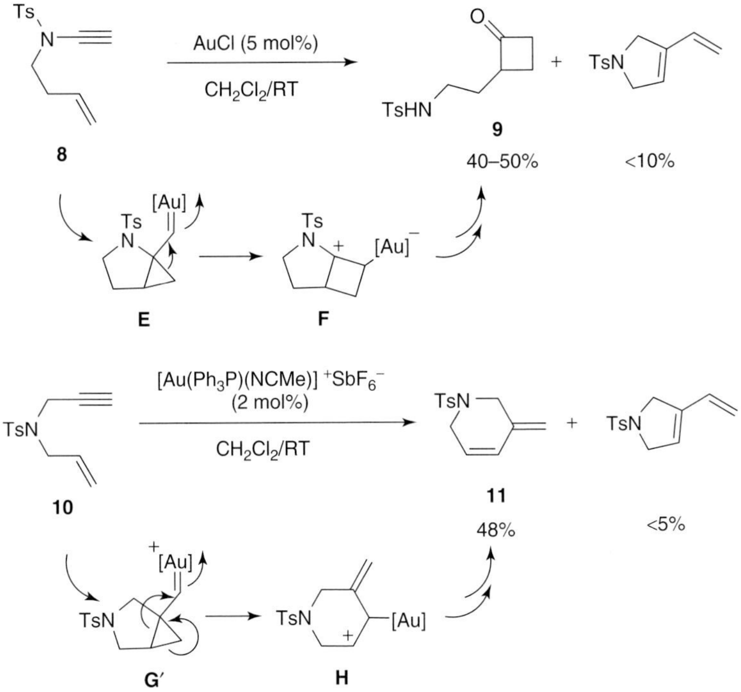

Scheme 2.3 Gold-catalyzed reactions and proposed mechanisms for the cycloisomerization of ene-ynamide **8** and enyne **10**.

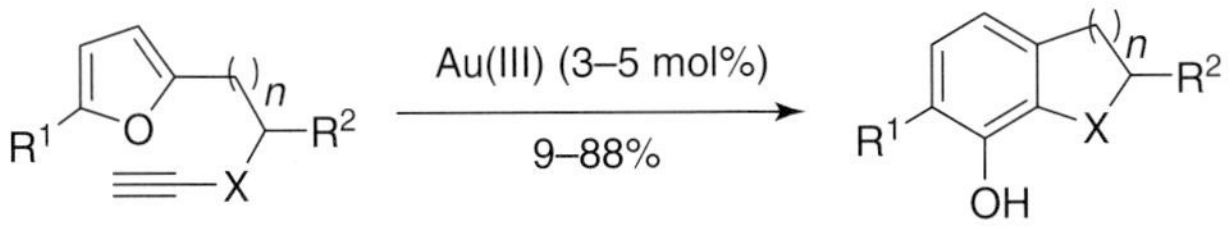

Scheme 2.4 Gold-catalyzed synthesis of dihydroindole, dihydrobenzofuran, chroman, and tetrahydroquinolines.

2.2.3
Gold-Catalyzed Cycloaromatization of Push–Pull Dienyne Acids: Synthesis of 2,3-Disubstituted Phenols

All the transformations depicted in the previous section reveal the high influence of the electronic nature of the substitution in the alkyne terminus of the enyne in these catalyzed processes. In this sense, we have recently described a highly efficient and simple procedure for the synthesis of push–pull dienynes (by reaction of Fischer alkynyl carbene complexes with commercially available 2-methoxy- and

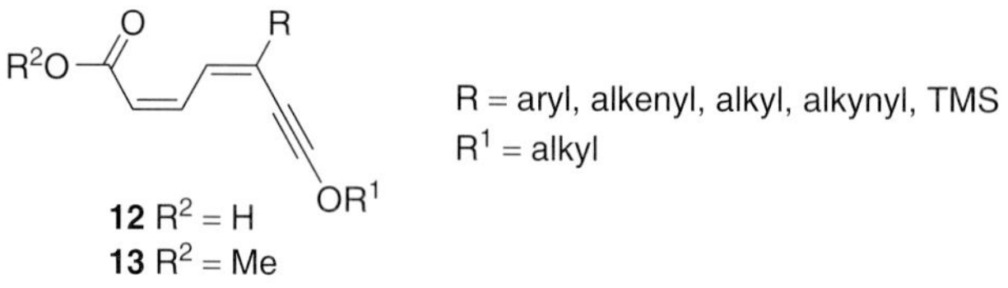

Figure 2.3 Push–pull 2,4-dien-6-yne acids and esters.

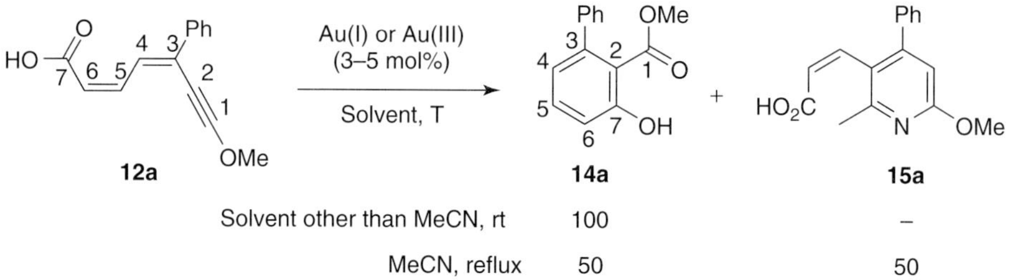

Scheme 2.5 Gold-catalyzed reactions of push–pull dienyne carboxylic acid **12a**.

	14a	15a
Solvent other than MeCN, rt	100	–
MeCN, reflux	50	50

2-trimethylsilyloxyfuran), which present an alkoxy group directly attached to the triple bond and a conjugated acid or ester functionality in the other side of the molecule (Figure 2.3) [17].

Therefore, we envisioned that these new activated conjugated dienynes **12** and **13** could be appropriate substrates for the development of new gold-catalyzed processes.

Reaction of 2,4-dien-6-yne carboxylic acid **12a** in the presence of gold(I) and gold(III) catalysts occurs at room temperature in various solvents to afford 2,3-disubstituted phenol **14a** (Scheme 2.5). The cycloaromatization of **12a** to form **14a** follows a novel reaction pattern where the new carbon–carbon bond is formed between carbons 2 and 7 of the π-conjugated system. Remarkably, this novel rearrangement is very different to the classical cyclization reactions of conjugated polyenynes that involve 6π electrons, such as Bergman, Saito-Myers, or Moore cycloaromatizations, where the new bond is created between carbons 1 and 6 [18]. Interestingly, reaction in acetonitrile at reflux provides an equimolecular mixture of the expected phenol **14a** and pyridine **15a** that result from the formal [4+2] cycloaddition of the dienyne and the nitrile (Scheme 2.5). This intermolecular transformation is disclosed in the following section.

Captodative dienyne acid **12a** was selected as a model system to assess the catalyst activity and to determine the optimum reaction conditions. Thus, several complexes derived from gold and platinum were able to catalyze the novel cycloaromatization. Gold(I) cationic complexes, generated *in situ* with silver salts, are superior catalysts than neutral gold or platinum complexes. Among them, the system AuClP(p-CF$_3$-C$_6$H$_4$)$_3$/AgSbF$_6$ exhibited the best results. Notably, two crucial factors to achieve good reaction yields were (i) premixing the gold catalyst

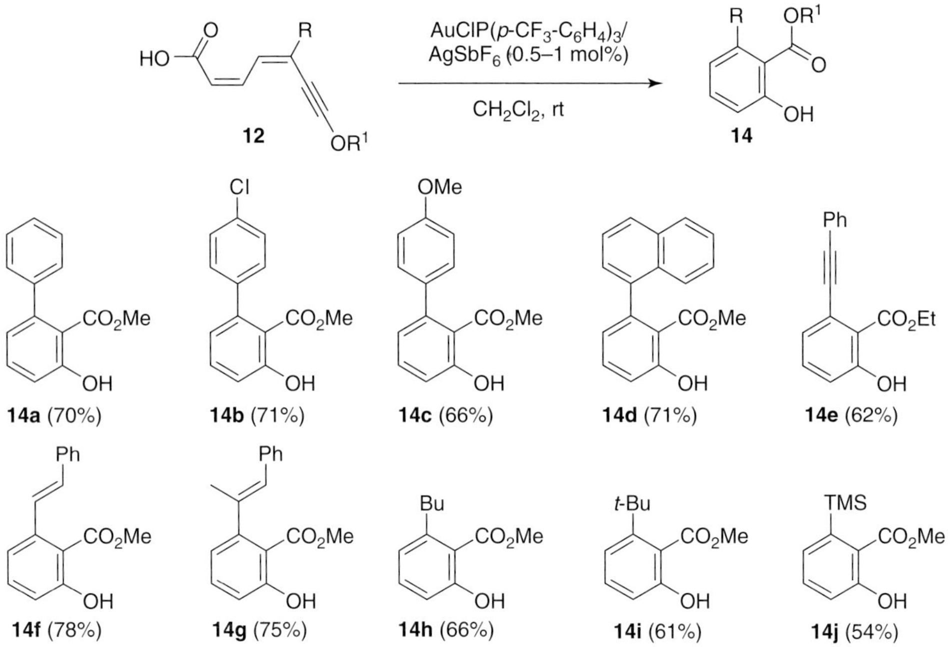

Scheme 2.6 Gold-catalyzed synthesis of 2,3-disubstituted phenols: scope of the cyclization.

with the silver salt and (ii) the premixing time – an optimum 30 minutes time leads to the highest yield observed.

It is worth to point out the high activity of the catalytic system for this transformation that allows the use of low catalyst loadings and mild reaction conditions. Therefore, the catalyst loading could be reduced to 0.5 mol% affording phenol **14a** in nearly the same yield than reactions in the presence of 3 mol% catalyst. Such catalyst loading is one order of magnitude lower than the amount employed in most of the gold-catalyzed processes.

The 2,7-cycloaromatization of push–pull dienyne carboxylic acids proved to be general. Thus, the reaction works nicely for R being an aromatic ring, with both electron-withdrawing and electron-donating substituents. It also leads to good yields for alkynyl, alkenyl, linear or branched alkyl, or silyl-substituted dienyne carboxylic acids (Scheme 2.6) [19]. Notably, the cycloadducts **14** obtained are, in fact, five-substituted salicylic esters that are usually difficult to prepare due to steric hindrance and are present in several natural or synthetic compounds with relevant biological activity.

On the other hand, and in contrast to gold-catalyzed cycloisomerization of nonheteroatom-substituted acetylenic carboxylic acids [20], total regioselectivity was observed in this novel 2,7-cycloaromatization. Further, the mild conditions (rt and 0.5–1.0 mol% catalyst) and the short reaction times (5–30 minutes) are especially noteworthy compared to related cyclizations of nonconjugated alkynyl ketones, which require 2.0–5.0 mol% catalyst and heating at 100 °C [21].

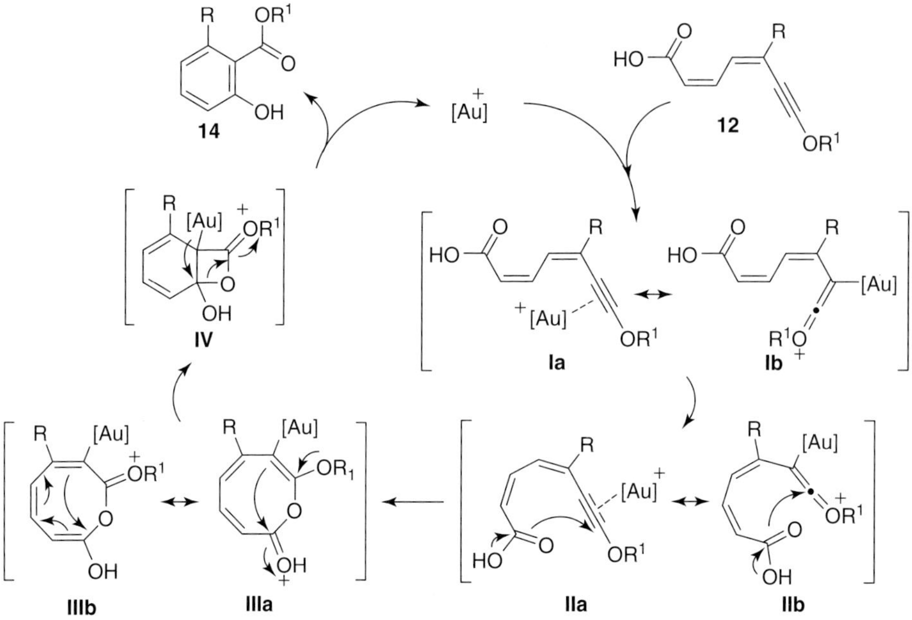

Scheme 2.7 Proposed mechanism for the 2,7-cycloaromatization of captodative dienyne carboxylic acids.

A mechanism that would explain the formation of phenols **14** is depicted in Scheme 2.7. Initial gold catalyst coordination to the triple bond to form intermediate **Ia**, which is stabilized by the electron-donating ability of the alkoxy group through resonance structure **Ib**, is followed by an *s-trans-s-cis* isomerization, leading to intermediates **IIa** or **IIb**. An intramolecular regioselective nucleophilic attack of the carboxylic group to the more electrophilic of the two sp carbons should take place leading to eight-membered cyclic intermediates **III**. It must be noticed the role of the alkoxy group in promoting such cyclization as for dienyne acid **12e**, which exhibits two triple bonds of different electronic nature, the nucleophilic attack of the carboxylic group occurs chemoselectively to the ethoxy-substituted triple bond. The alkoxy group would also promote the evolution of intermediate **IIIa** by an intramolecular attack to the activated carbonyl group, forming bicyclic species **IV**. Alternatively, a 6π-electrocyclization through intermediate **IIIb** may take place, leading to **IV**. A final aromatization would trigger the ring opening of the four-membered ring in intermediate **IV** to give the final product **14**, and would regenerate the catalyst, which may be incorporated into a new cycle. The catalytic cycle in Scheme 2.7 represents the first transition metal–catalyzed cycloaromatization reaction of dienyne carboxylic acids.

In order to analyze the electronic requirements of the transformation, dienyne carboxylic acids **16** and **17** that present a *p*-methoxyphenyl and a phenyl groups, respectively, linked to the triple bond, instead of an alkoxy group, were prepared

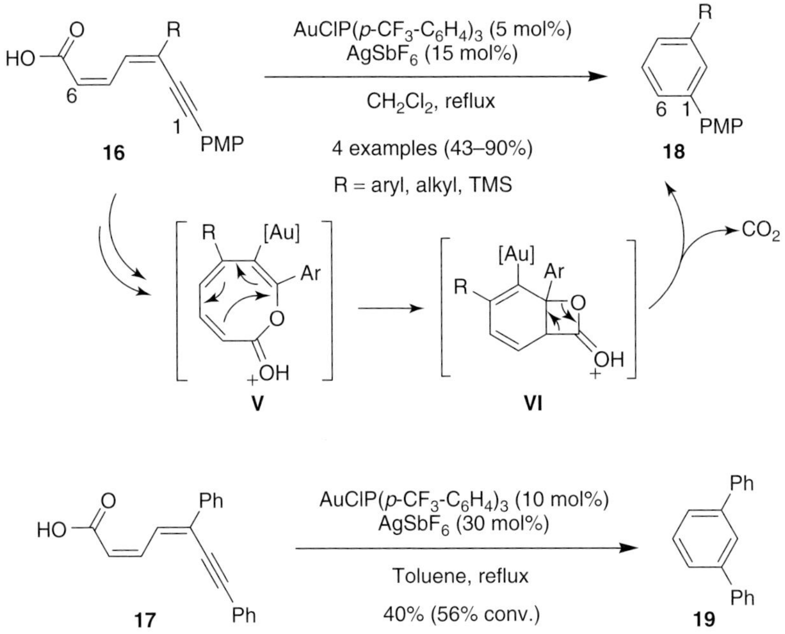

Scheme 2.8 Gold-catalyzed 1,6-cycloaromatization of dienyne acids and proposed mechanism.

(Scheme 2.8). The strong electron-donating effect displayed by the methoxy group in comparison with the *p*-methoxyphenyl and the phenyl groups is clearly pointed out by the ^{13}C NMR shifts of the β-alkyne carbon in compounds **12** (33.9–39.5 ppm) versus **16** (84.3–87.3 ppm) and **17** (85.8 ppm).

When conjugated dienyne acids **16** were treated under the typical reaction conditions described in Scheme 2.6, no traces of phenol derivatives were detected. Instead, the formation of unsymmetrical biphenyls or *m*-terphenyls **18** was observed. These cycloadducts result from a sequence involving cyclization, by creation of a bond between carbons 1 and 6, and decarboxylation. Consequently, the 2,7-cycloaromatization is absolutely dependent on the electronic nature of the dienyne acid, a strong electron-donating group linked to the triple bond appears to be a requisite to promote the transformation.

The 1,6-cyclization–decarboxylation sequence also proved to be general, although higher catalyst loadings [AuClP(*p*-CF$_3$-C$_6$H$_4$)$_3$ (5 mol%)/AgSbF$_6$ (15 mol%)] and temperature (refluxing CH$_2$Cl$_2$) were required. Accordingly, unsymmetrical *m*-terphenyls or biphenyls were obtained in moderate to high yields (Scheme 2.8). This transformation takes place even with a nonactivated substrate. Thus, phenyl-substituted dienyne carboxylic acid **17** reacts to form *m*-terphenyl **19**, although it required higher temperatures (refluxing toluene) and catalyst loading [AuClP(*p*-CF$_3$-C$_6$H$_4$)$_3$ (10 mol%)/AgSbF$_6$ (30 mol%)], and the conversion did not proceed beyond 56%, probably due to catalyst thermal decomposition.

The formation of biphenyls and terphenyls **18** would probably follow a similar pathway to the one proposed for phenols **14** at the earlier stages of the mechanism (metal coordination, *s-trans-s-cis* isomerization, and carboxylic acid nucleophilic attack). However, due to the absence of a strong electron-donating group, intermediate **V**, which in fact is an analog of **IIIa** (Scheme 2.7), reacts through an electrocyclic ring closure between positions 1 and 6 to give bicyclic intermediates **VI**. Aromatization should then occur to form **18** by CO_2 extrusion and regeneration of the gold catalyst (Scheme 2.8).

2.3
Gold-Catalyzed Intermolecular Cycloadditions

2.3.1
Cycloadditions of Enynes, Propargyl Acetylenes, and Alkynyl Cyclopropanes

In contrast to the intramolecular cycloisomerization of enynes, the development of gold-catalyzed intermolecular cycloadditions of these substrates has been less fruitful. In this regard, Echavarren described the first gold-catalyzed intermolecular cyclopropanation of 1,6-enynes with alkenes (Scheme 2.9) [22]. In addition, Yamamoto, Asao, and coworkers reported [4+2] benzannulation reactions of aromatic enynal or enynone substrates with several 2π-systems such as acetylenes, enol ethers, and carbonyl compounds [23].

On the other hand, the use of propargyl esters **20** in gold-catalyzed intermolecular reactions has been more successful. These unsaturated systems undergo [1,2]- or [1,3]-acyl migrations, mostly depending on the substitution of the alkyne, leading to the formation of gold-carbene **J** or allene **K** intermediates (Figure 2.4) [24].

In particular, Toste and coworkers have taken advantage of the reactivity of carbene type intermediates **J**, formed by [1,2]-migration, in a variety of intermolecular cycloaddition transformations (Scheme 2.10) [25]. Thus, cyclopropanation reactions of this intermediate to form alkenyl cyclopropanes **21** were achieved in the presence of olefins [25a]. Significantly, an asymmetric version of the reaction has also been developed using gold complexes bearing chiral phosphines [25a]. However, reactions of propargyl esters containing diynes produced functionalized benzonorcaradienes **22** in a formal [4+3] annulation reaction [25b]. This process involves the generation of terminal carbene species by rearrangement of intermediate **J**. Subsequent tandem cyclopropanation/hydroarylation afforded the observed adducts. Moreover, reaction of propargyl esters with 1,3-enynes, instead of simple olefins, allowed the synthesis of styrenes **23**, which formally result from a regioselective cross [4+2] reaction of two different enynes, or fluorenes **24** [25c]. Notably, either cycloadduct could be selectively obtained by simply changing the silver counterion of the gold(I) catalyst. Further, highly substituted azepines **25** were synthesized by gold(III)-catalyzed [4+3] cycloaddition of propargyl esters and 1-azadienes [25d].

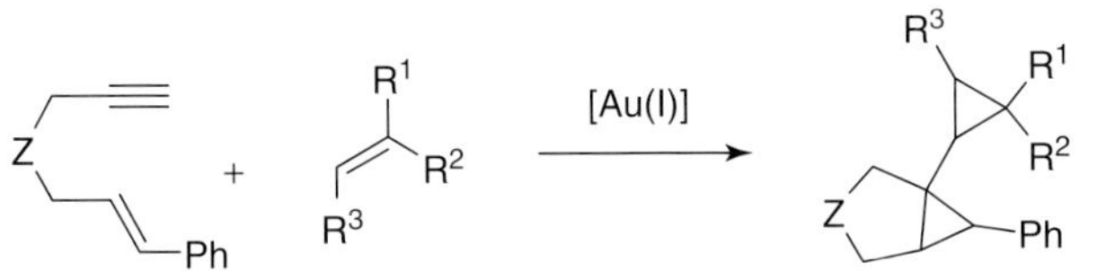

Scheme 2.9 Gold-catalyzed intermolecular cyclopropanation of 1,6-enynes.

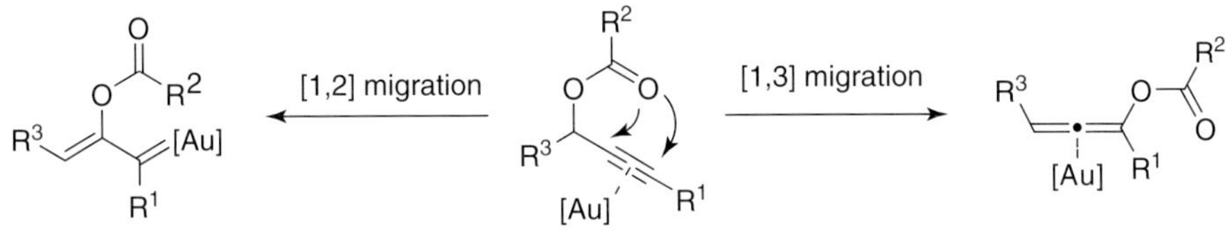

Figure 2.4 Gold-catalyzed [1,2] and [1,3] migrations of propargyl esters.

In addition, gold(I)-catalyzed [4+1] cycloaddition of propargyl tosilates with imines leading to cyclopent-2-enimines has been recently reported by González and coworkers [26]. This novel transformation was initiated by a 1,2-migration of the tosyl group in the starting material.

More recently, Zhang *et al.* described the generation of gold-containing all-carbon 1,4-dipole intermediates **M** from 1-(1-alkynylcyclopropyl)ketones **26** (Scheme 2.11, via *A*). These novel dipole species efficiently react in a formal [4+2] cycloaddition with a variety of dipolarophiles such as indoles, aldehydes, ketones, and imines affording highly functionalized six-membered carbo- and heterocycles (Scheme 2.11) [27a]. In contrast, reaction with enol ethers as dipolarophiles produced bicyclo[3.2.0]heptanes **27** by a [3+2] cycloaddition reaction via 1,3-dipole intermediate **N** (Scheme 2.11, via *B*) [27b].

On the other hand, Toste *et al.* reported an intermolecular enantioselective gold-catalyzed 1,3-dipolar cycloaddition of Münchnones with electron-deficient olefins [28].

2.3.2
Gold-Catalyzed Intermolecular Hetero-Dehydro-Diels–Alder Cycloaddition of Push–Pull Dienynes with Nonactivated Nitriles: Regioselective Synthesis of Pyridines

As depicted in Scheme 2.5, reaction of captodative dyenine carboxylic acid **12a** in acetonitrile in the presence of gold(I) or gold(III) catalysts afforded a mixture of phenol **14a** and pyridine **15a**. The latter adduct results from a formal intermolecular [4+2] cycloaddition of the nitrile with the enyne without apparent participation of the carboxylic acid functionality. Remarkably, this novel regioselective approach to tetrasubstituted pyridines is the first example of a catalyzed intermolecular hetero-dehydro-Diels–Alder reaction (HDDAR).

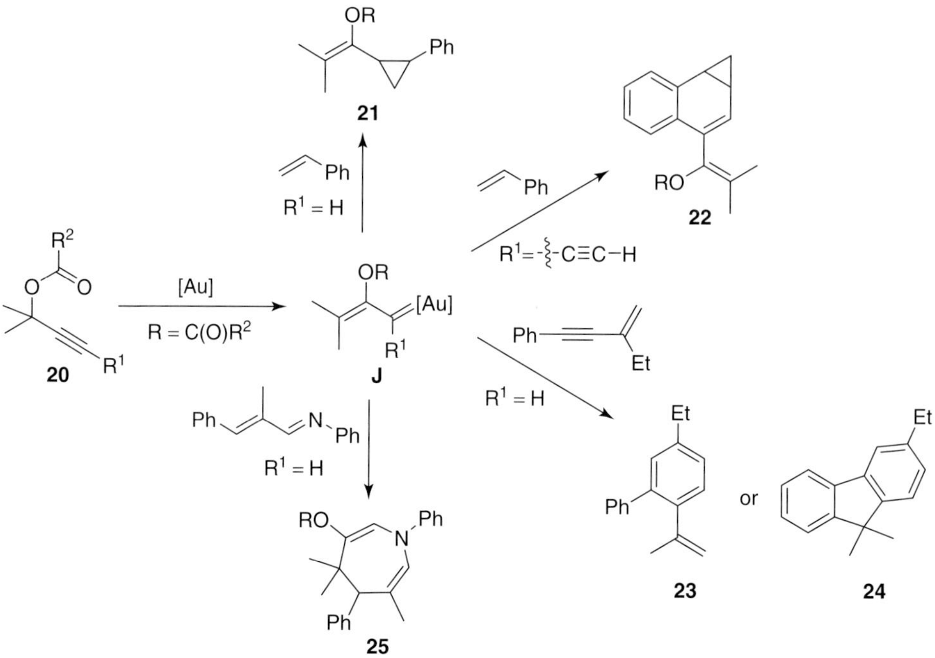

Scheme 2.10 Gold-catalyzed intermolecular cycloadditions of propargyl esters with olefins, 1,3-enynes and 1-azadienes.

The proposed mechanism for the formation of phenol **14a** involves a nucleophilic attack of the carboxylic group (Scheme 2.7). Having this in mind, we performed the reaction with 2,4-dien-6-yne ester **13a**. Consequently, the phenol formation could be suppressed and the corresponding pyridine **28a** was selectively obtained in good yield (Scheme 2.12) [29].

Reactions in the absence of catalyst or with metal complexes derived from Pt, Cu, Ag, or, Pd produced the dimerization of the starting dyenine or the formation of complex reaction mixtures. However, the [4+2]-cycloaddition reaction could be achieved with different gold(I) and gold(III) catalysts. Among them, the system AuClPEt$_3$/AgSbF$_6$ exhibited the best result. Interestingly, the reaction yield was further improved by the employment of 1,2-dichloroethane (DCE) as solvent, and the amount of acetonitrile could be reduced to 20 equiv.

The scope of this new intermolecular cycloaddition was next examined. Thus, a wide variety of dienynes **13** and a set of commercially available nonactivated nitriles were tested (Scheme 2.13). Substitution tolerated at the 3-position of the dienyne includes phenyl groups; aromatic rings with electron-withdrawing or electron-donating groups; and alkenyl, linear, and branched alkyl groups. Significantly, contrary to most Diels–Alder reactions, where the nitrile paucity is limited to activated nitriles such as trichloroacetonitrile or tosyl cyanide, our methodology allows the use of nonactivated nitriles. Therefore, besides acetonitrile, primary, secondary, and tertiary alkyl, aromatic, olefinic, and heteroaromatic nitriles

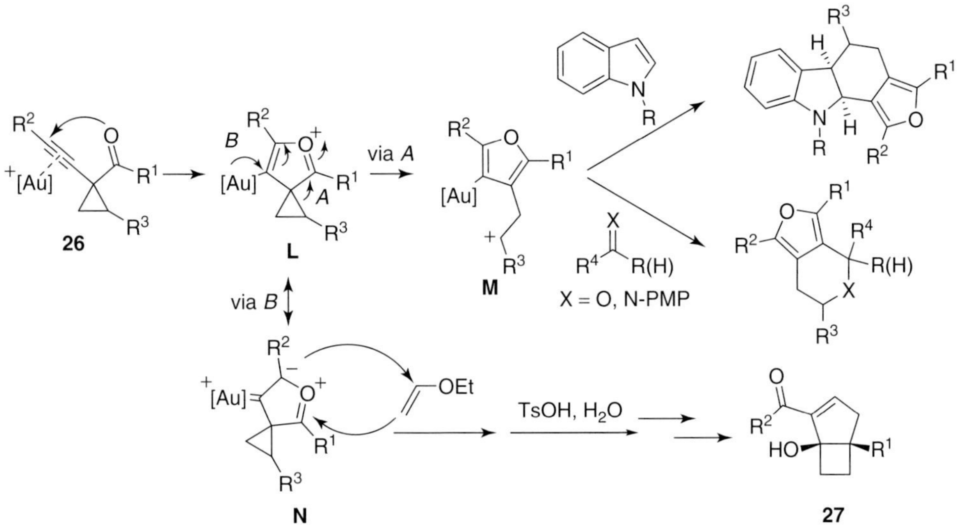

Scheme 2.11 Dipolar [4+2] and [3+2] intermolecular cycloadditions.

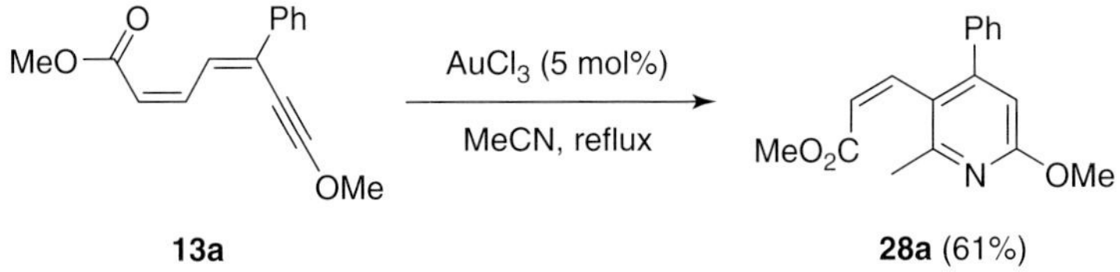

Scheme 2.12 Gold-catalyzed intermolecular HDDAR of push–pull dyenine **13a** with acetonitrile.

reacted with dienynes **13**. In all cases, the reaction proceeded satisfactorily yielding the expected tetrasubstituted pyridines **28** as single regioisomers.

As described for the 2,7-cycloaromatization of captodative dienyne acids **12** (Section 2.2.3), the electronic nature of the conjugated system is crucial for the [4+2] intermolecular cycloaddition to occur and, therefore, a push–pull system is required. This hypothesis is supported by the fact that for dienyne **29**, which presents two triple bonds of different electronic nature, only the electron-rich one partakes in the intermolecular cycloaddition (Scheme 2.14).

To further prove this assumption, simpler substrates **31–34** were prepared (Figure 2.5) and tested. A neutral substrate such as (E)-1-phenyloct-1-en-3-yne **31** did not evolve under the optimized reaction conditions. Moreover, electron-deficient enynes or dienynes such as (Z)-methyl nona-2-en-4-ynoate **32** and (2Z, 4Z)-methyl 5,7-diphenylhepta-2,4-dien-6-ynoate **33** just underwent partial double bond isomerization. Furthermore, (E)-1-ethoxy-oct-3-en-1-yne **34**, an enyne that bears an electron-donating substituent, did polymerize with all the tested catalysts, even at low temperatures ($-20\,^\circ$C).

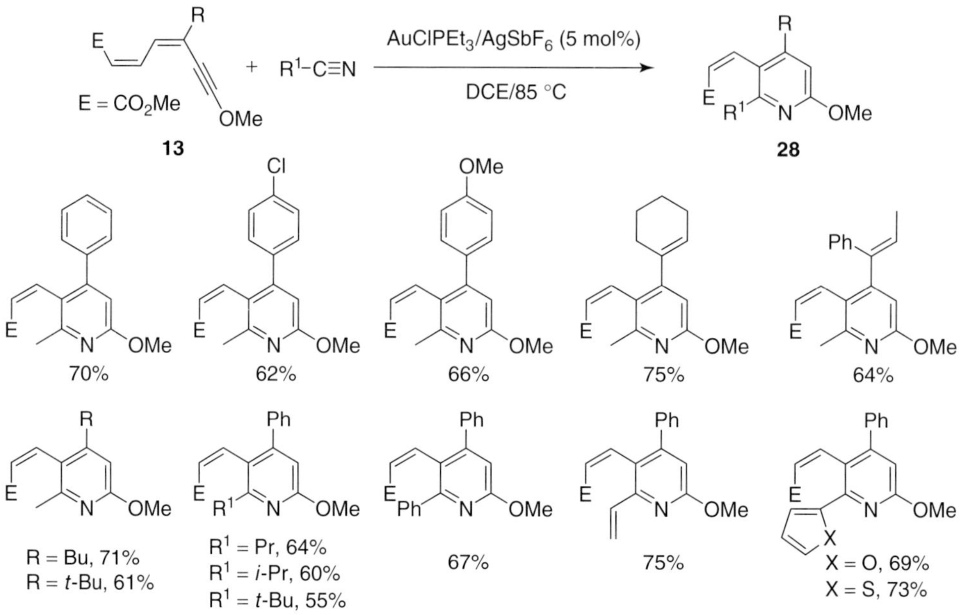

Scheme 2.13 Scope of the gold-catalyzed intermolecular HDDAR of push–pull dyenines **13** with nonactivated nitriles.

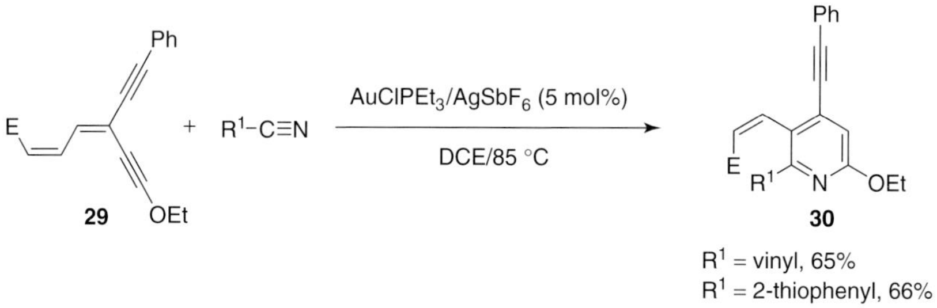

Scheme 2.14 Gold-catalyzed intermolecular HDDAR of push–pull dyenine **29** with nitriles.

A mechanism that explains the formation of pyridines **28** and **30** is depicted in Scheme 2.15. Initial coordination of the triple bond to the gold catalyst would take place to form intermediate **VIIa**, which presents a resonance structure **VIIb**, due to the electron-donating group directly attached to the triple bond. The push–pull substitution on the dienyne would facilitate the regioselective nucleophilic attack of the nitrile, leading to the formation of species **VIIIa**. Both steps, triple bond activation and subsequent nucleophilic attack, are well documented for gold-mediated transformations. A cyclization may then occur through resonance structure **VIIIb** or, alternatively, by intramolecular nucleophilic attack through structure **VIIIc**. This last option would involve the electron-withdrawing ester group in the reaction

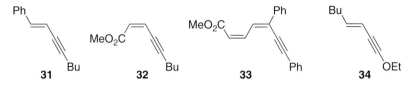

Figure 2.5 Enynes that do not participate in the HDDAR.

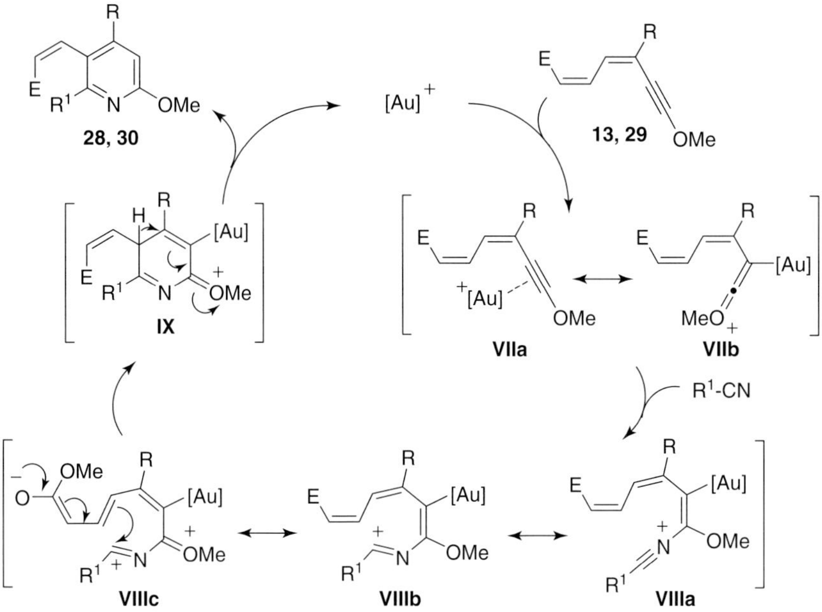

Scheme 2.15 Proposed mechanism for the gold-catalyzed intermolecular HDDAR of push–pull dyenines with nitriles.

mechanism and would explain why the reaction with electron-rich enyne **34** evolves by other routes. Consequently, dihydropyridine **IX** would be formed and a final protodemetalation would render the reaction products **28** or **30** and allow the incorporation of the gold catalyst into a new cycle.

To expand the scope of this novel gold-catalyzed intermolecular cycloaddition, reactions with other unsaturated nucleophiles are currently being investigated. To our delight, reaction with imines occurs under similar reaction conditions to afford dihydropyridones **35**. This intermolecular transformation is also general and allows the synthesis of a broad range of highly substituted dihydropyridones in moderate to good yields and, importantly, with complete regio- and diastereoselection (Scheme 2.16) (Fernández-Rodríguez et al., 2009, manuscript in preparation).

The mechanism of this transformation is similar to the one proposed for the reaction of nitriles depicted in Scheme 2.15.

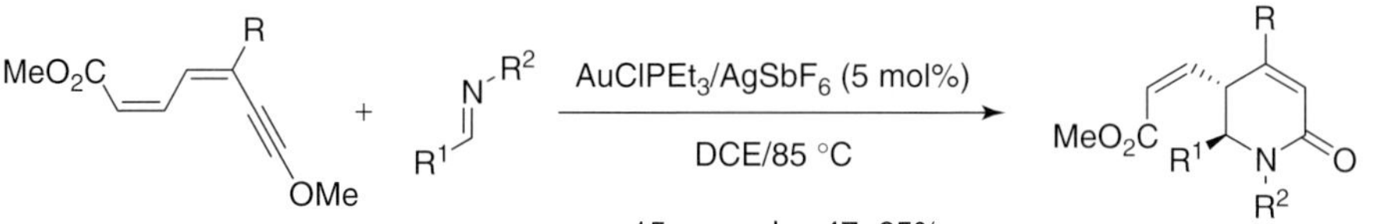

Scheme 2.16 Gold-catalyzed intermolecular HDDAR of push–pull dyenines with imines.

2.4
Conclusions and Future Perspectives

Gold-catalyzed cycloisomerizations and intermolecular cycloadditions of enynes are highly influenced by the electronic nature of the substitution in the alkyne terminus of the substrate. In this sense, easily and efficiently accessible push–pull 2,4-dien-6-ynes, which present an alkoxy group directly attached to the triple bond and a conjugated acid or ester functionality in the other side of the molecule, proved to be appropriate substrates for the development of new gold-catalyzed intra- and intermolecular transformations.

In particular, captodative dienyne acids react under mild reaction conditions to form 2,3-disubstituted phenols by a novel regioselective gold-catalyzed 2,7-cycloaromatization reaction. This process is absolutely dependent on the electronic properties of the dienyne acid; if a strong electron-donating group is not directly linked to the triple bond, a regioselective 1,6-cyclization–decarboxylation sequence takes place upon warming.

On the other hand, a regioselective intermolecular HDDAR occurs between the corresponding push–pull 2,4-dien-6-yne esters and nonactivated nitriles leading to the formation of tetrasubstituted pyridines. Moreover, a related [4+2] cycloaddition takes place with imines as unsaturated nucleophiles to afford highly substituted dihydropyridones in a regio- and diastereoselective manner. A captodative system is required for both processes to occur. Importantly, these sequences promoted by gold catalysts represent the first examples of catalyzed HDDAR.

The use of these push–pull dienynes and other related highly activated substrates in novel metal-catalyzed intermolecular cycloadditions ([4+2], [4+3], etc.) with a variety of unsaturated nucleophiles is currently under investigation.

Acknowledgments

I would like to thank Prof. Barluenga, Prof. Aguilar, and all the researchers whose work has been presented in this chapter, particularly Dra. Patricia García-García for her inputs and critical discussions during the writing of this chapter. Special thanks to Prof. Pignataro for giving the opportunity to write this chapter and MICINN (Spain) for funding by "Juan de la Cierva" and "Ramón y Cajal" Programs.

References

1. See for example: (a) de Meijere, A. and Diederich, F.D. (eds) (2004) *Metal-catalyzed Cross-Coupling Reactions*, Wiley-VCH Verlag GmbH, Weinheim; (b) Beller, M. and Bolm, C. (eds) (2004) *Transition Metals for Organic Synthesis*, Wiley-VCH Verlag GmbH, Weinheim; (c) Hegedus, L.S. (1999) *Transition Metals in the Synthesis of Complex Organic Molecules*, University Science Books, Sausalito.
2. Nolan, S.P. (2007) *Nature*, **445**, 496–497.
3. For general recent reviews see: (a) Kirsch, S.F. (2008) *Synthesis*, 3183–3204; (b) Shen, H.C. (2008) *Tetrahedron*, **64**, 7847–7870; (c) Shen, H.C. (2008) *Tetrahedron*, **64**, 3885–3903; (d) Arcadi, A. (2008) *Chem. Rev.*, **108**, 3266–3325; (e) Li, Z., Brouwer, C., and He, C. (2008) *Chem. Rev.*, **108**, 3239–3365; (f) Hashmi, A.S.K. (2007) *Chem. Rev.*, **107**, 3180–3211; (g) Jiménez-Núñez, E. and Echavarren, A.M. (2007) *Chem. Commun.*, 333–346.
4. For reviews regarding the effects of ligands in homogeneous gold catalysis see: (a) Gorin, D.J., Sherry, B.D., and Toste, F.D. (2008) *Chem. Rev.*, **108**, 3351–3378; (b) Marion, N. and Nolan, S.P. (2008) *Chem. Soc. Rev.*, **37**, 1776–1782.
5. For applications of gold catalysis in total synthesis see: Hashmi, A.S.K. and Rudolph, M. (2008) *Chem. Soc. Rev.*, **37**, 1766–1775.
6. Gorin, D.J. and Toste, F.D. (2007) *Nature*, **446**, 395–403.
7. Fürstner, A. and Davies, P.W. (2007) *Angew. Chem. Int. Ed.*, **46**, 3410–3449.
8. For reviews of metal-catalyzed cycloisomerization of enynes see: (a) Lee, S.I. and Chatani, N. (2009) *Chem. Commun.*, 371–384; (b) Michelet, V., Toullec, P.Y., and Genêt, J.P. (2008) *Angew. Chem. Int. Ed.*, **47**, 4268–4315.
9. For reviews on gold-catalyzed cycloisomerizations including mechanistic perspectives see: (a) Jiménez-Núñez, E. and Echavarren, A.M. (2008) *Chem. Rev.*, **108**, 3326–3350; (b) Zhang, L., Sun, J., and Kozmin, S.A. (2006) *Adv. Synth. Catal.*, **348**, 2271–2296.
10. For reactions with enol-ethers see: (a) Sherry, B., Maus, L., Laforteza, B.N., and Toste, F.D. (2006) *J. Am. Chem. Soc.*, **128**, 8132–8133; (b) Nieto-Oberhuber, C., Muñoz, M.P., Buñuel, E., Nevado, C., Cárdenas, C.J., and Echavarren, A.M. (2004) *Angew. Chem. Int. Ed.*, **43**, 2402–2406; For silylenol-ethers see: (c) Lee, K. and Lee, P.H. (2007) *Adv. Synth. Catal.*, **349**, 2092–2096; (d) Linghu, X., Kennedy-Smith, J.J., and Toste, F.D. (2007) *Angew. Chem. Int. Ed.*, **46**, 7671–7673; (e) Staben, S.T., Kennedy-Smith, J.J., Huang, D., Corkey, B.K., LaLonde, R.L., and Toste, F.D. (2006) *Angew. Chem. Int. Ed.*, **45**, 5991–5994; For enamines see: (f) Binder, J.T., Crone, B., Haug, T.T., Menz, H., and Kirsch, S.F. (2008) *Org. Lett.*, **10**, 1025–1028; For furans see: (g) Hashmi, A.S.K., Rudolph, M., Siehl, H.-U., Tanaka, M., Bats, J.W., and Frey, W. (2008) *Chem. Eur. J.*, **14**, 3703–3708 and references therein; For indoles see: (h) Ferrer, C., Amijs, C.H.M., and Echavarren, A.M. (2007) *Chem. Eur. J.*, **13**, 1358–1373, references cited therein.
11. Zhang, L. and Kozmin, S.A. (2004) *J. Am. Chem. Soc.*, **126**, 11806–11807.
12. Sun, J., Conley, M.P., Zhang, L., and Kozmin, S.A. (2006) *J. Am. Chem. Soc.*, **128**, 9705–9710.
13. Similar cycloisomerizations have been reported. See for example: (a) Luzung, M.R., Markham, J.P., and Toste, F.D. (2004) *J. Am. Chem. Soc.*, **126**, 10858–10859; (b) Mamane, V., Gress, T., Krause, H., and Fürstner, A. (2004) *J. Am. Chem. Soc.*, **126**, 8654–8655; (c) Harrak, Y., Blaszykowski, C., Bernard, M., Cariou, K., Mainetti, E., Mouires, V., Dhiname, A.L., Fensterbank, L., Malacria, M. (2004) *J. Am. Chem. Soc.*, **126**, 8656–8657; (d) see also Ref. 10b.
14. Couty, S., Meyer, C., and Cossy, J. (2006) *Angew. Chem. Int. Ed.*, **45**, 6726–6730.
15. Nieto-Oberhuber, C., Muñoz, M.P., López, S., Jiménez-Núñez, E., Nevado, C., Herrero-Gómez, E., Raducan, M.,

and Echavarren, A.M. (2006) *Chem. Eur. J.*, **12**, 1677–1693.

16. Hashmi, A.S.K., Rudolph, M., Bats, J.W., Frey, W., Rominger, F., and Oeser, T. (2008) *Chem. Eur. J.*, **14**, 6672–6678.

17. Barluenga, J., García-García, P., de Sáa, D., Fernández-Rodríguez, M.A., Bernardo de la Rúa, R., Ballesteros, A., Aguilar, E., and Tomás, M. (2007) *Angew. Chem. Int. Ed.*, **46**, 2610–2612.

18. For recent reviews on the Bergman, Saito-Myers or Moore cyclizations see: (a) Kar, M. and Basak, A. (2007) *Chem. Rev.*, **107**, 2861–2890; (b) Klein, M., Walenzyk, T., and König, B. (2004) *Collect. Czech. Chem. Commun.*, **69**, 945–965; (c) Wang, K.K. (1996) *Chem. Rev.*, **96**, 207–222; (d) Grissom, J.W., Gunawardena, G.U., Klingberg, D., and Huang, D. (1996) *Tetrahedron*, **52**, 6453–6518.

19. García-García, P., Fernández-Rodríguez, M.A., and Aguilar, E. (2009) *Angew. Chem. Int. Ed.*, **48**, 5534–5537.

20. Genin, E., Toullec, P.Y., Antoniotti, S., Brancour, C., Gent, J.-P., and Michelet, V. (2006) *J. Am. Chem. Soc.*, **128**, 3112, 3113.

21. Jin, T. and Yamamoto, Y. (2007) *Org. Lett.*, **9**, 5259–5262.

22. López, S., Herrero-Gómez, E., Pérez-Galán, P., Nieto-Oberhuber, C., and Echavarren, A.M. (2006) *Angew. Chem. Int. Ed.*, **45**, 6029–6032.

23. For an account including gold- and copper-catalyzed [4+2] benzannulation reactions see: (a) Asao, N. (2006) *Synlett*, 1645–1656; See also: (b) Asao, N., Aikawa, H., and Yamamoto, Y. (2004) *J. Am. Chem. Soc.*, **126**, 7458–7459; (c) Asao, N., Kasahara, T., and Yamamoto, Y. (2003) *Angew. Chem. Int. Ed.*, **42**, 3504–3506; (d) Asao, N., Takahashi, K., Lee, S., Kasahara, T., and Yamamoto, Y. (2002) *J. Am. Chem. Soc.*, **124**, 12650–11265.

24. For reviews see: (a) Marco-Contelles, J. and Soriano, E. (2007) *Chem. Eur. J.*, **13**, 1350–1357; (b) Marion, N. and Nolan, S.P. (2007) *Angew. Chem. Int. Ed.*, **46**, 2750–2752.

25. (a) Johansson, M.J., Gorin, D.J., Staben, S.T., and Toste, F.D. (2005) *J. Am. Chem. Soc.*, **127**, 18002–18003; (b) Gorin, D.J., Dubé, P., and Toste, F.D. (2006) *J. Am. Chem. Soc.*, **128**, 14480–14481; (c) Gorin, D.J., Watson, I.D.G., and Toste, F.D. (2008) *J. Am. Chem. Soc.*, **130**, 3736–3737; (d) Shapiro, N.D. and Toste, F.D. (2008) *J. Am. Chem. Soc.*, **130**, 9244–9245.

26. Suárez-Pantiga, S., Rubio, E., Alvarez-Rúa, C., and González, J.M. (2009) *Org. Lett.*, **11**, 13–16.

27. (a) Zhang, G., Huang, X., Li, G., and Zhang, L. (2008) *J. Am. Chem. Soc.*, **130**, 1814–1815; (b) Li, G., Huang, X., and Zhang, L. (2008) *J. Am. Chem. Soc.*, **130**, 6944–6945.

28. Melhado, A.D., Luparia, M., and Toste, F.D. (2007) *J. Am. Chem. Soc.*, **129**, 12638–12639.

29. Barluenga, J., Fernández-Rodríguez, M.A., García-García, P., and Aguilar, E. (2008) *J. Am. Chem. Soc.*, **130**, 2764–2765.

3
N-Heterocyclic Carbenes in Copper-Catalyzed Reactions
Silvia Díez-González

3.1
Introduction

N-Heterocyclic carbenes (NHCs) were first reported by Wanzlick in the 1960s and NHC–transition metal complexes have been known since 1968 [1, 2]. In 1988, Bertrand and coworkers succeeded in isolating the first stable carbene [3, 4]. Unfortunately, the reported (phosphino)(silyl)carbene did not show any coordination ability for transition metals. The isolation of a free imidazol-2-ylidene by Arduengo *et al.* in 1991 provided access to free isolable carbenes prepared in a single-step synthesis from imidazolium salts [5]. In consequence, a new and exciting field of research unfolded for chemists.

Now widely employed also as organocatalysts [6], NHCs were first considered as simple phosphine mimics in organometallic chemistry [7]. However, increasing experimental data show that NHC–metal catalysts can sometimes surpass their phosphine-based counterparts, and, more interestingly, display a complementary activity and scope. Mainly known for their major breakthroughs in palladium- and ruthenium-catalyzed reactions [8, 9], we intend here to give an overview of the contribution made by NHC ligands to the field of copper-catalyzed transformations [10]. First, the different synthetic strategies available for the preparation of well-defined [(NHC)Cu] complexes is presented (Section 3.2). Then, the state of the art of their applications in homogeneous catalysis is covered (Section 3.3) with a particular focus on our own contribution to the field (Sections 3.4 and 3.5).

3.2
Preparation of NHC-Containing Copper Complexes

Arduengo and coworkers reported the first NHC–copper complex: a bis-NHC cationic copper(I) compound prepared from copper(I) triflate and 2 equivalents of an imidazol-2-ylidene [11]. Soon after, neutral monocarbene copper(I) complexes were synthesized by Raubenheimer and coworkers via the alkylation of thiazolyl or imidazolyl-cuprates (Scheme 3.1) [12]. Six years later, Danopoulos and coworkers

Ideas in Chemistry and Molecular Sciences: Advances in Synthetic Chemistry. Edited by Bruno Pignataro
Copyright © 2010 WILEY-VCH Verlag GmbH & Co. KGaA, Weinheim
ISBN: 978-3-527-32539-9

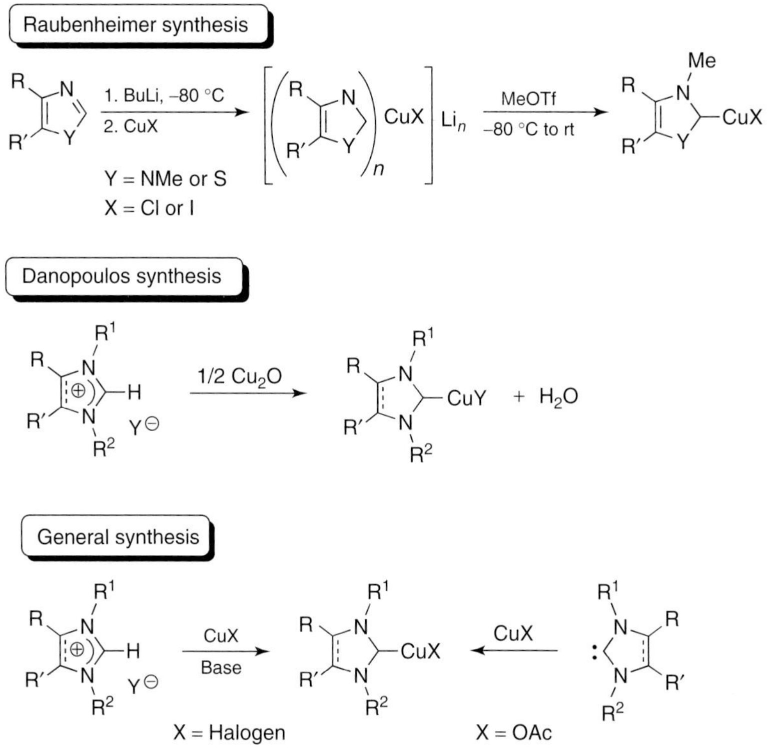

Scheme 3.1 Preparation of [(NHC)CuX] complexes.

reported the first monomeric copper(I) imidazolylidene complex by deprotonation of the starting salt by copper(I) oxide [13]. This reaction notably obviates the need for strong bases, and water is the only generated by-product. Nevertheless, this approach remains largely underexploited [14] and these complexes are more generally prepared by making a copper salt (CuCl, CuBr, CuI, or CuOAc) react with a free carbene, either isolated or *in situ* generated (Scheme 3.1) [15].

Most of these [(NHC)CuX] complexes are indefinitely water- and oxygen-stable and have been used as convenient precursors of often more unstable related compounds. These transformations, or activation steps, can be divided in two main categories: reaction with a *tert*-butoxide salt or with an organometallic reagent. Thus, from halogenated complexes, the corresponding *tert*-butoxide analogs can be prepared in high yields (Scheme 3.2) [16]. These activated species can then be readily transformed into the corresponding hydrides [16], boryl [17], or dibenzoylmethanoate (DBM) [18] derivatives.

Alternatively, the reaction of [(NHC)CuX] species (X = halogen or Ac) with aluminum- or magnesium-based organometallic reagents has led to the isolation of alkyl derivatives, which can also be regarded as "activated" compounds [19, 20]. Such alkyl complexes can indeed be transformed into their corresponding anilido, alkoxide, acetylide [21], and thiolate [22] analogs [23] (Scheme 3.3).

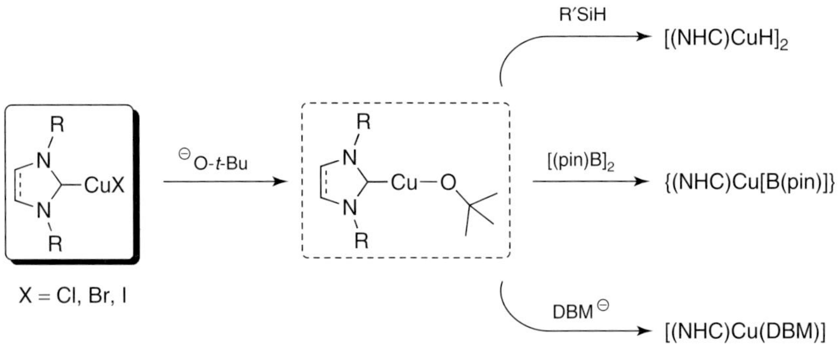

3.2 Preparation of NHC-Containing Copper Complexes

Scheme 3.2 Derivatization of [(NHC)CuX] complexes via an alkoxide complex.

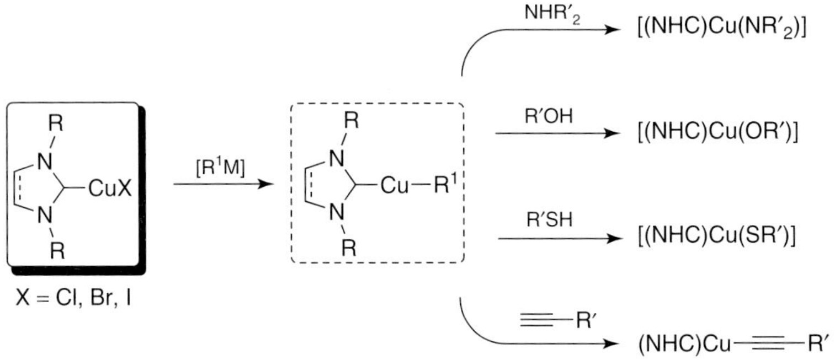

Scheme 3.3 Derivatization of [(NHC)CuX] complexes via an alkyl complex.

NHC-containing copper(I) complexes can alternatively be prepared by carbene transfer from the corresponding NHC/silver(I) reagents [24], a frequent approach for the preparation of NHC complexes of late transition metals [25]. Another well-established strategy for preparing NHC-containing complexes, such as phosphine displacement, has notably allowed for the preparation of ketiminate-containing complexes [26] (Scheme 3.4).

Even if so far efforts have been mainly focused on copper(I) species, some NHC–copper(II) complexes are also known in the literature. The first example was prepared by Meyer and coworkers by oxidation of its copper(I) analog [27].

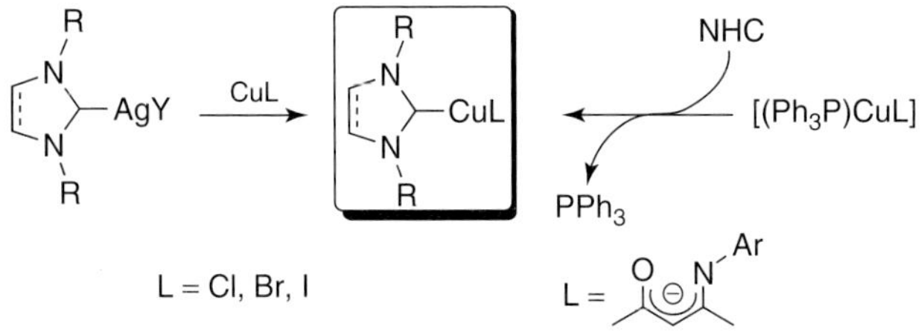

Scheme 3.4 Further approaches for the synthesis of [(NHC)Cu] complexes.

Other copper(II) complexes have been synthesized by carbene transfer from silver species [28], or alternatively from the corresponding NHC–lithium adducts [29], or by direct reaction with a free carbene [30].

3.3 Main Applications of [(NHC)Cu] Complexes in Catalysis

Since the preparation of the first reported NHC–copper derivatives in 1993 [11], an increasing number of organic transformations have benefited of their application. Their straightforward preparation, along with their (in most cases) remarkable stability toward oxygen and moisture, has certainly influenced the ever-rising popularity of these complexes in organic chemistry [31].

In particular, NHCs are ligands of choice for the copper-catalyzed allylic alkylation reaction. Not only the reactions are usually very regioselective [32] but also outstanding asymmetric inductions have been achieved with the binaphthol-based NHCs developed by Hoveyda and co workers [33]. Thus, a dimeric NHC–silver(I) complex in combination with air-stable copper(II) salts allowed for the highly selective formation of quaternary stereogenic centers with a diversity of zinc reagents (Scheme 3.5). Of note, the derived copper(II) complexes were also synthesized and used to perform this transformation.

The use of NHC–oxy ligands represent indeed one of the most general methods for the use of hard metal alkyls in this reaction and they have been further employed in the preparation of enantiomerically pure allylsilanes via the allylic alkylation of vinylsilanes [34] or in the addition of vinylaluminum species [35].

Related biphenol-based NHC ligands have also shown a remarkable activity in the conjugate addition of zinc reagents onto cyclic enones [36]. The addition of diethylzinc to enones was actually the first reported catalytic application of [(NHC)Cu] complexes [37], and ever since this has been a very active field of research [38]. Nowadays, even all carbon quaternary stereogenic centers can be enantioselectively formed with NHC/copper-based catalytic systems by using zinc or Grignard reagents [39].

Conjugate addition reactions in the presence of [(NHC)Cu] systems are not restricted to C–C bond forming transformations. Anilido, alkoxide [40], and thiolate [22] NHC-containing complexes have been reported as efficient catalysts for the N–H, O–H, and S–H addition to electron-deficient olefins, respectively. In these reactions, the corresponding anti-Markonikov products were formed

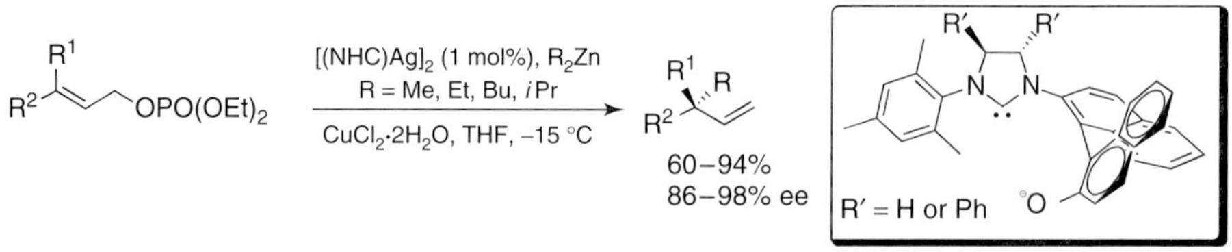

Scheme 3.5 [(NHC)Cu]-catalyzed allylic alkylation.

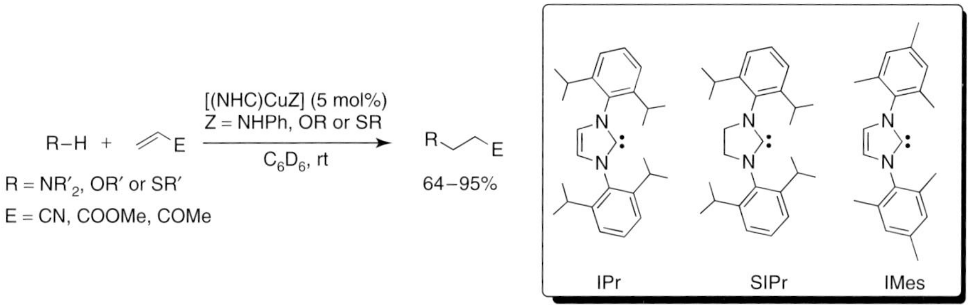

Scheme 3.6 [(NHC)Cu]-catalyzed conjugate addition to electron-deficient olefins.

(Scheme 3.6). These catalysts present the advantages of being broad in scope and being active under smooth reaction conditions. Furthermore, readily available NHC ligands such as IPr (IPr = N,N'-bis(2,6-diisopropylphenyl)imidazol-2-ylidene), SIPr (SIPr = N,N'-bis(2,6-diisopropylphenyl)imidazolin-2-ylidene), and IMes (IMes = N,N'-bis(2,4,6-trimethylphenyl)imidazol-2-ylidene) were used in this context.

Another family of copper-catalyzed transformations that have been widely explored with NHC ligands is the carbene transfer reactions. In particular, [(IPr)CuCl] displayed an outstanding catalytic activity in the cyclopropanation of alkenes and the related insertion reactions into the X–H bonds of amines and alcohols employing ethyl diazoacetate (EDA) as carbene source (Scheme 3.7) [41].

The most remarkable feature of this catalyst is the total suppression of the diazo homocoupling, a general drawback of this methodology. Indeed, [(IPr)CuCl],

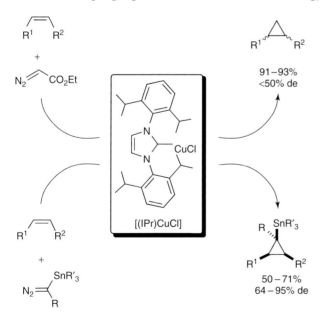

Scheme 3.7 [(IPr)CuCl]-catalyzed cyclopropanation reactions.

unlike other copper-based catalytic systems, does not react with EDA even in the absence of substrate. On the other hand, the moderate diastereoselectivity obtained in these cyclopropanation reactions could be substantially increased employing stannyldiazoacetate esters as carbene source (Scheme 3.7) [42]. Noteworthy, even if somewhat harsher reaction conditions (80 °C instead of room temperature) were required in these cases, the suppression of diazo homocoupling remained effective.

Another IPr-containing complex, [(IPr)Cu(DBM)], was applied to the related aziridination of olefins [43]. The most remarkable features of this catalyst are that the nitrene was generated *in situ* and that no large excess of the starting material was required.

Additionally, methylenation reactions have also been explored in this context. The combined use of trimethylsilyldiazomethane as reagent and a NHC–copper complex led to the efficient methylenation of a number of carbonyl compounds [44]. Although [(IPr)CuCl] and [(IMes)CuCl] performed equally well for aromatic or aliphatic aldehydes, [(IPr)CuCl] was the optimal catalyst for ketone methylenation (Scheme 3.8). For these reactions, ligandless systems were also examined and even if similar yields were obtained in some cases, an overall loss of efficiency was evidenced.

This strategy is complementary to the classic Wittig reaction since even base-sensitive- or electron-deficient substrates could conveniently be transformed into the corresponding alkenes in good yields. It is important to note that phosphines are not suitable ligands for this reaction due to the inefficient formation of phosphorus ylides under copper catalysis conditions [45].

Finally, diboration reactions have also focused a particular attention on NHC/copper systems. Sadighi and coworkers first reported in 2006 that [(ICy)Cu(O-*t*-Bu)] (ICy = N,N'-bis(cyclohexyl)imidazol-2-ylidene) was a good catalyst for the 1,2-diboration of aldehydes [46]. The actual active species, [(ICy)Cu(boryl)], was shown to be formed under these catalytic conditions (Scheme 3.9). Density functional theory (DFT) calculations pointed toward aldehyde insertion into the Cu–B bond to form a Cu–O–C species as initial step of the catalytic cycle [47]. This intermediate would undergo a σ-bond metathesis with the diborane to generate the reaction product and close the cycle. Boration-type reactions with NHC/copper systems have been successfully extended to aldimines [48], α,β-unsaturated esters [49], and alkenes [50].

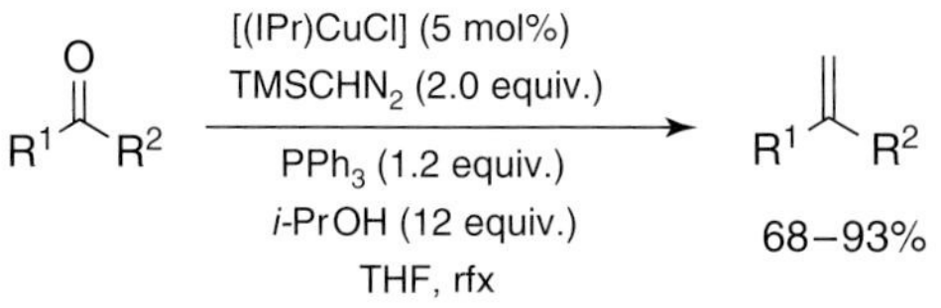

Scheme 3.8 [(IPr)CuCl]-catalyzed methylenation of ketones.

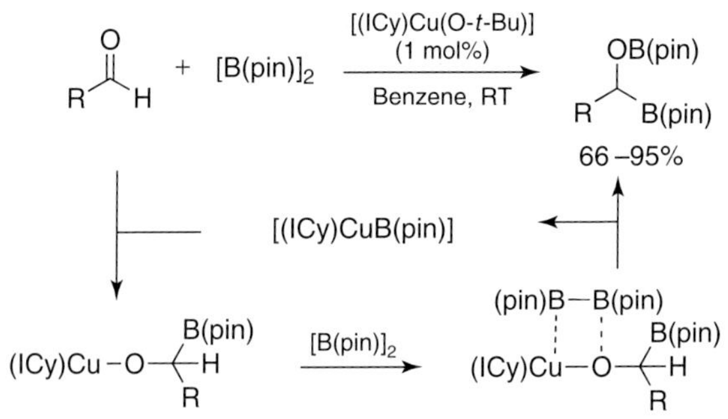

Scheme 3.9 [(ICy)Cu]-catalyzed 1,2-diboration of aldehydes.

3.4
Copper Hydride-Mediated Reactions

Reduction of carbonyl and related functions, such as imines or hydrazones, represents a fundamental protocol in organic synthesis [51]. Transition metal catalysis has been successfully applied to the reduction of olefins, alkynes, and many carbonyl compounds via hydrogenation or hydrosilylation [52]. The use by Stryker of a stabilized form of copper hydride, the hexameric [(Ph$_3$P)CuH]$_6$, represented a breakthrough in copper-catalyzed reduction reactions [53]. Regrettably, this complex is mostly effective as a stoichiometric reducing agent, which prevents its general applicability. The combination of Stryker's catalyst with a hydrosilane as a hydride source and/or with additional ligands was the starting point for the development of copper-mediated reduction reactions [54].

3.4.1
Hydrosilylation of Carbonyl Compounds

[(IPr)CuCl] was first reported as catalyst in conjugate reduction reactions of α,β-unsaturated esters and cyclic enones [55]. At room temperature, the desired reduced products were efficiently obtained using a hydrosilane as hydride source, as shown in Scheme 3.10. Soon after, the same complex was applied to the hydrosilylation of simple ketones to form the corresponding silyl ethers in excellent yields [56].

For more challenging ketones, ICy was found the most effective NHC ligand [57]. A number of NHC ligands were screened in this context, including the very bulky N,N'-bis(adamantyl)imidazol-2-ylidene (IAd) and N,N'-bis(tert-butyl) imidazol-2-ylidene (ItBu). These ligands, bearing such bulky groups on the nitrogen atoms of the NHC, were also found effective in these reactions, implying that not only steric but also electronic effects are relevant in this reaction [58].

Thus, in the presence of [(ICy)CuCl], a number of ketones with varying congestion around the carbonyl function could be efficiently reduced: this encompasses alkyl, aromatic, aliphatic, cyclic, and bicyclic ketones (Scheme 3.11). Even extremely hindered ketones reacted under these conditions to afford the corresponding silyl ethers in high yields and in short reaction times. This catalytic system

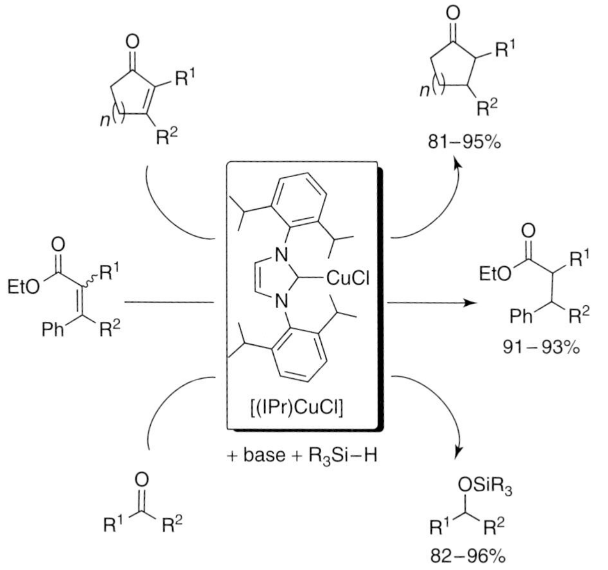

Scheme 3.10 [(IPr)CuCl]-catalyzed hydrosilylation of carbonyl compounds.

is very broad in scope, and diversely functionalized substrates (amino-, oxo-, haloketones, etc.) were smoothly converted in the presence of [(ICy)CuCl]. Of note, despite the broad applicability of this catalyst, another NHC ligand (SIMes = N,N'-bis(2,4,6-trimethylphenyl)imidazolin-2-ylidene) was found optimal for the hydrosilylation of heteroaromatic ketones.

Related available catalytic systems include copper complexes bearing a tetrahydropyrimidin-2-ylidene ligand, applied to the hydrosilylation of ketones

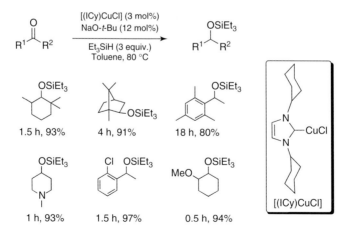

Scheme 3.11 [(ICy)CuCl]-catalyzed hydrosilylation of hindered and functionalized ketones.

and aldehydes [59]. Furthermore, Yun et al. showed that a [(NHC)Cu^{II}] can also mediate the hydrosilylation of ketones [30]. However, the nature of the active species in this case (a Cu^I or a Cu^{II} hydride) remains elusive so far.

Another family of NHC-containing complexes, [(NHC)$_2$Cu]X (X = PF$_6^-$ or BF$_4^-$), has recently been studied in this context [60]. The activity of these cationic bis-NHC complexes in the hydrosilylation of ketones was found to be strongly dependent on both the ligand and the counterion around the copper center. Although the ligand influence could not be simplified in pure steric or electronic arguments [58], complexes with a BF$_4^-$ counterion were systematically superior to their PF$_6^-$ analogs.

When compared with [(NHC)CuCl] complexes, these cationic complexes present the advantages of requiring lower reaction temperatures and smaller excess of hydrosilanes, while exhibiting better catalytic performances (Scheme 3.12). Cyclohexanone was indeed more efficiently reduced in the presence of [(IPr)$_2$Cu]BF$_4$ than with [(IPr)CuCl]. For more hindered ketones, the activity of [(ICy)CuCl] was compared with [(ICy)$_2$Cu]BF$_4$ in the reduction of dicyclohexyl ketone. In this case, a faster reaction was observed with the neutral complex under more forcing conditions though. However, using comparable reaction conditions (55 °C, 2 equivalents of hydride source), the cationic complex was found again to be the best catalyst.

3.4.2
Mechanistic Considerations

The proposed mechanism for the [(NHC)CuCl]-catalyzed hydrosilylation of ketones is depicted in Scheme 3.13. First, formation of [(NHC)Cu(O-t-Bu)] would take place. Then, the actual active catalyst, a NHC–copper hydride species, is believed to be formed via a σ-bond metathesis between the copper alkoxide and the hydrosilane. This proposition is supported by the isolation and characterization of both intermediates [16]. Addition of the copper hydride to the carbonyl would result in a copper alkoxide that would lead to the expected silyl ether after another σ-bond metathesis with the hydrosilane, regenerating the active catalyst [61].

This mechanism is very similar to the one proposed for the phosphine–copper catalytic systems, supported by significant experimental evidence [62]. However, NHC-based catalytic systems generally require the presence of an excess of base, which role remains uncertain. Since hydrosilanes are prone to nucleophilic attack,

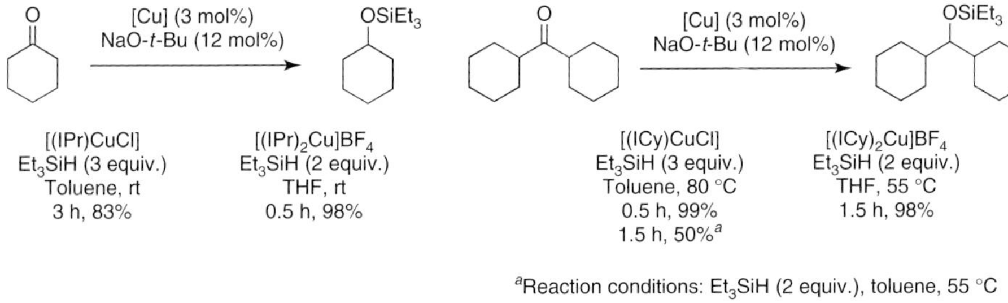

[(IPr)CuCl]
Et$_3$SiH (3 equiv.)
Toluene, rt
3 h, 83%

[(IPr)$_2$Cu]BF$_4$
Et$_3$SiH (2 equiv.)
THF, rt
0.5 h, 98%

[(ICy)CuCl]
Et$_3$SiH (3 equiv.)
Toluene, 80 °C
0.5 h, 99%
1.5 h, 50%^{a}

[(ICy)$_2$Cu]BF$_4$
Et$_3$SiH (2 equiv.)
THF, 55 °C
1.5 h, 98%

^{a}Reaction conditions: Et$_3$SiH (2 equiv.), toluene, 55 °C

Scheme 3.12 Comparison of hydrosilylation catalysts.

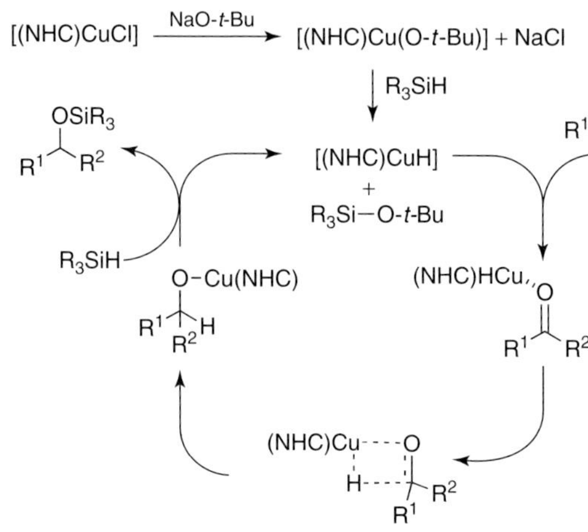

Scheme 3.13 Proposed mechanism for the [(NHC)Cu]-catalyzed hydrosilylation of ketones.

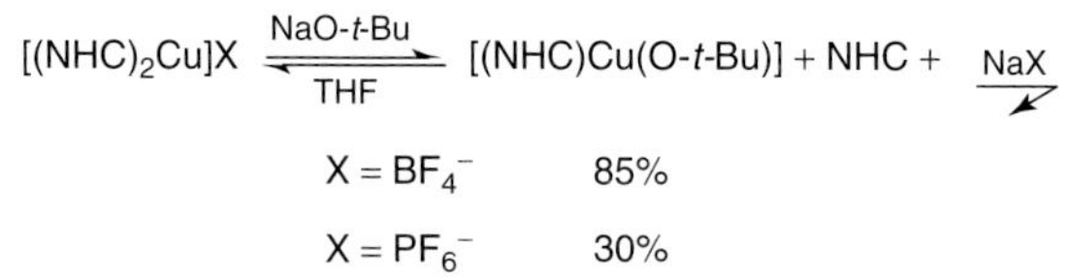

Scheme 3.14 Activation of [(NHC)$_2$Cu]X complexes toward hydrosilylation.

it was proposed that such excess of base could interact with the hydrosilane to facilitate the second σ-bond metathesis of the catalytic cycle [61].

As for the cationic bis-NHC complexes, it was found that the activation of the starting complex, rather than the catalytic cycle itself, is different. In this case, one of the two NHC ligands is displaced by t-BuO$^-$ in the presence of NaO-t-Bu, producing the neutral [(NHC)Cu(O-t-Bu)], direct precursor of the active species. The released NHC, being nucleophilic, could then facilitate the σ-bond metathesis leading to the formation of the silyl ether [6, 63]. This could explain the better catalytic performance of [(NHC)$_2$Cu]X complexes when compared with the parent [(NHC)CuX].

Moreover, the previously mentioned counterion effect in the catalytic activity was rationalized as a consequence of the difference in solubility in the reaction solvent of the inorganic salts formed during the reaction. As shown in Scheme 3.14, the more insoluble NaX is in tetrahydrofuran (THF), the more efficient the activation step will be.

3.4.3
Related Transformations

When using a copper hydride as reducing agent in a conjugate reduction, the copper enolate intermediate can be engaged in further reactions rather than quenched. Hence, the conjugate reduction/aldol condensation tandem reaction

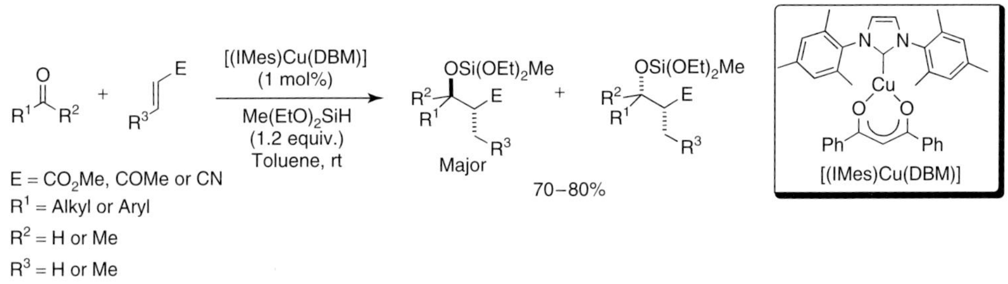

Scheme 3.15 [(NHC)Cu]-catalyzed reductive aldol condensation.

was first explored with Stryker's reagent as a hydride source [64]. The use of other ligands, mainly diphosphines, subsequently allowed for the generalization of this methodology [65].

To date, only a single example involving NHC ligands in this tandem reaction is available [18]. With an IMes ligand, the direct reduction of the electrophile (aldehyde or ketone) was minimized and good yields were obtained from a number of electrophilic C=C double bonds (Scheme 3.15). Furthermore, a reasonable *anti* diastereoselectivity was obtained under these reaction conditions.

Finally, the activity of NHC-bearing copper hydrides toward alkynes or alkenes remains greatly unexplored. Sadighi and coworkers reported the hydrocupration of 3-hexyne by an isolated dimeric [(NHC)CuH]$_2$ complex [16]. Nevertheless, only propargyl oxiranes has been properly examined to date. This family of alkynes yielded diversely functionalized α-hydroxyallenes diastereoselectively [66].

3.5
[3+2] Cycloaddition of Azides and Alkynes

3.5.1
Click Chemistry

In 2001, Sharpless and coworkers defined the concept of "Click chemistry" and the criteria for a transformation to be considered as Click [67]. Inspired by nature, the objective is to rapidly create molecular diversity through the use of reactive modular building blocks and only benign reaction conditions, simple work-up, and purification procedures.

After the discovery in 2002 of copper(I) as efficient and regiospecific catalyst for the reaction of azides and alkynes yielding 1,2,3-triazoles [68] (1,3-dipolar Huisgen cycloaddition [69]), this transformation has become the best Click reaction to date. Catalytic systems for this reaction most often consist of a copper(II) salt and a reducing agent, but diverse family of ligands have also been shown to protect Cu^I centers during this reaction [70].

Screening of a set of [(NHC)CuX] complexes under standard cycloaddition conditions showed that [(SIMes)CuBr] was the best catalyst for this transformation [71].

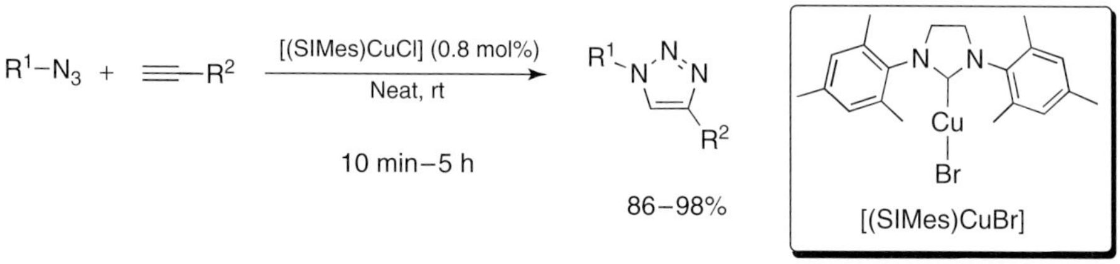

Scheme 3.16 [(SIMes)CuBr]-catalyzed formation of triazoles.

Although poor conversions were obtained in organic solvents, a strong acceleration rate was observed in water. Furthermore, neat reactions proceeded smoothly with no detectable formation of undesired by-products and the catalyst loading could be lowered to 0.8 mol% with no loss of activity, ensuring straightforward reaction work-ups (Scheme 3.16). This transformation is broad in scope and triazoles were isolated in excellent yield and high purity after simple filtration or extraction.

Pleasantly, azides generated *in situ* from the corresponding halides and NaN$_3$ reacted at room temperature to efficiently yield triazoles. This catalytic system represents an impressive improvement when compared to previously reported ligandless conditions [72]. Again, water was the best solvent for this transformation. To ensure short reaction times and minimize any decomposition of the starting materials, 5 mol% of copper complex was used. However, at lower catalyst loadings the reaction still proceeded to completion after extended reaction times. No competitive formation of N–H triazoles, resulting from the addition of an inorganic azide to the alkyne, was detected under these conditions.

To date, [(SIMes)CuBr], or its unsaturated analog, has been successfully used for the preparation of triazoles-containing carbanucleosides [73], porphyrins [74], or platinum-based anticancer drugs [75].

The ability of NHC ligands to modulate the copper center reactivity has alternatively been applied to the development of a latent catalyst for this cycloaddition reaction [76]. A screening of different [(NHC)CuCl] complexes showed that [(SIPr)CuCl] was the least active complex of the series. The worst activity was observed in DMSO as solvent and under these conditions a number of azides and alkynes did not react under ambient conditions, even after a week of stirring. However, the cycloaddition reaction proceeded smoothly and in high yields at 60 °C upon addition of water to the reaction mixture (Scheme 3.17). This approach was applied to a variety of substrates, although highly activated substrates were unsuitable since they reacted during the latency period. This study represents a proof-of-concept and it should be a useful starting point to broaden the applications of this transformation, especially in biology and material science.

Alternatively, the bis-NHC-containing family of complexes that we showed superior to neutral [(NHC)CuX] in hydrosilylation reactions (Section 3.3), has displayed an outstanding activity in cycloaddition reactions even at very low catalyst loadings [77]. With only 40–100 ppm of [(ICy)$_2$Cu]PF$_6$, an array of triazoles could be efficiently prepared (Figure 3.1).

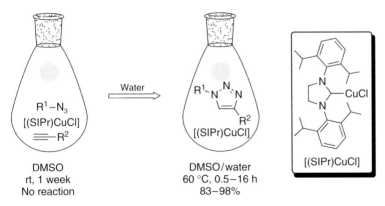

Scheme 3.17 Latent catalyst for [3+2] cycloaddition reactions.

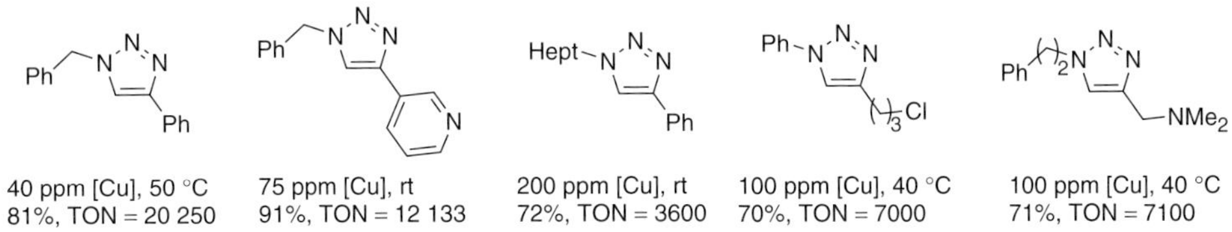

Ph–triazole–Ph	Ph–triazole–(pyridyl)	Hept–triazole–Ph	Ph–triazole–(CH₂)₃Cl	Ph–CH₂–triazole–NMe₂
40 ppm [Cu], 50 °C	75 ppm [Cu], rt	200 ppm [Cu], rt	100 ppm [Cu], 40 °C	100 ppm [Cu], 40 °C
81%, TON = 20 250	91%, TON = 12 133	72%, TON = 3600	70%, TON = 7000	71%, TON = 7100

Figure 3.1 [(ICy)₂Cu]PF₆-catalyzed preparation of triazoles.

Mechanistic studies showed that one of the NHC ligand on the copper center acts as a base, deprotonating the starting alkyne to generate a copper acetylide and start the catalytic cycle. The enhanced catalytic activity of this system, when compared to mono-NHC catalysts, was attributed to the efficiency of the formed azolium salt to protonate the copper triazolide intermediate, closing the catalytic cycle (Scheme 3.18).

3.5.2
Use of Internal Alkynes: Mechanistic Implications

Traditionally, the starting point of the catalytic cycle for the copper-catalyzed Huisgen reaction is considered to be the formation of a copper–acetylide intermediate, which precludes internal alkynes as cycloaddition partners. Moreover, a hypothetical activation toward cycloaddition via π-coordination of copper(I) onto the alkyne (without deprotonation) has also been ruled out since the calculated activation barrier for this process exceeds that of the uncatalyzed process [78].

However, in the presence of [(SIMes)CuBr], 4,5-disubstituted triazoles could be isolated in fair to good yields after heating at 70 °C for 48 hours (Scheme 3.19) [79]. Optimization studies showed that both the copper salt and the NHC ligand were essential for this transformation.

In contrast to terminal alkynes, no acceleration rate was observed when the reactions with internal alkynes were carried out in the presence of water. This

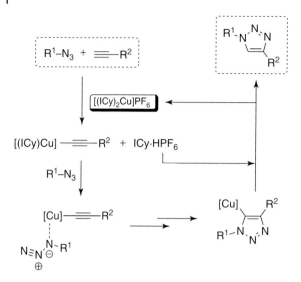

Scheme 3.18 Mechanistic proposal for the [(ICy)$_2$Cu]X-catalyzed [3+2] cycloaddition reactions.

observation suggests distinct mechanistic pathways for mono- and disubstituted alkynes. Although the copper ion is generally considered a poor π-back-donating ion, the ancillary ligands on the metal center play an essential role in its coordination to alkynes [80]. In fact, DFT calculations indicated that π-coordination of EtC≡CEt to [(SIMes)Cu]$^+$ is favored by almost 20 kcal mol^{-1} relative to coordination to [(MeCN)$_2$Cu]$^+$. These results led to the proposition that the beneficial effect of the NHC allows for the activation of disubstituted alkynes to proceed via a π-alkyne complex (Scheme 3.20). Of note, the widely accepted reaction pathway for terminal alkynes would still be applicable with this system. The recent isolation of an intermediate copper(I) triazolide complex **A** bearing a SIPr ligand strongly supports this proposition [81].

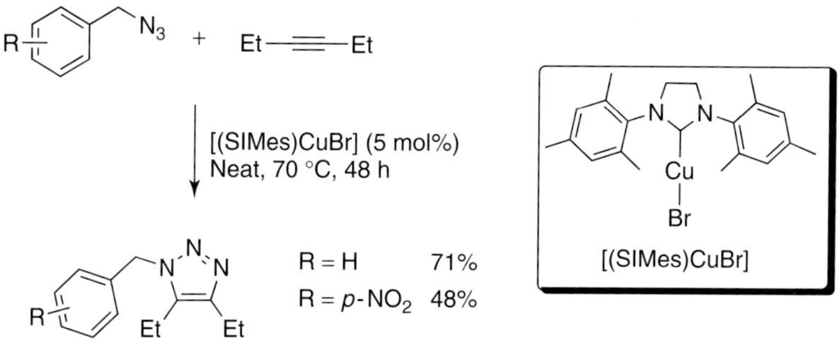

Scheme 3.19 [(SIMes)CuBr]-catalyzed cycloaddition reaction of benzyl azides and 3-hexyne.

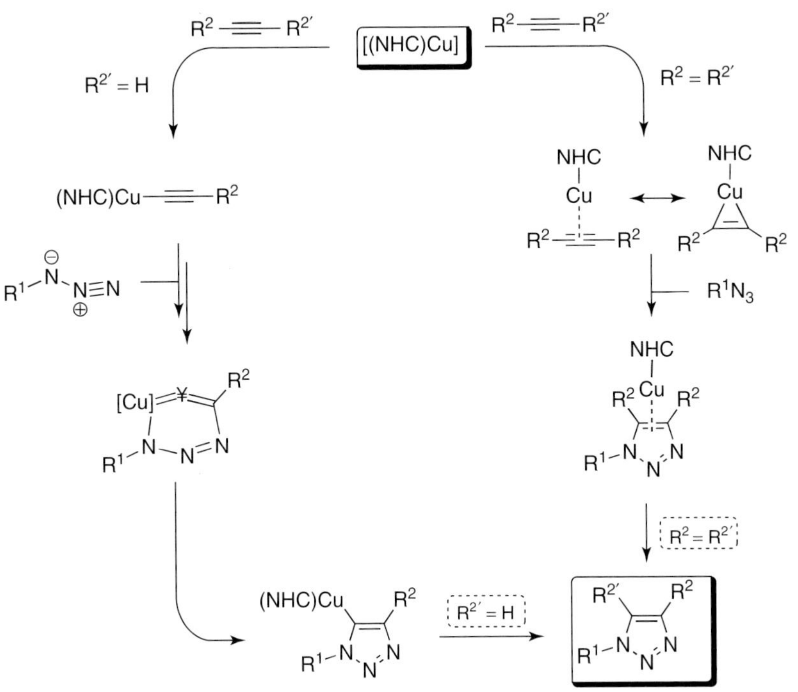

Scheme 3.20 Postulated mechanisms for the [(NHC)Cu]-catalyzed [3+2] cycloaddition of azides and alkynes.

3.6
Concluding Remarks

It is now well established that NHC ligands protect the copper(I) center and, more importantly, tune its reactivity. Their straightforward preparation, along with their stability and ease of manipulation, has certainly played an essential role in the increasing interest in these species.

Nevertheless, these complexes are of interest not only in the context of catalysis. For instance, NHC–copper complexes have shown remarkable toxicity against human cancer cells [82], good activity in processes of substantial industrial interest such as reduction of CO_2 to CO [83], and hydrogen storage [84]. Alternatively, these complexes possess an important organometallic interest and have notably revealed the existence of non-negligible π-interaction between copper (and other Group 11 metals) and NHC ligands [85], which, at the time, shattered the general assumption that NHC ligands are pure σ-donors.

Despite their success, NHC/copper systems are still underexploited, even in well-established copper-catalyzed reactions. For instance, only two reports in the literature deal with C–X bond forming cross-coupling reactions in the presence of these species [86]. However, the most important challenge for [(NHC)Cu] complexes at this point is not to find application to every single known copper-catalyzed reaction, but to display novel reactivities, complementary to known methodologies. In this important task, the study of novel architectures and/or activation modes will certainly be crucial for succeeding.

Acknowledgments

The ICIQ Foundation and the Spanish Ministerio de Ciencia e Innovación are gratefully acknowledged for financial support. Prof. Steven P. Nolan (ICIQ/University of St Andrews) is sincerely acknowledged for his great trust and support in the last few years.

References

1. (a) Wanzlick, H.-W. (1962) Nucleophile carben-chemie. *Angew. Chem.*, **74** (4), 129–134;   (b) Wanzlick, H.W., Esser, F., and Kleiner, H.J. (1963) Nucleophile carben-chemie, III. Neue verbindungen vom typ des bis-[1.3-diphenyl-imidazolidinylidens-(2)]. *Chem. Ber.*, **96** (5), 1208–1212;   (c) Wanzlick, H.-W. and Schönherr, H.-J. (1968) Direct synthesis of a mercury salt-carbene complex. *Angew. Chem. Int. Ed. Engl.*, **7** (2), 141–142.
2. (a) Öfele, K. (1968) 1,3-Dimethyl-4-imidazolinyliden-(2)-pentacarbonylchrom ein neuer übergangsmetall-carben-komplex. *J. Organomet. Chem.*, **12** (3), P42–P43;   (b) Öfele, K. (1970) Tetracarbonylbis(1,3-dimethyl-4-imidazolin-2-ylidene)chromium(0). *Angew. Chem. Int. Ed. Engl.*, **9** (9), 739–740;   (c) Öfele, K. (1970) Dichlor(2,3-diphenylcyclopropenyliden)-palladium(II). *J. Organomet. Chem.*, **22** (1), C9–C11.
3. Igau, A., Grutzmacher, H., Baceiredo, A., and Bertrand, G. (1988) Analogous α,α'-bis-carbenoid, triply bonded species: synthesis of a stable λ^3-phosphino carbene-λ^5-phosphaacetylene. *J. Am. Chem. Soc.*, **110** (19), 6463–6466.
4. For a very recent and comprehensive review on carbenes, see: de Frémont, P., Marion, N., and Nolan, S.P. (2009) Carbenes: synthesis, properties, and organometallic chemistry. *Coord. Chem. Rev.*, **253** (7–8), 862–892.
5. Arduengo, A.J., III, Harlow, R.L., and Kline, M.A. (1991) A stable crystalline carbene. *J. Am. Chem. Soc.*, **113** (1), 361–363.
6. (a) Enders, D., Niemeier, O., and Henseler, A. (2007) Organocatalysis by N-heterocyclic carbenes. *Chem. Rev.*, **107** (12), 5606–5655;   (b) Marion, N., Díez-González , S., and Nolan, S.P. (2007) N-Heterocyclic carbenes as organocatalysts. *Angew. Chem. Int. Ed.*, **46** (17), 2988–3000.
7. Green, J.C., Scurs, R.G., Arnold, P.L., and Cloke, G.N. (1997) An experimental and theoretical investigation of the electronic structure of Pd and Pt bis(carbene) complexes. *Chem. Commun.*, (20), 1963–1964.
8. (a) Marion, N. and Nolan, S.P. (2008) Well-defined N-heterocyclic carbenes–palladium(II) precatalysts for cross-coupling reactions. *Acc. Chem. Res.*, **41** (11), 1440–1449;   (b) Díez-González, S. and Nolan, S.P. (2007) Palladium-catalyzed reactions using NHC ligands. *Top. Organomet. Chem.*, **21**, 47–82;   (c) Kantchev, E.A.B., O'Brien, C.J., and Organ, M.G. (2007) Palladium complexes of N-heterocyclic carbenes as catalysts for cross-coupling reactions–a synthetic chemist's perspective. *Angew. Chem. Int. Ed.*, **46** (16), 2768–2813.

9. Grubbs, R.H. (ed.) (2003) *Handbook of Metathesis*, Wiley-VCH Verlag GmbH, Weinheim.
10. For related reviews, see: (a) Lin, J.C.Y., Huang, R.T.W., Lee, C.S., Bhattacharyya, A., Hwang, W.S., and Lin, J.B. (2009) Coinage metal-*N*-heterocyclic carbene complexes. *Chem.* **109** (8), 3561–3598; (b) Arnold, P. (2002) Organometallic chemistry of silver and copper N- heterocyclic complexes. *Heteroat. Chem.*, **13** (6), 534–539.
11. Arduengo, A.J.III, Dias, H.V.R., Calabrese, J.C., and Davidson, F. (1993) Homoleptic carbene-silver(I) and carbene-copper(I) complexes. *Organometallics*, **12** (9), 3405–3409.
12. (a) Raubenheimer, H.G., Cronje, S., van Rooyen, P.H., Olivier, P.J., and Toerien, J.G. (1994) Synthesis and crystal structure of a monocarbene complex of copper. *Angew. Chem. Int. Ed. Engl.*, **33** (6), 672–673; (b) Raubenheimer, H.G., Cronje, S., and Olivier, P.J. (1995) Synthesis and characterization of mono(carbene) complexes of copper and crystal structure of a linear thiazolinylidene compound. *J. Chem. Soc., Dalton Trans.* **2**, 313–316.
13. Tulloch, A.A.D., Danopoulos, A.A., Kleinhenz, S., Light, M.E., Hursthouse, M.B., and Eastham, G. (2001) Structural diversity in pyridine-*N*-functionalized carbene copper(I) complexes. *Organometallics*, **20** (10), 2027–2031.
14. (a) McKie, R., Murphy, J.A., Park, S.R., Spicer, M.D., and Zhou, S.-z. (2007) Homoleptic crown N-heterocyclic carbene complexes. *Angew. Chem. Int. Ed.*, **46** (34), 6525–6528; (b) Simonovic, S., Whitwood, A.C., Clegg, W., Harrington, R.W., Hursthouse, M.B., Male, L., and Douthwaite, R.E. (2009) Synthesis of copper(I) complexes of N-heterocyclic carbene-phenoxyimine/amine ligands: Structures of mononuclear copper(II), mixed-valence copper(I)/(II), and copper(II) cluster complexes. *Eur. J. Inorg. Chem.*, (13), 1786–1795.
15. For selected examples, see: (a) Mankad, N.P., Gray, T.G., Laitar, D.S., and Sadighi, J.P. (2004) Synthesis, structure, and CO_2 reactivity of a two-coordinate (carbene)copper(I) methyl complex. *Organometallics*, **23** (6), 1191–1193; (b) Schneider, N., César, V., Bellemin-Laponnaz, S., and Gade, L.H. (2005) Synthesis and structural chemistry of oxazolinyl-carbene copper(I) complexes. *J. Organomet. Chem.*, **690** (24-25), 5556–5561; (c) Michon, C., Ellern, A., and Angelici, R.J. (2006) Chiral tetradentate amine and tridentate aminocarbene ligands: synthesis, reactivity and X-ray structural characterizations. *Inorg. Chim. Acta*, **359** (14), 4549–4556.
16. Mankad, N.P., Laitar, D.S., and Sadighi, J.P. (2004) Synthesis, structure, and alkyne reactivity of a dimeric (carbene)copper(I) hydride. *Organometallics*, **23** (14), 3369–3371.
17. Laitar, D.S., Müller, P., and Sadighi, J.P. (2005) Efficient homogeneous catalysis in the reduction of CO_2 to CO. *J. Am. Chem. Soc.*, **127** (49), 17196–17197.
18. Welle, A., Díez-González, S., Tinant, B., Nolan, S.P., and Riant, O. (2006) A three-component tandem reductive aldol reaction catalyzed by N-heterocyclic carbene–copper complexes. *Org. Lett.*, **8** (26), 6059–6062.
19. (a) Goj, L.A., Blue, E.D., Delp, S.A., Gunnoe, T.B., Cundari, T.R., Pierpont, A.W., Petersen, J.L., and Boyle, P.D. (2006) Chemistry surrounding monomeric copper(I) methyl, phenyl, anilido, ethoxide, and phenoxide complexes supported by N-heterocyclic carbene ligands: reactivity consistent with both early and late transition metal systems. *Inorg. Chem.*, **45** (22), 9032–9045; (b) Goj, L.A., Blue, E.D., Delp, S.A., Gunnoe, T.B., Cundari, T.R., and Petersen, J.L. (2006) Single-electron oxidation of monomeric copper(I) alkyl complexes: evidence for reductive elimination through bimolecular formation of alkanes. *Organometallics*, **25** (17), 4097–4104; See also ref. 15a. For an alternative synthesis of [(NHC)Cu(aryl)] complexes, see: (c) Niemeyer, M. (2003) Reaktion von kupferarylen mit imidazol-2-ylidenen oder triphenylphosphan-bildung von 1:1-addukten mit zweifach koordinierten

kupferatomen. *Z. Anorg. Allg. Chem.*, **629** (9), 1535–1540.

20. Cp-containing [(NHC)Cu] complexes are a particular case, since despite a similar synthesis, they cannot be considered as activated, see: Ren, H., Zhao, X., Xu, S., Song, H., and Wang, B. (2006) Synthesis and structures of cyclopentadienyl N-heterocyclic carbene copper(I) complexes. *J. Organomet. Chem.*, **691** (19), 4109–4113.

21. Goj, L.A., Blue, E.D., Munro-Leighton, C., Gunnoe, B., and Petersen, J.L. (2005) Cleavage of X-H bonds (X = N, O, or C) by copper(I) alkyl complexes to form monomeric two-coordinate copper(I) systems. *Inorg. Chem.*, **44** (24), 8647–8649. See also, ref. 19a.

22. Delp, S.A., Munro-Leighton, C., Goj, L.A., Ramírez, M.A., Gunnoe, T.B., Petersen, J.L., and Boyle, P.D. (2007) Addition of S–H bonds across electron-deficient olefins catalyzed by well-defined copper(I) thiolate complexes. *Inorg. Chem.*, **46** (7), 2365–2367.

23. For a review on these family of complexes, see: Gunnoe, T.B. (2007) Reactivity of ruthenium(II) and copper(I) complexes that possess anionic heteroatomic ligands: synthetic exploitation of nucleophilicity and basicity of amido, hydroxido, alkoxo, and aryloxo ligands fort he activation of substrates that possess polarbonds as well as nonpolar C–H and H–H bonds. *Eur. J. Inorg. Chem.*, (9), 1185–1203.

24. For selected examples, see: (a) Arnold, P.L., Scarisbrick, A.C., Blake, A.J., and Wilson, C. (2001) Chelating alkoxy-N-heterocyclic carbene complexes of silver and copper. *Chem. Commun.*, (22), 2340–2341; (b) Wan, X.-J., Xu, F.-B., Li, Q.-S., Song, H.-B., and Zhang, Z.-Z. (2005) Synthesis and crystal structure of metal (M = Ag, Cu) crown ether with N-heterocyclic carbene linkage. *Inorg. Chem. Commun.*, **8** (11), 1053–1055; (c) Winkelmann, O., Näther, C., and Lüning, U. (2008) Bimacrocyclic NHC transition metal complexes. *J. Organomet. Chem.*, **693** (6), 923–932.

25. Lin, I.J.B. and Vasam, C.S. (2007) Preparation and application of N-heterocyclic carbene complexes of Ag(I). *Coord. Chem. Rev.*, **251** (5–6), 642–670, references therein.

26. Hsu, S.-H., Li, C.-Y., Chiu, Y.-W., Chiu, M.-C., Lien, Y.-L., Kuo, P.-C., Lee, H.M., Huang, J.-H., and Cheng, C.-P. (2007) Synthesis and characterization of Cu(I) and Cu(II) complexes containing ketiminate ligands. *J. Organomet. Chem.*, **692** (24), 5421–5428.

27. Hu, X., Castro-Rodriguez, I., and Meyer, K. (2003) Copper complexes of nitrogen-anchored tripodal N-heterocyclic carbene ligands. *J. Am. Chem. Soc.*, **125** (40), 12237–12241.

28. Larsen, A.O., Leu, W., Nieto Oberhuber, C., Campbell, J.E., and Hoveyda, A.H. (2004) Bidentate NHC-based chiral ligands for efficient Cu-catalyzed enantioselective allylic alkylations: structure and activity of an air-stable chiral Cu complex. *J. Am. Chem. Soc.*, **126** (36), 11130–11131.

29. Arnold, P.L., Rodden, M., Davis, K.M., Scarisbrick, A.C., Blake, A.J., and Wilson, C. (2004) Asymmetric lithium(I) and copper(II) alkoxy-N-heterocyclic carbene complexes; crystallographic characterisation and Lewis acid catalysis. *Chem. Commun.*, (14), 1612–1613.

30. Yun, J., Kim, D., and Yun, H. (2005) A new alternative to Stryker's reagent in hydrosilylation: synthesis, structure, and reactivity of a well-defined carbene–copper(II) acetate complex. *Chem. Commun.*, (41), 5181–5183.

31. For comprehensive reviews on catalytic applications of NHC/copper systems, see: (a) Díez-González, S. and Nolan, S.P. (2008) N-Heterocyclic carbene–copper complexes: synthesis and applications in catalysis. *Aldrichimica Acta*, **41** (2), 43–51; (b) Díez-González, S. and Nolan, S.P. (2008) N-Heterocyclic carbene–copper(I) complexes in homogeneous catalysis. *Synlett*, (14), 2158–2167.

32. (a) Tominaga, S., Oi, Y., Kato, T., An, D.K., and Okamoto, S. (2004) γ-Selective allylic substitution reaction with Grignard reagents catalyzed by copper N-heterocyclic carbene complexes and its application to

enantioselective synthesis. *Tetrahedron Lett.*, **45** (29), 5585–5588; (b) Okamoto, S., Tominaga, S., Saino, N., Kase, K., and Shimoda, K. (2005) Allylic substitution reactions with Grignard reagents catalyzed by imidazolium and 4,5-dihydroimidazolium carbene–CuCl complexes. *J. Organomet. Chem.*, **690** (24-25), 6001–6007; (c) Seo, H., Hirsch-Weil, D., Abboud, K.A., and Hong, S. (2008) Development of biisoquinoline-based chiral diaminocarbene ligands: Enantioselective $S_{N}2$ allylic alkylation catalyzed by copper-carbene complexes. *J. Org. Chem.*, **73** (5), 1983–1986.

33. Van Veldhuizen, J.J., Campbell, J.E., Giudici, R.E., and Hoveyda, A.H. (2005) A readily available chiral Ag-based N-heterocyclic carbene complex for use in efficient and highly enantioselective Ru-catalyzed olefin metathesis and Cu-catalyzed allylic alkylation reaction. *J. Am. Chem. Soc.*, **127** (18), 6877–6882. See also ref. 28.

34. Kacprzynski, M.A., May, T.L., Kazane, S.A., and Hoveyda, A.H. (2007) Enantioselective synthesis of allylsilanes bearing tertiary and quaternary Si-substituted carbons through Cu-catalyzed allylic alkylations with alkylzinc and arylzinc reagents. *Angew. Chem. Int. Ed.*, **46** (24), 4554–4558.

35. Lee, Y., Akiyama, K., Gilligham, D.G., Brown, M.K., and Hoveyda, A.H. (2008) Highly site- and enantioselective Cu-catalyzed allylic alkylation reactions with easily accessible vinylaluminum reagents. *J. Am. Chem. Soc.*, **130** (2), 446, 447.

36. Brown, M.K., May, T.L., Baxter, C.A., and Hoveyda, A.H. (2007) All-carbon quaternary stereogenic centers by enantioselective Cu-catalyzed conjugate additions promoted by a chiral N-heterocyclic carbene. *Angew. Chem. Int. Ed.*, **46** (7), 1097–1100.

37. (a) Fraser, P.K., and Woodward, S. (2001) Strong ligand accelerated catalysis by an Arduengo-type carbene in copper-catalysed conjugate addition. *Tetrahedron Lett.*, **42** (14), 2747–2749; (b) Pytkowicz, J., Roland, S., and Mangeney, P. (2001) Enantioselective conjugate addition of diethylzinc using catalytic silver(I) diaminocarbenes and $Cu(OTf)_2$. *Tetrahedron: Asymmetry*, **12** (15) 2087–2089; (c) Guillen, F., Winn, C.L., and Alexakis, A. (2001) Enantioselective conjugate addition of diethylzinc using catalytic silver(I) diaminocarbenes and $Cu(OTf)_2$. *Tetrahedron: Asymmetry*, **12** (15) 2083–2086.

38. For selected references, see: (a) Winn, C.L., Guillen, F., Pytkowicz, J., Roland, S., Mangeney, P., and Alexakis, A. (2005) Enantioselective copper catalysed 1,4-conjugate addition reactions using chiral N-heterocyclic carbenes. *J. Organomet. Chem.*, **690** (24-25), 5672–5695; (b) Clavier, H., Coutable, L., Toupet, L., Guillemin, J.-C., and Mauduit, M. (2005) Design and synthesis of new bidentate alkoxy-NHC ligands for enantioselective copper-catalyzed conjugate addition. *J. Organomet. Chem.*, **690** (23), 5237–5254; (c) Moore, T., Merzouk, M., and Williams, N. (2008) Synthesis of amidoalkyl imidazol-2-ylidene ligands and their application to enantioselective copper-catalysed conjugate Addition. *Synlett* (1), 21–24; (d) Brown, M.K. and Hoveyda, A.H. (2008) Enantioselective total synthesis of Clavirolide C. Applications of Cu-catalyzed asymmetric conjugate additions and Ru-catalyzed ring-closing metathesis. *J. Am. Chem. Soc.*, **130** (39), 12904–12906.

39. Martin, D., Kehrli, S., d'Augustin, M., Clavier, H., Mauduit, M., and Alexakis, A. (2006) Copper-catalyzed asymmetric conjugate addition of Grignard reagents to trisubstituted enones. Construction of all-carbon quaternary chiral centers. *J. Am. Chem. Soc.*, **128** (26), 8416–8417. See also ref. 36.

40. (a) Munro-Leighton, C., Blue, E.D., and Gunnoe, T.B. (2006) Anti-Markovnikov N-H and O-H additions to electron-deficient olefins catalyzed by well-defined Cu(I) anilido, ethoxide, and phenoxide systems. *J. Am. Chem. Soc.*, **128** (5), 1446–1447; (b) Munro-Leighton, C., Delp, S.A., Blue, E.D., and Gunnoe, T.B. (2007) Addition of N-H and O-H bonds of amines and

alcohols to electron-deficient olefins catalyzed by monomeric copper(I) systems: reaction scope, mechanistic details, and comparison of catalyst efficiency. *Organometallics*, **26** (6), 1483–1493.

41. Fructos, M.R., Belderrain, T.R., Nicasio, M.C., Nolan, S.P., Kaur, H., Díaz-Requejo, M.M., and Pérez, P.J. (2004) Complete control of the chemoselectivity in catalytic carbene transfer reactions from ethyl diazoacetate: an N-heterocyclic carbene-Cu system that suppresses diazo coupling. *J. Am. Chem. Soc.*, **126** (35), 10846–10847.

42. Gawley, R.E. and Narayan, S. (2005) Stannyl cyclopropanes by diastereoselective cyclopropanations with (tributylstannyl)-diazoacetate esters catalyzed by Cu(I) N-heterocyclic carbene. *Chem. Commun.*, (40), 5109–5111.

43. (a) Xu, Q. and Appella, D.H. (2008) Aziridination of aliphatic alkenes catalyzed by N-heterocyclic carbene copper complexes. *Org. Lett.*, **10** (7), 1497–1500; For applications in total synthesis of [(IPr)CuCl] as aziridination catalyst, see: (b) Trost, B.M. and Dong, G. (2006) New class of nucleophiles for palladium-catalyzed asymmetric allylic alkylation. Total synthesis of Agelastatin A. *J. Am. Chem. Soc.*, **128** (18), 6054–6055; (c) Liu, R., Herron, S.R., and Fleming, S.A. (2007) Copper-catalyzed tethered aziridination of unsaturated N-tosyloxy carbamates. *J. Org. Chem.*, **72** (15), 5587–5591.

44. (a) Lebel, H.H., Davi, M.M., Díez-González, S.S., and Nolan, S.P. (2007) Copper-carbene complexes as catalysts in the synthesis of functionalized styrenes and aliphatic alkenes. *J. Org. Chem.*, **72** (1), 144–149; For application in multicatalytic one-pot processes, see: (b) Lebel, H., Ladjel, C., and Bréthous, L. (2007) Palladium-catalyzed cross-coupling reactions in one-pot multicatalytic processes. *J. Am. Chem. Soc.*, **129** (43), 13321–13326; For application in total synthesis, see: (c) Lebel, H.H. and Parmentier, M.M. (2007) Copper-catalyzed methylenation reaction: total synthesis of (+)-Desoxygaliellalactone. *Org. Lett.*, **9** (18), 3563–3566.

45. (a) Liao, Y. and Huang, Y.Z. (1990) A novel olefination of diazo-compounds with carbonyl compounds mediated by tributylstibine and catalytic amount of Cu(I)I. *Tetrahedron Lett.*, **31** (41), 5897–5900; (b) Zhou, Z.L., Huang, Y.Z., and Shi, L.L. (1993) Reactions of diazocompounds with carbonyl compounds mediated by diorganyl telluride and catalytic amount of CuI compounds: conversion of aldehydes to a. *Tetrahedron*, **49** (31), 6821–6830.

46. Laitar, D.S., Tsui, E.Y., and Sadighi, J.P. (2006) Catalytic diboration of aldehydes via insertion into the copper-boron bond. *J. Am. Chem. Soc.*, **128** (34), 11036–11037.

47. Zhao, H., Dang, L., Marder, T.B., and Lin, Z. (2008) DFT studies on the mechanism of the diboration of aldehydes catalyzed by copper(I) boryl complexes. *J. Am. Chem. Soc.*, **130** (16), 5586–5594.

48. Beenen, M.A., An, C., and Ellman, J.A. (2008) Asymmetric copper-catalyzed synthesis of α-amino boronate esters from N-*tert*-butanesulfinyl aldimines. *J. Am. Chem. Soc.*, **130** (22), 6910–6911.

49. (a) Lillo, V., Prieto, A., Bonet, A., Díaz-Requejo, M.M., Ramírez, J., Pérez, P.J., and Fernández, E. (2009) Asymmetric β-boration of α,β-unsaturated esters with chiral (NHC)Cu catalysts. *Organometallics*, **28** (2), 659–662. (b) Bonet, A., Lillo, V., Ramírez, J., Díaz-Requejo, M.M., and Fernández, E. (2009) The selective catalytic formation of β-boryl aldehydes through a base-free approach. *Org. Biomol. Chem.*, **7** (8), 1533–1535.

50. (a) Laitar, D.S., Tsui, E.Y., and Sadighi, J.P. (2006) Copper(I) β-boroalkyls from alkene insertion: isolation and rearrangement. *Organometallics*, **25** (10), 2405–2408; (b) Lillo, V., Fructos, M.R., Ramírez, J., Braga, A.A.C., Maseras, F., Díaz-Requejo, M.M., Pérez, P.J., and Fernández, E. (2007) A valuable, inexpensive Cu/N-heterocyclic carbene catalyst for the selective diboration of styrene. *Chem. Eur. J.*, **13** (9), 2614–2621; (c) Dang, L., Zhao, H., Lin, Z., and Marder, T.B. (2008) Understanding the higher reactivity of $B(cat)_2$ versus $B(pin)_2$

in copper(I)-catalyzed alkene diboration reactions. *Organometallics*, **27** (6), 1178–1186; For the hydroboration of alkenes, see: (d) Lee, Y. and Hoveyda, A.H. (2009) Efficient boron–copper additions to aryl-substituted alkenes promoted by NHC–based catalysts. Enantioselective Cu-catalyzed hydroboration reactions. *J. Am. Chem. Soc.*, **131** (9), 3160–3161.
51. (a) Díez-González, S. and Nolan, S.P. (2007) Transition-metal catalyzed hydrosilylation of carbonyl compounds and imines. A review. *Org. Prep. Proc. Int.*, **36** (6), 523–559; (b) Smith, M.B. and March, J. (eds) (2001) *Advanced Organic Chemistry: Reactions, Mechanism, and Structure*, 5th edn, Wiley-Interscience, New York, pp. 1544–1604.
52. (a) Ojima, I., Li, Z., Zhu, J., and Apeloig, Y., (1998) Recent advances in the hydrosilylation and related reactions, in *Chemistry of Organic Silicon Compounds*, vol. 2 (eds Z. Rappaport and Y. Apeloig), John Wiley & Sons, Inc., New York, pp. 1687–1792; (b) Riant, O., Mostefaï, N., and Courmarcel, J. (2004) Recent advances in the asymmetric hydrosilylation of ketones, imines and electrophilic double bonds. *Synthesis*, (18), 2943–2958.
53. Mahoney, W.S., Brestensky, D.M., and Stryker, J.M. (1988) Selective hydride-mediated conjugate reduction of α,β-unsaturated carbonyl compounds using $[(Ph_3P)CuH]_6$. *J. Am. Chem. Soc.*, **110** (1), 291–293.
54. For recent reviews, see: (a) Lipshutz, B.H. (2009) Rediscovering organocopper chemistry through copper hydride. It's all about the ligand. *Synlett*, (4), 509–524; (b) Díez-González, S. and Nolan, S.P. (2008) Copper, silver, and gold complexes in hydrosilylation reactions. *Acc. Chem. Res.*, **41** (2), 349–358; (c) Rendler, S. and Oestreich, M. (2007) Polishing a diamond in the rough: CuH catalysis with silanes. *Angew. Chem. Int. Ed.*, **46** (4), 498–504.
55. Jurkauskas, V., Sadighi, J.P., and Buchwald, S.L. (2003) Conjugate reduction of α,β-unsaturated carbonyl compounds catalyzed by a copper carbene complex. *Org. Lett.*, **5** (14), 2417–2420.
56. Kaur, H., Zinn, F.K., Stevens, E.D., and Nolan, S.P. (2004) (NHC)Cu (NHC = N-heterocyclic carbene) complexes as efficient catalysts for the reduction of carbonyl compound. *Organometallics*, **23** (5), 1157–1160.
57. Díez-González, S., Kaur, H., Zinn, F.K., Stevens, E.D., and Nolan, S.P. (2005) A simple and efficient copper-catalyzed procedure for the hydrosilylation of hindered and functionalized ketones. *J. Org. Chem.*, **70** (12), 4784–4796.
58. (a) Jacobsen, H., Correa, A., Poater, A., Costabile, C., and Cavallo, L. (2009) Understanding the M–NHC (NHC = N-heterocyclic carbene) bond. *Coord. Chem. Rev.*, **253** (5-6), 687–703; (b) Díez-González, S. and Nolan, S.P. (2007) Stereoelectronic parameters associated with N-heterocyclic carbene (NHC) ligands: a quest for understanding. *Coord. Chem. Rev.*, **251** (5-6), 874–883.
59. Bantu, B., Wang, D., Wurst, K., and Buchmeiser, M.R. (2005) Copper(I) 1,3-R_2-3,4,5,6-tetrahydropyrimidin-2-ylidenes (R=mesityl, 2-propyl): Synthesis, X-ray structures, immobilization and catalytic activity. *Tetrahedron*, **61** (51), 12145–11215.
60. (a) Díez-González, S., Scott, N.M., and Nolan, S.P. (2006) Cationic copper(I) complexes as efficient precatalysts for the hydrosilylation of carbonyl compounds. *Organometallics*, **25** (9), 2355–2358; (b) Díez-González, S., Stevens, E.D., Scott, N.M., Petersen, J.L., and Nolan, S.P. (2008) Synthesis and characterization of $[Cu(NHC)_2]X$ complexes: Catalytic and mechanistic studies of hydrosilylation reactions. *Chem. Eur. J.*, **14** (1), 158–168.
61. Lorenz, C. and Schubert, U. (1995) An efficient catalyst for the conversion of hydrosilanes to alkoxysilanes. *Chem. Ber.*, **128** (12) 1267–1269.
62. (a) Moritami, Y., Appella, D.H., Jurkauskas, V., and Buchwald, S.L. (2000) Synthesis of β-alkyl cyclopentanones in high enantiomeric excess via copper-catalyzed asymmetric conjugate reduction. *J. Am. Chem. Soc.*, **122** (28), 6797–6798; (b) Yun, J. and Buchwald,

S.L. (2001) One-pot synthesis of enantiomerically enriched 2,3-disubstituted cyclopentanones via copper-catalyzed 1,4-reduction and alkylation. *Org. Lett.*, **3** (8), 1129–1131.

63. For an example of NHC–Si interaction see: Bonnette, F., Kato, T., Destarac, M., Mignani, G., Cossío, F.P.M., and Baceiredo, A. (2007) Encapsulated N-heterocyclic carbenes in silicones without reactivity modification. *Angew. Chem. Int. Ed.*, **46** (45), 8632–8635.

64. (a) Chiu, P., Chen, B., and Cheng, K.F. (1998) A conjugate reduction-intramolecular aldol strategy toward the synthesis of pseudolaric acid A. *Tetrahedron Lett.*, **39** (50), 9229–9232; (b) Chiu, P., Szeto, C.-P., Geng, Z., and Cheng, K.-F. (2001) Tandem conjugate reduction-aldol cyclization using Stryker's reagent. *Org. Lett.*, **3** (12), 1901–1903; (c) Chiu, P., Szeto, C.P., Geng, Z., and Cheng, K.F. (2001) Application of the tandem Stryker reduction–aldol cyclization strategy to the asymmetric synthesis of lucinone. *Tetrahedron Lett.*, **42** (24), 4091–4093; (d) Chiu, P., Leung, S.K. (2004) Stoichiometric and catalytic reductive aldol cyclizations of alkynediones induced by Stryker's reagent. *Chem. Commun.*, (20) 2308–2309; (e) Chiu, P. (2004) Organosilanes in copper-mediated conjugate reductions and reductive aldol reactions. *Synthesis*, (13) 2210–2215.

65. (a) Lam, H.W. and Joensuu, P.M. (2005) Cu(I)-Catalyzed reductive aldol cyclizations: diastereo- and enantioselective synthesis of β-hydroxylactones. *Org. Lett.*, **7** (19), 4225–4228; (b) Deschamp, J., Chuzel, O., Hannedouche, J., and Riant, O. (2006) Highly diastereo- and enantioselective copper-catalyzed domino reduction/aldol reaction of ketones with methyl acrylate. *Angew. Chem. Int. Ed.*, **45** (8), 1292–1297; (c) Zhao, D.B., Oisaki, K., Kanai, M., and Shibasaki, M. (2006) Catalytic enantioselective intermolecular reductive aldol reaction to ketones. *Tetrahedron Lett.*, **47** (9), 1403–1407; (d) Chuzel, O., Deschamp, J., Chausteur, C., and Riant, O. (2006) Copper(I)-catalyzed enantio- and diastereoselective tandem reductive aldol reaction. *Org. Lett.*, **8** (26), 5943–5946.

66. Deutsch, C., Lipshutz, B.H., and Krause, N. (2007) Small but effective: copper hydride catalyzed synthesis of α-hydroxyallenes. *Angew. Chem. Int. Ed.*, **46** (10), 1650–1653. See also, Deutsch, C., Lipshutz, B.H., and Krause, N. (2009) (NHC) Cutt-catalyzed to allenes via propargylic carbonate S_{N^2} Reductions. *Org. Lett.*, **11** (21), 5010–5012.

67. Kolb, H.C., Finn, M.G., and Sharpless, K.B. (2001) Click chemistry: diverse chemical function from a few good reactions. *Angew. Chem. Int. Ed.*, **40** (11), 2004–2021.

68. (a) Tornøe, C.W., Christensen, C., and Meldal, M. (2002) Peptidotriazoles on solid phase: [1,2,3]-Triazoles by regiospecific copper(I)-catalyzed 1,3-dipolar cycloadditions of terminal alkynes to azides. *J. Org. Chem.*, **67** (9), 3057–3064; (b) Rostovtsev, V.V., Green, L.G., Fokin, V.V., and Sharpless, K.B. (2002) A stepwise Huisgen cycloaddition process: Copper(I)-catalyzed regioselective ligation of azides and terminal alkynes. *Angew. Chem. Int. Ed.*, **41** (14), 2596–2599.

69. (a) Huisgen, R. (1989) Kinetics and reaction mechanisms: selected examples from the experience of forty years. *Pure Appl. Chem.*, **61** (4), 613–628; (b) Padwa, A. (1991) Nonpolar additions to alkenes and alkynes, in *Comprehensive Organic Synthesis*, Vol. 4 (eds B.M.Trost and I. Fleming), Pergamon Press, Oxford, pp. 1069–1109.

70. (a) Pérez-Balderas, F., Ortega-Muñoz, M., Morales-Sanfrutos, J., Hernández-Mateo, F., Calvo-Flores, F.G., Calvo-Asín, J.A., Isac-García, J., and Santoyo-González, F. (2003) Multivalent neoglycoconjugates by regiospecific cycloaddition of alkynes and azides using organic-soluble copper catalysts. *Org. Lett.*, **5** (11), 1951–1954; (b) Chan, T.R., Hilgraf, R., Sharpless, K.B., and Fokin, V.V. (2004) Polytriazoles as copper(I)-stabilizing ligands in catalysis. *Org. Lett.*, **6** (17), 2853–2855; (c) Gerard, B., Ryan, J., Beeler, A.B., and Porco, J.A., Jr. (2006) Synthesis of 1,4,5-trisubstituted-1,2,3-triazoles by

copper-catalyzed cycloaddition-coupling of azides and terminal alkynes. *Tetrahedron*, **62** (26), 6405–6411; (d) Candelon, N., Lastécoutères, D., Diallo, A.K., Ruiz Aranzaes, J., Astruc, D., and Vincent, J.-M. (2008) A highly active and reusable copper(I)-tren catalyst for the 'Click' 1,3-dipolar cycloaddition of azides and alkynes. *Chem. Commun.*, (6), 741–743.

71. Díez-González, S., Correa, A., Cavallo, L., and Nolan, S.P. (2006) (NHC)Copper(I)-catalyzed [3+2] cycloaddition of azides and mono- or disubstituted alkynes. *Chem. Eur. J.*, **12** (29), 7558–7564.

72. (a) Appukkuttan, P., Dehaen, W., Fokin, V.V., and Van der Rycken, E. (2004) A microwave-assisted Click chemistry synthesis of 1,4-disubstituted 1,2,3-triazoles via a copper(I)-catalyzed three-component reaction. *Org. Lett.*, **6** (23), 4223–4225; (b) Feldman, A.K., Colasson, B., and Fokin, V.V. (2004) One-pot synthesis of 1,4-disubstituted 1,2,3-triazoles from in situ generated azides. *Org. Lett.*, **6** (22), 3897–3899.

73. Broggi, J., Díez-González, S., Petersen, J.L., Berteina-Raboin, S., Nolan, S.P., and Agrofoglio, L.A. (2008) Study of copper(I) catalysts for the synthesis of carbanucleosides via azide-alkyne 1,3-dipolar cycloaddition. *Synthesis*, (1), 141–148.

74. Séverac, M., Le Pleux, L., Scarpaci, A., Blart, E., and Odobel, F. (2007) Synthesis of new azido porphyrins and their reactivity in copper(I)-catalyzed Huisgen 1,3-dipolar cycloaddition reaction with alkynes. *Tetrahedron Lett.*, **48** (37), 6518–6522.

75. Maisonial, A., Serafin, P., Traïkia, M., Debiton, E., Théry, V., Aitken, D.J., Lemoine, P., Voissat, B., and Gautier, A. (2008) Click chelators for platinum-based anticancer drugs. *Eur. J. Inorg. Chem.*, (2), 298–305.

76. Díez-González, S., Stevens, E.D., and Nolan, S.P. (2008) A [(NHC)CuCl] complex as a latent Click catalyst. *Chem. Commun.*, (39), 4747–4749.

77. Díez-González, S. and Nolan, S.P. (2008) [(NHC)$_2$Cu]X complexes as efficient catalysts for azide-alkyne Click chemistry at low catalyst loadings. *Angew. Chem. Int. Ed.*, **47** (46), 8881–8884.

78. Himo, F., Lovell, T., Hilgraf, R., Rostovtsev, V.V., Noodleman, L., Sharpless, K.B., and Fokin, V.V. (2005) Copper(I)-catalyzed synthesis of azoles. DFT study predicts unprecedented reactivity and intermediates. *J. Am. Chem. Soc.*, **127** (1), 210–216.

79. For the first example of a copper-catalyzed cycloaddition of an internal alkyne, see ref. 72. See also ref. 71d.

80. Thompson, J.S., Bradley, A.Z., Park, K.-H., Dobbs, K.D., and Marshall, W. (2006) Copper(I) complexes with bis(trimethylsilyl)acetylene: role of ancillary ligands in determining π back-bonding interactions. *Organometallics*, **25** (11), 2712–2714.

81. Nolte, C., Mayer, P., and Straub, B.F. (2007) Isolation of a copper(I) triazolide: a click intermediate. *Angew. Chem. Int. Ed.*, **46** (12), 2101–2103.

82. Teyssot, M.-L., Jarrousse, A.-S., Chevry, A., De Haze, A., Beaudoin, C., Manin, M., Nolan, S.P., Díez-Gonzále, S., Morel, L., and Gautier, A. (2009) Toxicity of copper(I)-NHC complexes against human tumor cells: Induction of cell cycle arrest, apoptosis, and DNA cleavage. *Chem. Eur. J.*, **15** (2), 314–318.

83. Zhao, H., Lin, Z., and Marder, T.B. (2006) Density functional theory studies on the mechanism of the reduction of CO2 to CO catalyzed by copper(I) boryl complexes. *J. Am. Chem. Soc.*, **128** (49), 15637–15643 See also ref. 17.

84. Keaton, R.J., Blacquiere, J.M., and Baker, R.T. (2007) Base metal catalyzed dehydrogenation of ammonia-borane for chemical hydrogen storage. *J. Am. Chem. Soc.*, **129** (7), 1844–1855.

85. (a) Hu, X., Castro-Rodríguez, I., Olsen, K., and Meyer, K. (2004) Group 11 metal complexes of N-heterocyclic carbene ligands: nature of the metal–carbene bond. *Organometallics*, **23** (4), 755–764; (b) Nemcsok, D., Wichmann, K., and Frenking, G. (2004) The significance of π interactions in group 11 complexes with N-heterocyclic carbenes. *Organometallics*, **23** (15), 3640–3646; (c) Kausamo, A., Tuononen, H.M.,

Krahulic, K.E., and Roesler, R. (2008) N-Heterocyclic carbenes with inorganic backbones: electronic structures and ligand properties. *Inorg. Chem.*, **47** (3), 1145–1154.

86. (a) Haider, J., Kunz, K., and Scholz, U. (2004) Highly selective copper-catalyzed monoarylation of aniline. *Adv. Synth. Catal.*, **346** (7), 717–722; (b) Yan, J.-C., Zhou, L., and Wang, L. (2008) Amination reactions of aryl halides with nitrogen-containing reagents catalyzed by CuI in ionic liquid. *Chin. J. Chem.*, **26** (1), 165–169.

4
Supported Organocatalysts as a Powerful Tool in Organic Synthesis

Francesco Giacalone, Michelangelo Gruttadauria, and Renato Noto

4.1
Introduction

Enantiomerically pure substances, important both in the fields of pharmaceuticals, agrochemicals, and fine chemicals as well as key intermediates in the organic synthesis, until recently have been obtained mainly by employing two classes of catalysts: transition metal complexes and enzymes. However, while the latter are very expensive, rather unstable, and condition dependent, the former are often moisture- and oxygen-sensitive and require demanding reaction conditions such as absolute solvents, low temperature, inert atmosphere, and so on. Moreover, transition metal complexes give rise to the metal leaching phenomenon, which is detrimental for pharmaceutical purposes, since it leads to pollution, waste treatments, and product contamination [1].

Very recently a third class is pushing up and growing in importance: organocatalysts. These are small organic molecules composed of carbon, nitrogen, oxygen, sulfur, phosphorus, and fluorine, which do not contain a metal element in their structure and they are able, in substoichiometric amounts, to promote an acceleration in chemical reactions (see Figure 4.1 for some example). In addition, organocatalysts are robust and low-cost compounds, nontoxic with high resistance to air and moisture. All these advantages have had a fundamental importance in the birth and paramount growing of a new branch of chemistry: *organocatalysis*.

Although during the end of the nineteenth and early and mid-twentieth centuries some organically catalyzed reactions have been described, it was in the third millennium when organocatalysis has been applied to enantioselective synthesis, emerging as a new field in organic synthesis [2]. In fact, in 2000, the proline-catalyzed aldol reaction [3], which was previously reported by Hajos and Parrish [4] and Eder *et al.* [5], was rediscovered, and in the same year, MacMillan used the term *organocatalysis* [6].

Enantioselective organocatalytic reactions can be carried out in several ways by secondary or primary amine catalysis via enamine or via iminium ions [7], by phase transfer catalysis [8], by Brønsted acids catalysis [9], or by H-bonding catalysis [10]. Covalent binding of substrate normally requires high catalyst loading

Ideas in Chemistry and Molecular Sciences: Advances in Synthetic Chemistry. Edited by Bruno Pignataro
Copyright © 2010 WILEY-VCH Verlag GmbH & Co. KGaA, Weinheim
ISBN: 978-3-527-32539-9

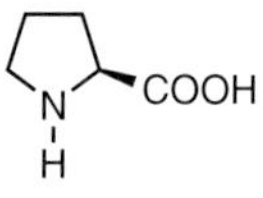

Figure 4.1 Selection of typical organocatalysts.

1 (L-Proline)
2 (Imidazolidinone derivative) 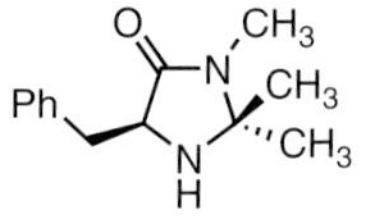
3 (Quinine) 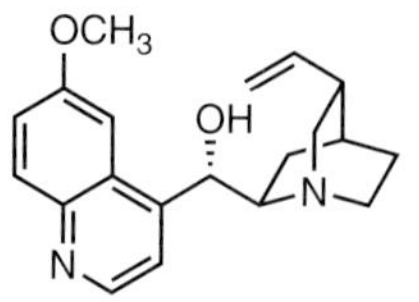
4 (BINAP) 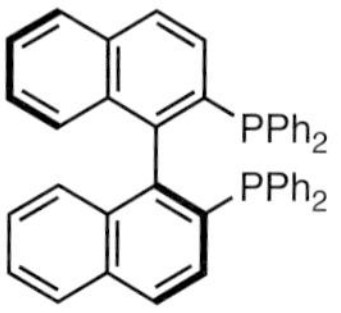
5 (Thiourea derivative) 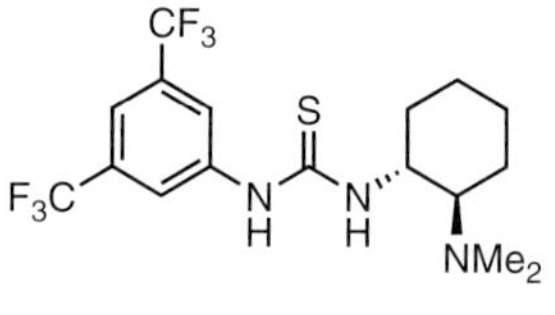
6 (D-Fructose derivative) 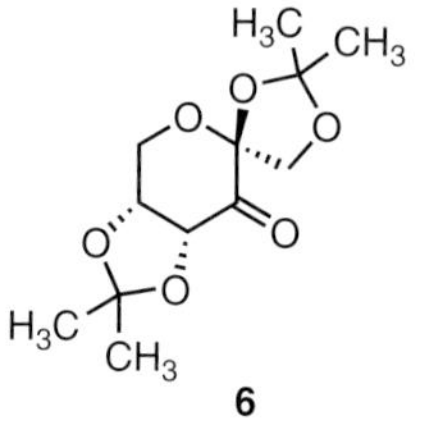

(for proline catalysis typically 20–30 mol%). Noncovalent interactions such as hydrogen-bonding facilitates low catalyst loadings (down to 0.001 mol%) [11].

A huge amount of data have been reported for secondary amine catalysts, mainly because the renaissance of this topic regarded the use of proline and MacMillan's imidazolidinones. L-Proline [12] can be meant as the simplest "enzyme" and, in addition to aldol reaction [13], it has been successfully applied to many other reactions such as Robinson annulation [4, 5, 14], Mannich reactions [15], Michael reactions [16], direct electrophilic α-aminations [17], Diels–Alder reactions [18], Baylis–Hillman reactions [19], aza-Morita–Baylis–Hillman reactions [20], α-selenenylation [21], oxidation [22], chlorination [23], and others [24].

Imidazolidinones have been employed for enantioselective organocatalytic Diels–Alder reactions [6], Friedel–Craft reactions [25] and, more recently, for enantioselective organo-SOMO (singly occupied molecular orbital) reactions such as α-allylation and α-vinylation of aldehydes [26].

Concerning proline-catalyzed reactions, many proline derivatives have been designed and synthesized with the aim of increasing reactivity and stereoselectivity. Parallelly, immobilization of organocatalysts has been growing in importance because the possibility of recycling a catalyst has an important impact on catalysts costs [27]. This importance is evident both when a less expensive catalyst such as simple proline is used in large quantities, or when more expensive catalysts prepared in multistep syntheses are used. Then, from an economical point of view, improved immobilization procedures that allow recovery and reuse of organocatalysts could be of still higher value, thereby increasing the greenness of the process. Moreover, morphological properties of heterogeneous supports, such as polystyrene or silica, may have a great influence on the outcome of the reactions. As a consequence, these materials can be modulated in such a way that high stereoselectivities can be achieved.

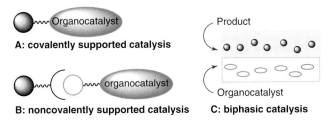

Figure 4.2 Approaches for organocatalyst immobilization.

The immobilization of the organocatalysts that need to be used under heterogeneous conditions can be achieved through different and generally applicable approaches (Figure 4.2):

1) **Covalently supported catalysts:** The organocatalyst is covalently anchored to a soluble (e.g., PEG, dendrimer) or insoluble (e.g., MCM-41, polystyrene, magnetite) support.
2) **Noncovalently supported catalysts:** The organocatalyst is adsorbed (e.g., onto (ILs) ionic liquids-modified SiO_2), dissolved (e.g., polyelectrolytes), included (e.g., β-Cyclodextrin, zeolites, clays) or linked by electrostatic interactions (e.g., PS-SO_3H, layered double hydroxides) in several supports.
3) **Biphasic catalysts:** The organocatalyst is dissolved and remains in ILs and, after the reaction, the product is separated by distillation, crystallization, or some other physical means. Ionic liquid–anchored organocatalysts can be considered as an advanced development of this approach since this simplifies the work-up and avoids extraction and phase separation.

In addition, immobilization of organocatalysts (e.g., proline) allows their use in aqueous media, which is of special interest because it is directly relevant to the class I aldolase-catalyzed reactions under physiological conditions. Indeed, another aspect of organocatalyzed reactions that is currently receiving great interest is the use of water as reaction medium both for supported and unsupported organocatalysts. The use of a high amount of water has been developed with the aim to obtain higher yields and stereoselectivities [28]. Its use is correlated with more hydrophobic catalysts, both in the presence and absence of acid additives. The higher hydrophobicity can be obtained by the proper modification of the organocatalyst structure in the case of unsupported catalysts, or by the nature of the support (polystyrene).

4.2
L-Proline and its Derivatives on Ionic Liquid-Modified Silica Gels

The asymmetric intermolecular aldol reaction affording β-hydroxy ketones, important building blocks for the synthesis of polyfunctional compounds and natural products [29], is one of the most important methods of C–C bond-formation in organic synthesis [30]. Proline and its derivatives have emerged as powerful catalysts for this reaction, operating often as a bifunctional catalysts and playing the role of simplified versions of the type I aldolase enzyme [31]. On the other hand, ionic liquids have become very useful alternative solvents for synthesis and catalysis since they are able to dissolve a great variety of catalysts [32]. The big attention ionic liquids have attracted, is due to their peculiar characteristics such as low vapor pressure, immiscibility with several solvents, and strong solubilizing ability. For all these advantages, they have found several applications in the immobilization and recycling of catalysts including the immobilization of L-proline [33]. An advance of this approach has been the use of ionic liquid–anchored organocatalysts. For instance, proline has been covalently linked to ionic liquid moieties such as in Figure 4.3. Such an approach allowed reaching of higher stereoselectivities and higher activities compared to the use of simple proline dissolved into ionic liquids [34]. Modification of the ionic liquid moiety also allowed the use in aqueous medium [35].

Recently, the new concept of supported ionic liquid catalysis, which involves the treatment of a monolayer of covalently attached ionic liquid on the surface of silica gel with additional ionic liquid [36] has been reported. This approach allows to solubilize the catalyst on the ionic liquid layer, while the solid support makes easier the reaction mixture separation and catalyst recovery. Moreover, since ionic liquids are expensive it is desirable to minimize the amount of ionic liquid used in usual biphasic reaction systems. In order to develop a new methodology for organocatalyst recycling using a supported ionic liquid approach, Gruttadauria and coworkers reported the first example of supported ionic liquid asymmetric catalysis. The case of proline-catalyzed aldol reaction was investigated. In this approach, the organocatalyst, proline, was immobilized by adsorption on the surface of modified silica gel with a monolayer of covalently attached ionic liquid [37]. To this covalently attached ionic liquid layer additional ionic liquid was also added by adsorption.

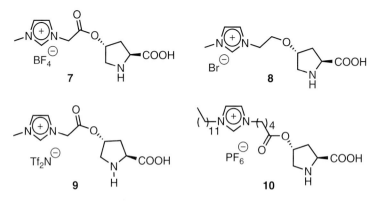

Figure 4.3 Structures of some ionic liquid–immobilized proline.

These layers serve as the reaction phase in which the homogeneous chiral catalyst is dissolved.

In a first attempt, proline was added from an acetonitrile/water solution on three different imidazolium-modified silica gels **14–16** with additional adsorbed ionic liquid (Scheme 4.1) [37a].

Data obtained showed that good results were achieved with BF_4-modified silica gel containing additional [bmim]BF_4 (bmim: 1-butyl-3-methylimidazole). Interestingly, ionic liquid adsorbed onto unmodified silica gel gave poor results showing the importance of ionic liquid modification of its surface.

Later, a deeper study on this subject was carried out [37b]. Several modified silica gels were prepared (Scheme 4.2). To some of these silica gels additional ionic liquids were adsorbed. Proline was added from an acetonitrile/water or methanol solution.

From the data obtained it was concluded that catalytic materials **17**/proline and **15**-[bmim]BF_4/proline were the most promising, whereas, silica gels modified with aliphatic ionic moiety (**18**) or without ionic moiety (**19**) gave very poor results. Such materials were employed in the aldol reaction between acetone and a set of aldehydes. Moreover, several recycling studies were also carried out. The best system was **15**/[bmim]BF_4/proline. After seven cycles the ee value was still good, however, a decrease in conversion was observed.

Then the support was regenerated washing the modified silica with methanol to remove exhausted proline and adsorbed ionic liquid and recharged with fresh [bmim]BF_4/proline. The regenerated material was successfully used for further six cycles with high yield and enantioselectivity.

This methodology was further developed by using other ionic liquid–modified silica gels.

Silica gels **15**, **20–23** (see Figure 4.4) were used and, in these cases, no additional ionic liquid was adsorbed onto their surfaces. Proline and the more expensive tripeptide H-L-Pro-L-Pro-L-Asp-NH_2 **24** were immobilized. Tripeptide **24** was previously found to be a powerful catalyst for the aldol reaction between acetone and aldehydes [38]. It was used in 1 mol% giving, in several cases, better ee values than proline under homogeneous conditions (Scheme 4.3).

Better enantioselectivities were obtained when the catalyst was adsorbed from a methanol solution rather than from aqueous solutions. Moreover, both enantioselectivities and yields were comparable or better than those obtained by using proline under homogeneous conditions. It is worthy to mention that this system gave higher enantioselectivities with respect to those obtained in pure acetone. These results indicated that **22**/pro material was superior to both **15**/pro and **20**/pro. Tripeptide H-Pro-Pro-Asp-NH_2 was supported on silica gel **22** because of the better overall performance of this support [39]. The use of chloride- or hexafluorophosphate-modified silica gels **21** and **23** gave higher enantioselectivities in the first reaction cycle. However, recycling studies further indicated that silica gel **22** was indeed the best support. Comparison between proline- and tripeptide-supported catalysts showed that the former material was more recyclable. Indeed, the L-proline-supported material **22**/pro was used, at least, up to

72 | *4 Supported Organocatalysts as a Powerful Tool in Organic Synthesis*

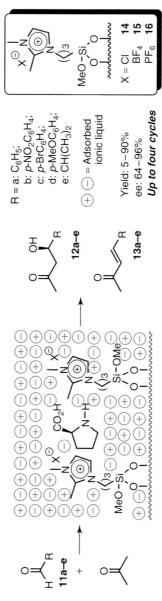

Scheme 4.1 Aldol reaction catalyzed by proline supported on ionic liquid–modified silica gels **14–16** with additional ionic liquid.

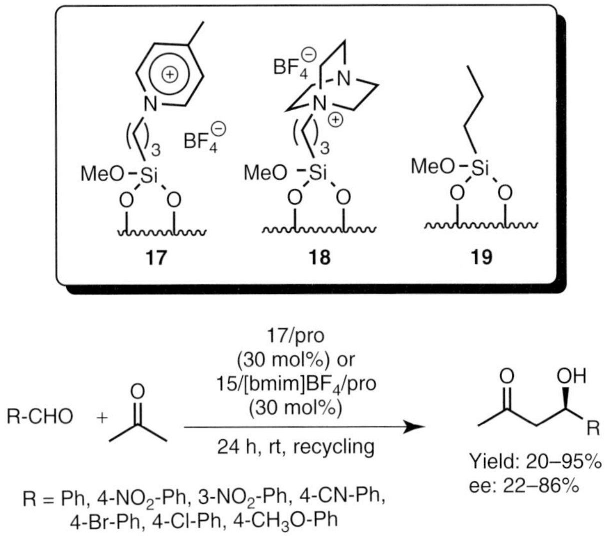

Scheme 4.2

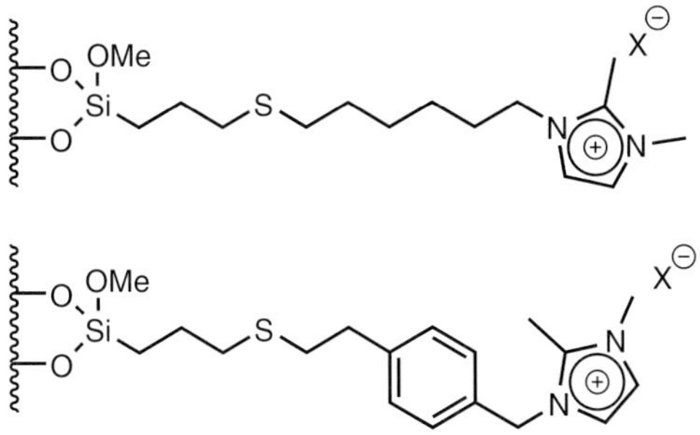

R-CHO + acetone → 17/pro (30 mol%) or 15/[bmim]BF$_4$/pro (30 mol%), 24 h, rt, recycling → aldol product

R = Ph, 4-NO$_2$-Ph, 3-NO$_2$-Ph, 4-CN-Ph, 4-Br-Ph, 4-Cl-Ph, 4-CH$_3$O-Ph

Yield: 20–95%
ee: 22–86%

Figure 4.4 Structures of silica-supported ionic liquids **20–23**.

nine times with unchanged yield and selectivity. Interestingly, comparison of these data with literature values showed that the monolayer of covalently attached ionic liquid behaved as a bulk ionic liquid.

4.3 Polystyrene-Supported Proline as a Versatile and Recyclable Organocatalyst

As stated above, since the first report in 2000 [3] on the use of L-proline as an organocatalyst, a lot of effort was devoted to its immobilization and recycling [27]. In fact, in 2001, Benaglia, Cozzi, and coworkers covalently anchored L-proline on PEG$_{500}$ monomethyl ether by means of a succinate spacer (**25** in Figure 4.5), and they used such soluble and recyclable catalysts in the enantioselective aldol

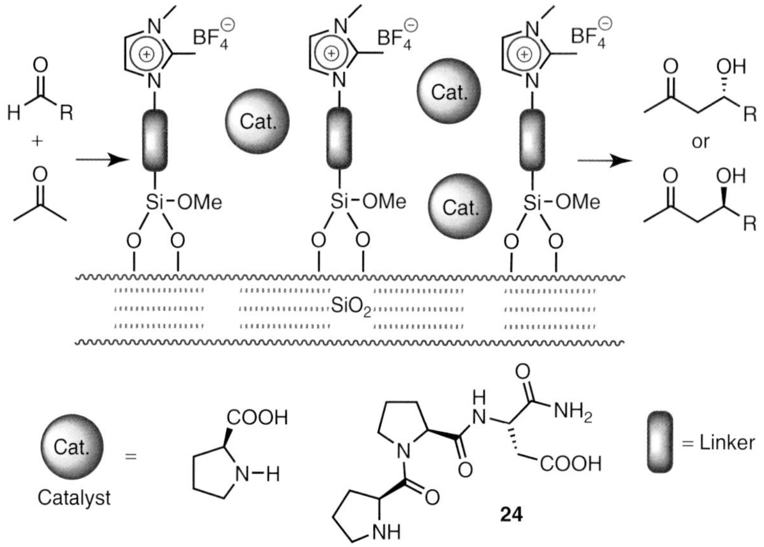

Scheme 4.3

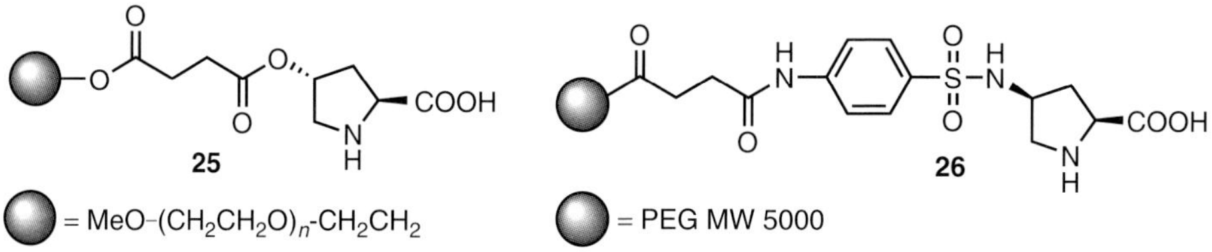

Figure 4.5 Chemical structures for catalysts **25** and **26**.

condensation between acetone or hydroxyacetone with several aldehydes, in the synthesis of the Wieland–Mischler ketone and in the Mannich reaction [40], as well as in the addition of ketones to β-nitrostyrene and in the addition of 2-nitropropane to cyclohexanone [41]. For this latter reaction also, PEG-resin **26** has been tested with fair results [42].

Since then, several groups have been engaged in the preparation and use of polymer-supported prolines and, to date, polystyrene-resins resulted to be the mostly employed and those showing best results. Pericás widely employed the so-called "click-chemistry" protocol in order to covalently bound proline to resins (Figure 4.6).

Catalysts **27–31** appeared to be very active and selective materials. Catalyst **27** was used in the aldol reaction between several ketones and arylaldehydes [43, 44]. The reaction worked nicely in water in the presence of a catalytic amount of DiMePEG in order to facilitate reactant diffusion to the resin. The same catalyst was active also in the α-aminoalkylation of aldehydes and ketones with good yield and high

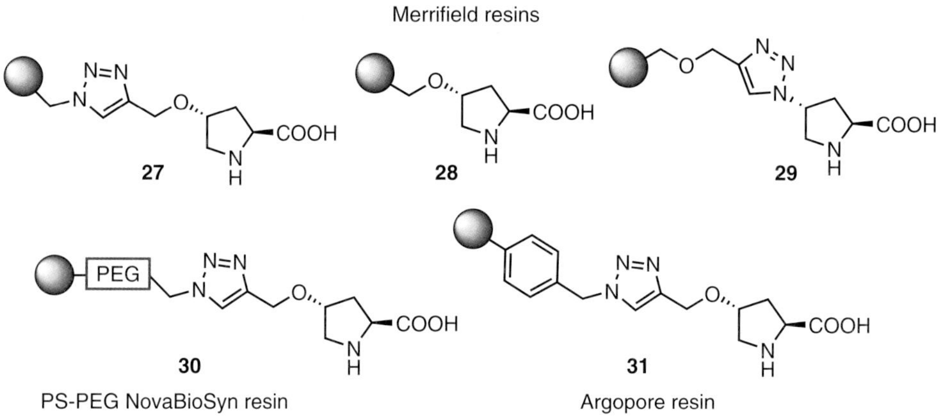

Figure 4.6 Chemical structures for polystyrene-supported catalysts **27–31**.

enantioselectivity [45]. Catalyst **29** swelled in water in spite of the hydrophobicity of the polymer backbone, therefore, behaving as an artificial aldolase [44]. It is noteworthy that in all cases, recycle and reuse of catalyst was accomplished with no losses in activity or stereoselectivity.

Beside polymeric materials, several other supports such as mesoporous MCM-41 silica [46], magnetite [47], ionic liquids [34] have been employed for the anchoring of proline and its derivatives.

In 2007, Gruttadauria and coworkers immobilized 4-hydroxy-L-proline on the commercially available mercaptomethyl polymer–bound (1% cross-linked with DVB, spherical beads, particle size 100–200 mesh, 2.5 mmol/g loading) [48]. The anchorage of L-proline was accomplished in two steps (Scheme 4.4): (i) synthesis of styrene derivative **33** of hydroxy-L-proline; (ii) radical reaction between the polystyrene and **33** or **34** followed by the deprotection of proline moiety.

This flexible synthetic strategy is straightforward, leading to high-loading catalysts **35** and **36** (proline content ca. 1.4 mmol/g) in high yields. Once prepared, catalysts were tested both in nonasymmetric and asymmetric reactions with the aim to disclose whether our supported proline is also as versatile as proline itself catalyzing several kind of reactions, with the additional value of a recyclable and reusable material.

4.3.1
Nonasymmetric Reactions

4.3.1.1 α-Selenenylation of Aldehydes
Synthesis of α-phenylseleno-derivatives of carbonyl compounds is highly important since they represent valuable and versatile intermediates in the organic synthesis [49]. However, most of the synthetic procedures developed for the preparation of such compounds are indirect methods involving the reaction between aldehyde or ketone enolate and a selenium-based electrophilic reactant. Recently, Wang and

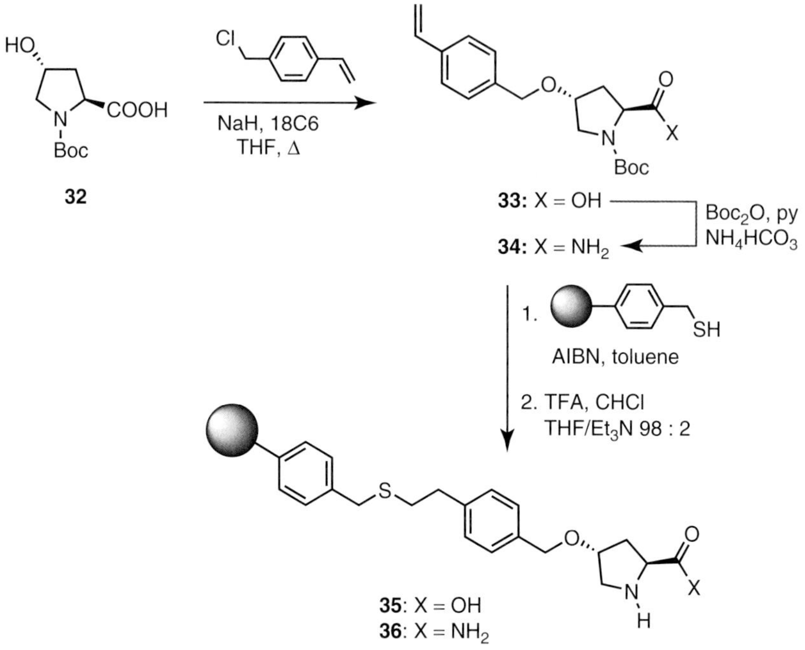

Scheme 4.4 Synthesis of polystyrene-supported proline and prolinamide **35** and **36**.

coworkers reported a new catalytic procedure for the direct α-selenenylation of aldehydes, in which the best results were observed when proline and prolinamide were used as catalysts in 30 mol%, the latter being the most active [21].

Hence, catalyst **35** (30 mol%) was tested in the α-selenenylation of aldehydes, employing as the selenenylating agent N-(phenylseleno)phthalimide (NPSP) in dichloromethane at room temperature (Table 4.1) [48].

Interestingly, resin **35** resulted to be an active and recyclable catalyst, giving rise to clean and high-yielding reactions. The proposed mechanism for the α-selenenylation of aldehydes in the presence of proline involves the formation of an enamminic intermediate, which then undergoes rate-limiting attack at the electrophilic selenium atom of NPSP [21] (Scheme 4.5). Formation of the Se–C bond takes place in concert with the departure of phthalimide anion to produce the α-selenoiminium ion intermediate, which in turn adds water to produce the corresponding carbinolamine, the precursor of the α-selenoaldehyde product.

This mechanism also applies in the case of prolinamide, which appeared to be a more active catalyst. In this sense, polystyrene-supported prolinamide **36** has also been tested for the title reaction (Table 4.2) [48]. It is worthy to note that even **36** works nicely with this reaction and, as observed under homogeneous conditions, it is more efficient than the corresponding proline-supported catalyst **35**. However, recycling studies revealed that, after three cycles, the catalytic activity falls down.

Table 4.1 α-Selenenylation of aldehydes catalyzed by polystyrene-supported proline **35** [48].

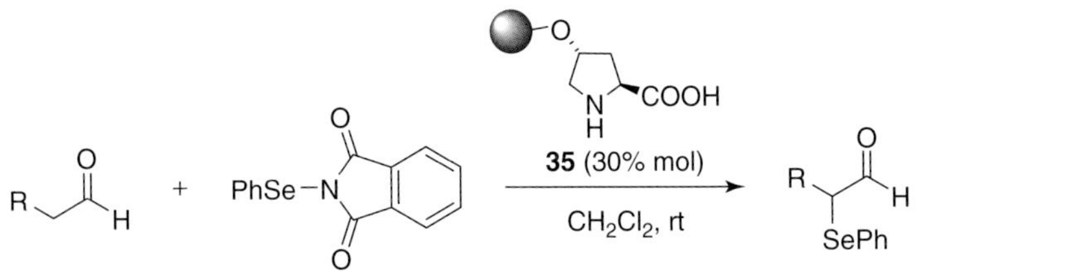

R–CHO + PhSe–N(phthalimide) →[**35** (30% mol)][CH$_2$Cl$_2$, rt] R–CH(SePh)–CHO

Entry	Cycle	Aldehyde	Product	Time (h)	Yield (%)
1	1–2	C$_9$H$_{19}$–CHO	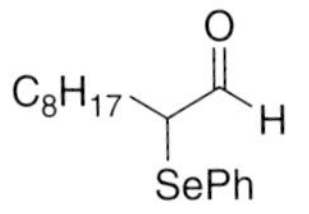 C$_8$H$_{17}$–CH(SePh)–CHO	1–2.5	65–85
2	1–3	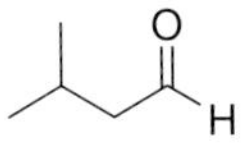 (CH$_3$)$_2$CHCH$_2$C(O)H (isovaleraldehyde-type)	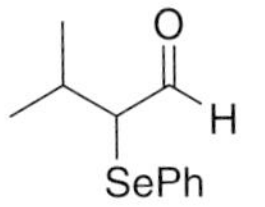 (CH$_3$)$_2$CH–CH(SePh)–CHO	2.5	88–90
3	1–4	C$_5$H$_{11}$–CHO	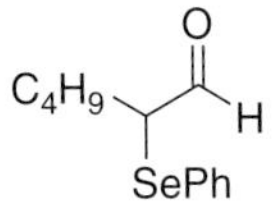 C$_4$H$_9$–CH(SePh)–CHO	2.5	96–97
4	1	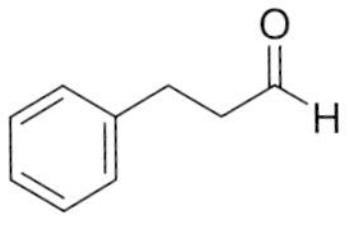 PhCH$_2$CH$_2$CHO	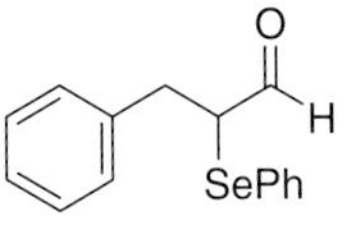 PhCH$_2$–CH(SePh)–CHO	2.5	85

So far, this study represents, to the best of our knowledge, the first examples of heterogeneous catalytic α-selenenylation of aldehydes.

4.3.1.2 Baylis–Hillman Reaction

The Baylis–Hillman reaction (B–H) [50] is, essentially, an atom-economical C–C bond-forming three-component reaction involving the coupling of the α-position of activated alkenes with carbon electrophiles in the presence of a tertiary amine. This reaction provides a simple and convenient methodology for the synthesis of multifunctionalized molecules as useful building blocks in organic synthesis [51]. Usually, as tertiary amine, DABCO, and imidazole are often used. In 2002, Shi reported that L-proline (30 mol%) in the presence of imidazole (30 mol%) is able to catalyze the B–H reaction between methyl vinyl ketone and aromatic aldehydes nonenantioselectively [52]. More recently, it has been reported that the above reaction can be carried out using a lower amount of catalyst (10 mol%) in a 9/1 dimethylformamide/water mixture, in which water plays a crucial role

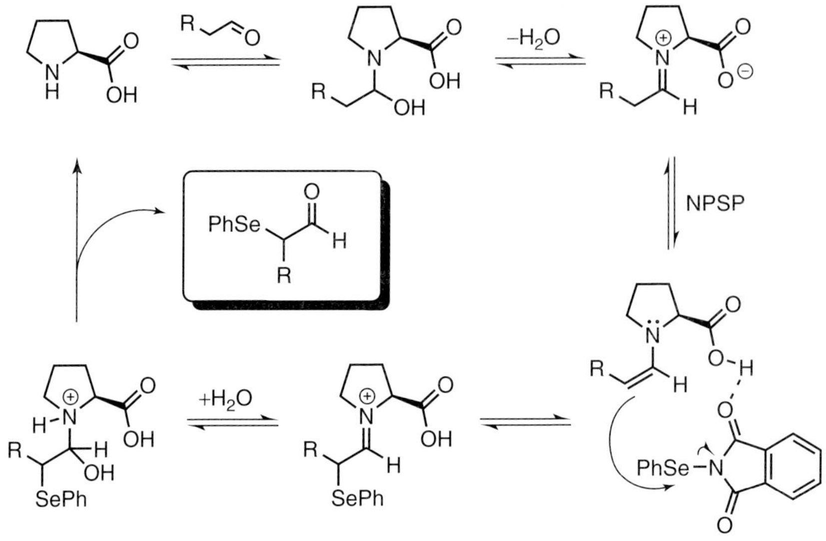

Scheme 4.5 Proposed mechanism for the α-selenenylation of aldehydes in the presence of L-proline.

in the acceleration of the reaction [53]. These findings led to the investigation of the use of supported proline **35** as a heterogeneous catalyst in the B–H reaction between alkyl vinyl ketones and substituted benzaldehydes. In fact, despite the great importance of such a reaction, only few reports with not outstanding results deal with heterogeneous B–H reaction [54]. After a quick screening of solvents, it was found that, in this case also, the best reaction medium was a 9/1 dimethylformamide/water mixture, employing both supported proline and imidazole in 10 mol% (Scheme 4.6) [55].

Reactions carried out with benzaldehydes having EWG groups in position 4 gave high isolated yields, while longer reaction times were needed and low yields were obtained with less reactive benzaldehydes such as 4-CH_3-, 4-CH_3O-benzaldehyde, and 2-naphthaldehyde. No enantioselectivity was observed under these reaction conditions. The proposed mechanism for the title reaction is shown in Scheme 4.7.

As in the case of the α-selenenylation reaction, the first step involves the formation of the intermediate α,β-unsaturated iminium ion by reaction between methyl vinyl ketone and the supported proline, which subsequently undergoes a nucleophilic attack by imidazole. The so-formed enamine thus reacts with the aldehyde to afford, after elimination of imidazole and hydrolysis of the iminium ion, the B–H adduct. Finally, recycling studies have been performed in this case also, revealing that, even for the B–H reaction, the catalyst can be used and recycled, with almost no losses in efficiency, for at least five times. The work reported here represents the first example about the use of supported proline as a heterogeneous catalyst in the B–H reaction between alkyl vinyl ketones and substituted benzaldehydes.

Table 4.2 α-Selenenylation of aldehydes catalyzed by polystyrene-supported prolinamide **36** [48].

Entry	Cycle	Product	4 (% mol)	Time (h)	Yield (%)
1	1	C₄H₉–CH(SePh)–CHO	5	0.08	52
2	1–4	C₄H₉–CH(SePh)–CHO	5	2.5	96–40
3	1	(CH₃)₂CH–C(SePh)–CHO	5	2.5	86
4	1	C₈H₁₇–CH(SePh)–CHO	5	2.5	84
5	1	PhCH₂–CH(SePh)–CHO	5	2.5	87

4.3.2
Asymmetric Reactions

4.3.2.1 Aldol Reaction

Although L-proline is able to promote aldol reaction with good stereocontrol in organic solvents like DMF or DMSO [3], no enantioselectivity was obtained for reactions carried out in water [56]. Nevertheless, as stated before, in such a medium, excellent results have been reported by Barbas and Hayashi when proline or pyrrolidine endowed with large apolar groups have been used [57]. In this regard, since polystyrene may be meant as a large apolar substituent, material, **35** may be useful for asymmetric aldol reaction in water [48, 58]. In this regard, it

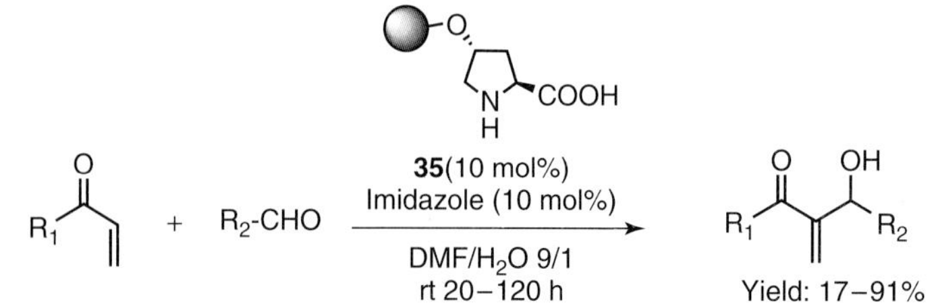

Scheme 4.6 Supported proline-catalyzed Baylis–Hillman reaction.

$R_1 = CH_3$ R_2 = 4-NO$_2$-Ph, 4-CN-Ph, 4-CF$_3$-Ph, 4-Br-Ph, 2-Cl-Ph, 3-Cl-Ph, 4-Cl-Ph, 3-NO$_2$-Ph, 3-Br-Ph, 2-CN-Ph, 2-F-Ph, 2-furyl, Ph, 4-CH$_3$-Ph, 4-CH$_3$O-Ph, 2-naphthyl, 2-Cl-5-NO$_2$-Ph

$R_1 = CH_2CH_3$ R_2 = 4-NO$_2$-Ph, 4-CN-Ph, 4-CF$_3$-Ph

Reaction conditions: 35 (10 mol%), Imidazole (10 mol%), DMF/H$_2$O 9/1, rt 20–120 h, Yield: 17–91%

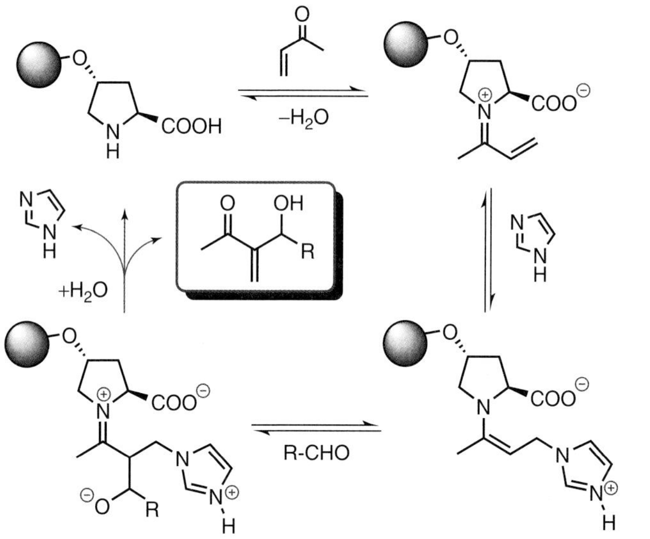

Scheme 4.7 Proposed mechanism for the supported proline/imidazole catalyzed Baylis–Hillman reaction.

was observed that, even if four phases (two solids: 4-nitrobenzaldehyde and resin **35**; two liquids: water and cyclohexanone) were present, after 22 hours of stirring the reaction was quantitative and with good stereoselectivity. Hence, the scope of the reaction was investigated rationally, first finding the right amount of water and screening several solvents. Interestingly, the reaction did not work either in polar aprotic solvents or in apolar solvents. On the contrary, polar protic solvents showed decreased conversion in the order CH$_3$OH > EtOH > i-PrOH. Then, several aromatic aldehydes were investigated in water giving excellent results, with reaction times ranging between 22 and 120 hours (Scheme 4.8).

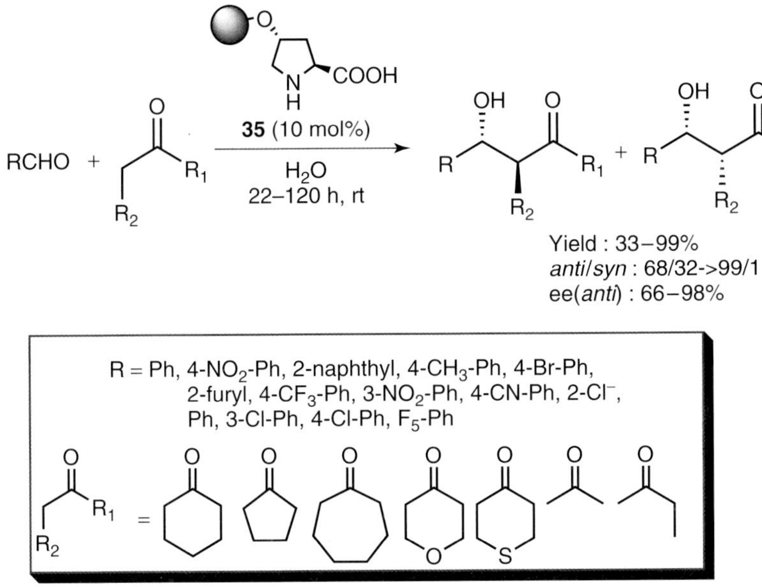

Scheme 4.8 Asymmetric aldol reaction catalyzed by supported proline **35**.

Good to excellent conversions, diastereoselectivities, and enantioselectivities were observed in almost all the examined substrates, and importantly, no additives in the reaction are required. Low levels of conversions were observed only for the more sterically hindered β-naphthaldehyde and the less reactive p-tolualdehyde, while aliphatic aldehydes did not react. In addition, a set of ketones has been also checked. Interestingly, the reaction between tetrahydro-4H-thiopyran-4-one and 4-nitrobenzaldehyde worked well in chloroform with a stoichiometric amount of water (7 μl) (yield: 55%, rd: 98:2; ee: 95%). Finally, recycling studies showed no decrease in yield and stereoselectivity after five cycles.

To explain the observed higher stereoselectivity, compared to those obtained using nonsupported proline, it has been hypothesized that hydrophilic proline moieties lie at the resin/H_2O interface, which facilitates the formation of a hydrophobic core on the inner surface of the resin (Figure 4.7). Water molecules, lying in the hydrophilic outer region, force hydrophobic aryl aldehydes into the restricted hydrophobic inner pocket. In this way, water promotes the reaction and increases the stereoselectivity. Thus, this material can be considered a better mimic of natural class I aldolase enzymes. Although immobilization of L-proline may be considered useless due to its low cost and availability in both enantiomeric forms, its immobilization has allowed to reach, in several cases including this [43, 44], higher performances compared to the native catalyst. Moreover, the linking strategy may be used as a model for the immobilization of more complex and expensive catalysts to lower the costs and avoid wastes, thereby improving the greenness of the process.

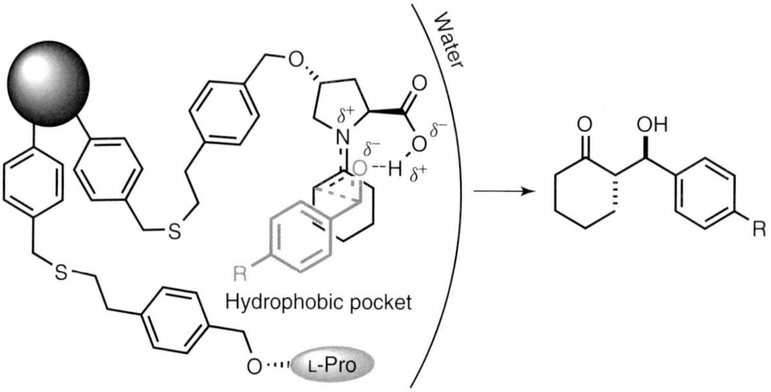

Figure 4.7 Proposed transition state model for aldol reaction using resin **35**.

4.4
Prolinamide-Supported Polystyrenes as Highly Stereoselective and Recyclable Organocatalysts for the Aldol Reaction

Recently, several substituted prolinamides have been found to be active and highly stereoselective catalysts for direct aldol reaction both in organic solvents [59] and in aqueous conditions [60]. Among them, catalysts with general structure **37** (Figure 4.8), showed enhanced performances, including excellent ee for aldols derived from acetone (catalyst loading: 5 mol% at −40 °C) [59a] or in brine (catalyst loading: 0.5 mol% at −5 °C) [60b].

Immobilization of more expensive organocatalysts, and hence their recovery and reuse, such as those substituted prolinamides, may be of higher interest from an economical point of view as well as in order to make more sustainable, the entire process. Interestingly, there are no reports in the literature about the use of immobilized organocatalysts giving high ee values (>90%) in the reaction between acetone and substituted benzaldehydes. Thus, it was considered possible to obtain high ee values for the above reaction using an immobilized organocatalyst. In this sense, the idea of using immobilized compounds **37** has been coupled with the anchoring procedure previously seen, with the aim to obtain a recyclable

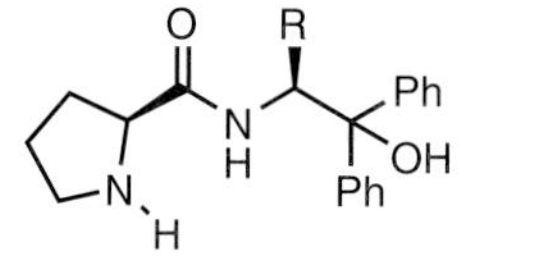

37a, R = H;   **37b**, R = Me
37c, R = i-Pr;   **37d**, R = i-Bu
37e, R = s-Bu;   **37f**, R = Bn
37g, R = Ph

Figure 4.8 Structure of catalysts **37a–g**.

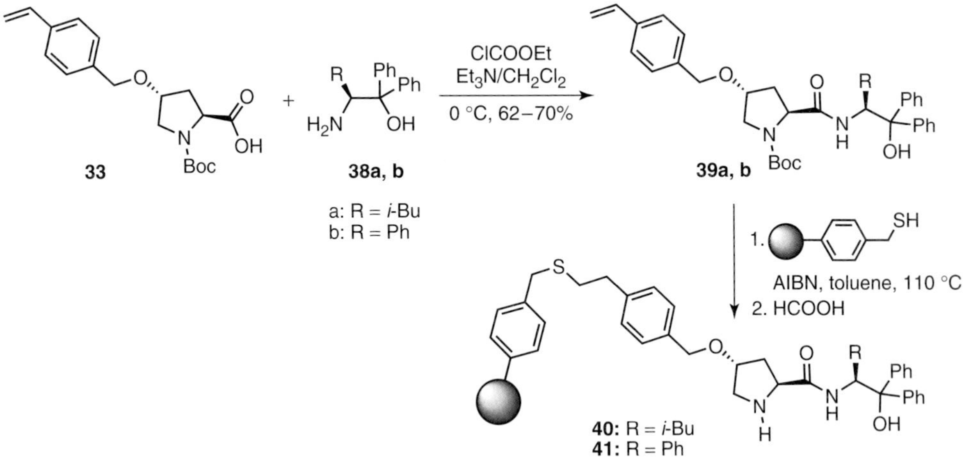

4.4 Prolinamide-Supported Polystyrenes as effective Organocatalysts | 83

Scheme 4.9 Synthetic of polystyrene-supported prolinamides **40** and **41**.

catalyst, trying to maintain its excellent behavior. In fact, although sometimes an immobilized catalyst could work better or "differently" than its unsupported version, often in the immobilization, one catalyst may partially or entirely lose the original catalytic activity [61].

Once again, the syntheses of resins **40** and **41** were achieved by following a straightforward procedure (Scheme 4.9) [62]. First, the carboxylic group of proline-derivative **33** was converted into the corresponding amide **39a, b** in good yield by reaction with amino alcohols **38a, b** in the presence of ethylchloroformate. Finally radical reactions were performed for the immobilization.

First, the best solvent system was determined and, in the mean time, recycling studies were performed. Water/chloroform 1 : 2 (v/v) mixture was found to be the best reaction medium. Resins were then recovered by filtration and reused in the next cycle, but dramatic drop in activity was observed. It was argued that a certain amount of catalyst was probably deactivated by an excess of the ketone with formation of the corresponding imidazolidinone (Figure 4.9) [63]. Therefore, catalysts were regenerated by treatment with formic acid in order to hydrolyze the imidazolidinone [64]. Remarkably, activity and enantioselectivity were restored and regeneration of resins was performed after each cycle.

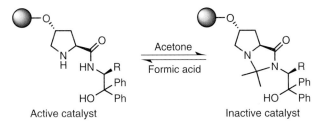

Figure 4.9 Inactivation and reactivation of resins **40** and **41**.

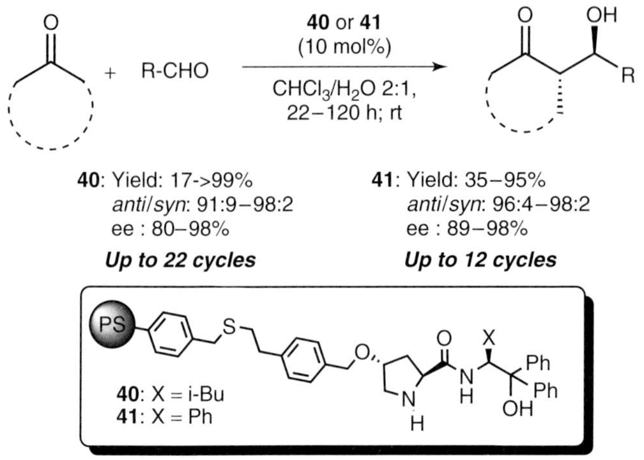
40 or 41
(10 mol%)
R-CHO
CHCl3/H2O 2:1,
22–120 h; rt
O
OH
R
40: Yield: 17->99%
anti/syn: 91:9–98:2
ee : 80–98%
Up to 22 cycles
41: Yield: 35–95%
anti/syn: 96:4–98:2
ee : 89–98%
Up to 12 cycles
PS
S
X
Ph
Ph
OH
40: X = i-Bu
41: X = Ph

Scheme 4.10

Next, catalysts **40** and **41** were used in the aldol reaction between acetone or cyclohexanone and several substituted benzaldehydes (Scheme 4.10). It is worth mentioning that both catalysts worked well with acetone, with resin **41** being the best one, showing good to excellent results both in terms of conversion (50–99%) and enantiomeric ratio (89–97%). Good diastereoselectivities were achieved with supported prolinamide **41** when cyclohexanone was the ketone of choice (96/4–98/2), as well as with other cyclic ketones.

Besides the astonishing recyclability of both materials, up to 12 cycles for **41** and up to 22 cycles for **40**, the performances obtained in some cases improved the values observed under homogeneous conditions or with the previously seen proline-based polystyrene **35**. Particularly, the enantioselectivities obtained employing acetone as ketone were, to the best of our knowledge, the highest achieved with a supported proline-derivative. It is worthy to note that these catalysts worked nicely at room temperature while the excellent results displayed by prolinamides **37a–g** under homogeneous condition were obtained on carrying out the reactions at −5 or −10 °C.

4.5
Outlook and Future Perspectives

Immobilization of proline and proline derivatives, covalently or not, has attracted much interest. It is fascinating how the immobilization of these simple molecules has stimulated the synthetic creativity of researchers [27]. Covalently linked organocatalysts make the recovery procedure very easy, avoiding leaching of the catalyst and simplifying the product isolation. This is true in the case of heterogeneous supports such as cross-linked polystyrene or silica. In these cases, the morphological properties of the support have a great influence on the outcome of the reactions. As a consequence, these materials may be less effective than their nonsupported homogeneous counterparts but, as just shown, in other cases they can be modulated in such a way that higher stereoselectivities can be achieved. In our opinion, studies for new highly active, stereoselective, and highly recyclable organocatalysts are always desirable. In fact, examples in which the organocatalyst is used in low percentages (<5 mol%) are scarcely known [11, 60b, 65]. Very recently, low-molecular inexpensive hydrophobic proline derivatives able to catalyze aldol reaction in a percentage as low as 0.5 mol% with high stereoselectivities have been reported [62a]. A proline covalently linked to an ionic liquid moiety has been employed in 0.1 mol% in the asymmetric aldol reaction displaying TON of up to 980 [66]. In this way one can reach the effectiveness obtained with catalytic organometallic complexes.

Complex and more expensive Organocatalysts may also be supported, such as the so-called Jørgensen catalyst [67] or, generally speaking, diarylprolinol ethers [68], which are able to promote several reactions [69] and cascades [70] affording natural products such as (+)-angustureine [71] or α-tocopherol [72].

As an alternative, other supports may be investigated in order to exploit their effect on enantioselective reactions. Some novel support may be represented by single- and multiwalled carbon nanotubes, which can be used as covalent or noncovalent supports [73]. In the latter case, pyrene-containing organocatalysts may represent good candidates for supramolecular supported catalysts. Also, nanostructured inorganic supports may be useful as reactant-confining nanoreactors in order to lead to improved conversions and selectivities. A novel approach in this sense may be represented by the hybrid metal-organic frameworks (MOFs) [74], crystalline materials assembled by the bonding of metal ions with polyfunctional organic ligands. MOFs, in some cases, may display catalytic activity by themselves [75] or may be modified in order to promote organic reactions [76]. To date, nothing has been reported on organocatalysts confined inside MOFs' structures or as a part of the constituting organic framework. This approach may lead to highly organized catalytic spaces or nanovessels, and may constitute a real step further in the field of organocatalysis.

Noticeably, no investigations have been reported yet about the use of continuous flow methods in organocatalysis. Indeed, a system in which the catalyst must not be removed from the reaction vessel is very attractive. We strongly believe that further interesting developments in this appealing field will appear soon.

References

1. Fubini, B. and Areán, L.O. (1999) Chemical aspects of the toxicity of inhaled mineral dusts. *Chem. Soc. Rev.*, **28**, 373–382.
2. (a) Dalko, P.I. and Moisan, L. (2001) Enantioselective organocatalysis. *Angew. Chem. Int. Ed.*, **40**, 3726–3748; (b) Berkessel, A. and Gröger, H. (2005) *Asymmetric Organocatalysis: From Biomimetic Concepts to Applications in Asymmetric Synthesis*, Wiley-VCH Verlag GmbH, Weinheim; (c) Benaglia, M., Puglisi, A., and Cozzi, F. (2003) Polymer-supported

organic catalysts. *Chem. Rev.*, **103**, 3401–3430; (d) (2004) Special issue: asymmetric organocatalysis. *Acc. Chem. Res.*, **37**, 487–631; (e) (2004) Special issue on organocatalysis. *Adv. Synth. Catal.*, **346**, 1007–1249; (f) (2006) Special issue on organocatalysis. *Tetrahedron*, **62**, 243–502; (g) (2007) Special issue on organocatalysis. *Chem. Rev.*, **107**, 5413–5883; (h) Pellissier, H. (2007) Asymmetric organocatalysis. *Tetrahedron*, **63**, 9267–9331; (i) Dondoni, A. and Massi, A. (2008) Asymmetric organocatalysis: from infancy to adolescence. *Angew. Chem. Int. Ed.*, **47**, 4638–4660; (j) Melchiorre, P., Marigo, M., Carlone, A., and Bartoli, G. (2008) Asymmetric aminocatalysis-Gold rush in organic chemistry. *Angew. Chem. Int. Ed.*, **47**, 6138–6171; (k) Guillena, G., Nájera, C., and Ramón, D.J. (2007) Enantioselective direct aldol reaction: the blossoming of modern organocatalysis. *Tetrahedron: Asymmetry*, **18**, 2249–2293.
3. List, B., Lerner, R.A., and Barbas, C.F. III (2000) Proline-catalyzed direct asymmetric aldol reactions. *J. Am. Chem. Soc.*, **122**, 2395–2396.
4. (a) Hajos, Z.G. and Parrish, D.R. (1971) Werkwijze voor de bereiding van 1,3-dioxycycloalkanen. German Patent 2102 623, filled Jan. 21, 1970 and issued Jul. 29, 1971; (b) Hajos, Z.G. and Parrish, D.R. (1976) Asymmetric synthesis of organic compounds. US Patent 3,975,440; filled Dec. 9, 1970 and issued Aug.17, 1976; (c) Hajos, Z.G. and Parrish, D.R. (1974) Asymmetric synthesis of bicyclic intermediates of natural product chemistry. *J. Org. Chem.*, **39**, 1615–1621.
5. (a) Eder, U., Sauer, G., and Wiechert, R. (1971) Verfahren zur herstellung optisch aktiver bicycloalkan derivate. German Patent 2,014,757, filled Mar. 20, 1970 and issued Oct. 7, 1971; (b) Eder, U., Sauer, G., and Wiechert, R. (1971) New type of asymmetric cyclization to optically active steroid cd partial structures. *Angew. Chem. Int. Ed.*, **10**, 496–497.
6. Ahrendt, K.A., Borths, C.J., and MacMillan, D.W.C. (2000) New strategies for organic catalysis: the first highly enantioselective organocatalytic Diels-Alder reaction. *J. Am. Chem. Soc.*, **122**, 4243–4244.
7. (a) Mukherjee, S., Woon Yang, Y., Hoffmann, S., and List, B. (2007) Asymmetric enamine catalysis. *Chem. Rev.*, **107**, 5471–5569; (b) Erkkilä, A., Majander, I., and Pihko, P.M. (2007) Iminium catalysis. *Chem. Rev.*, **107**, 5416–5470; (c) Peng, F. and Shao, Z. (2008) Advances in asymmetric organocatalytic reactions catalyzed by chiral primary amines. *J. Mol. Catal. A: Chem.*, **285**, 1–13; (d) Chen, Y.-C. (2008) The development of asymmetric primary amine catalysts based on cinchona alkaloids. *Synlett*, 1919–1930.
8. Hashimoto, T. and Maruoka, K. (2007) Recent development and application of chiral phase-transfer catalysts. *Chem. Rev.*, **107**, 5656–5682.
9. Akiyama, T. (2007) Stronger Brønsted acids. *Chem. Rev.*, **107**, 5744–5758.
10. Doyle, A.G. and Jacobsen, E.N. (2007) Small-molecule H-bond donors in asymmetric synthesis. *Chem. Rev.*, **107**, 5713–5743.
11. Kotke, M. and Schreiner, P.R. (2007) Generally applicable organocatalytic tetrahydropyranylation of hydroxy functionalities with very low catalyst loading. *Synthesis*, 779–790.
12. Review on proline-catalyzed reactions: (a) List, B. (2002) Proline-catalyzed asymmetric reactions. *Tetrahedron*, **58**, 5573–5590; (b) Jarvo, E.R. and Miller, S.J. (2002) Amino acids and peptides as asymmetric organocatalysts. *Tetrahedron*, **58**, 2481–2495; (c) Notz, W., Tanaka, F., and Barbas, C.F. III (2004) Enamine-based organocatalysis with proline and diamines: the development of direct catalytic asymmetric aldol, Mannich, Michael, and Diels-Alder reactions. *Acc. Chem. Res.*, **37**, 580–591.
13. Selected recent examples on proline-catalyzed aldol reactions: (a) Sakthivel, K., Notz, W., Bui, T., and Barbas, C.F. III (2001) Amino acid catalyzed direct asymmetric aldol reactions: a bioorganic approach to catalytic asymmetric carbon-carbon bond-forming reactions. *J. Am. Chem. Soc.*, **123**, 5260–5267; (b) Suri, J.T., Mitsumori, S., Albertshofer, K., Tanaka, F., and

Barbas, C.F. III (2006) Dihydroxyacetone variants in the organocatalytic construction of carbohydrates: mimicking tagatose and fuculose aldolases. *J. Org. Chem.*, **71**, 3822–3228; (c) Grondal, C. and Enders, D. (2006) Direct asymmetric organocatalytic de novo synthesis of carbohydrates. *Tetrahedron*, **62**, 329–337; (d) Suri, J.T., Ramachary, D.B., and Barbas, C.F. III (2005) Mimicking dihydroxy acetone phosphate-utilizing aldolases through organocatalysis: a facile route to carbohydrates and aminosugars. *Org. Lett.*, **7**, 1383–1385; (e) Ibrahem, I. and Córdova, A. (2005) Amino acid catalyzed direct enantioselective formation of carbohydrates: one-step de novo synthesis of ketoses. *Tetrahedron Lett.*, **46**, 3363–3367; (f) Storer, R.I. and MacMillan, D.W.C. (2004) Enantioselective organocatalytic aldehyde-aldehyde cross-aldol couplings. The broad utility of a-thioacetal aldehydes. *Tetrahedron*, **60**, 7705–7714; (g) Casas, J., Sundén, H., and Córdova, A. (2004) Direct organocatalytic asymmetric α-hydroxymethylation of ketones and aldehydes. *Tetrahedron Lett.*, **45**, 6117–6119; (h) Pan, Q., Zou, B., Wang, Y., and Ma, D. (2004) Diastereoselective aldol reaction of N,N-dibenzyl- α-amino aldehydes with ketones catalyzed by proline. *Org. Lett.*, **6**, 1009–1012; (i) Northrup, A.B., Mangion, I.K., Hettche, F., and MacMillan, D.W.C. (2004) Enantioselective organocatalytic direct aldol reaction of α-oxylaldehydes: step one in a two-step synthesis of carbohydrates. *Angew. Chem. Int. Ed.*, **43**, 2152–2154; (j) Allemann, C., Gordillo, R., Clemente, F.R., Cheong, P.H.-Y., and Houk, K.N. (2004) Theory of asymmetric organocatalysis of aldol and related reactions: rationalizations and predictions. *Acc. Chem. Res.*, **37**, 558–569; (k) Thayumanavan, R., Tanaka, F., and Barbas, C.F. III (2004) Direct organocatalytic asymmetric aldol reactions of α-amino aldehydes: expedient syntheses of highly enantiomerically enriched anti-β-hydroxy-α-amino acids. *Org. Lett.*, **6**, 3541–3544; (l) Pidathala, C., Hoang, L., Vignola, N., and List, B. (2003) Direct catalytic asymmetric enolexo aldolizations. *Angew. Chem. Int. Ed.*, **42**, 2785–2788.

14. Bui, T. and Barbas, C.F. III (2000) A proline-catalyzed asymmetric Robinson annulation reaction. *Tetrahedron Lett.*, **41**, 6951–6954.

15. (a) List, B. (2000) The direct catalytic asymmetric three-component Mannich reaction. *J. Am. Chem. Soc.*, **122**, 9336; (b) Córdova, A., Notz, W., Zhong, G., Betancort, J.M., and Barbas, C.F. III (2002) A highly enantioselective amino acid-catalyzed route to functionalized α-amino acids. *J. Am. Chem. Soc.*, **124**, 1842–1843; (c) List, B., Pojarliev, P., Biller, W.T., and Martin, H.J. (2002) The proline-catalyzed direct asymmetric three-component mannich reaction: scope, optimization, and application to the highly enantioselective synthesis of 1,2-amino alcohols. *J. Am. Chem. Soc.*, **124**, 827–833; (d) Notz, W., Tanaka, F., Watanabe, S., Chowdari, N.S., Turner, J.M., Thayumanavan, R., and Barbas, C.F. III (2003) The direct organocatalytic asymmetric Mannich reaction: unmodified aldehydes as nucleophiles. *J. Org. Chem.*, **68**, 9624–9634; (e) Chowdari, N.S., Suri, J.T., and Barbas, C.F. III (2004) Asymmetric synthesis of quaternary α- and β-amino acids and β-lactams via proline-catalyzed Mannich reactions with branched aldehyde donors. *Org. Lett.*, **6**, 2507–2510; (f) Hayashi, Y., Tsuboi, W., Ashimine, I., Urushima, T., Shoji, M., and Sakai, K. (2003) The direct and enantioselective, one-pot, three-component, cross-Mannich reaction of aldehydes. *Angew. Chem. Int. Ed.*, **42**, 3677–3680; (g) Córdova, A. (2003) Asymmetric tandem Mannich-Michael reaction. *Synlett*, 1651–1654.

16. (a) List, B., Pojarliev, P., and Martin, H.J. (2001) Efficient proline-catalyzed Michael additions of unmodified ketones to nitro olefins. *Org. Lett.*, **3**, 2423–2425; (b) Betancort, J.M. and Barbas, C.F. III (2001) Catalytic direct asymmetric Michael reactions: taming naked aldehyde donors. *Org. Lett.*, **3**, 3737–3740; (c) Enders, D. and Seki,

A. (2002) Proline-catalyzed enantioselective Michael additions of ketones to nitrostyrene. *Synlett*, 26–28; (d) Gryko, D. (2005) Organocatalytic transformation of 1,3-diketones into optically active cyclohexanones. *Tetrahedron: Asymmetry*, **16**, 1377–1383; (e) Mangion, I.K. and MacMillan, D.W.C. (2005) Total synthesis of Brasoside and Littoralisone. *J. Am. Chem. Soc.*, **127**, 3696–3697.

17. (a) Bøgevig, A., Juhl, K., Kumaragurubaran, N., Zhuang, W., and Jørgensen, K.A. (2002) Direct organo-catalytic asymmetric α-amination of aldehydes – a simple approach to optically active α-amino aldehydes, α-amino alcohols, and α-amino acids. *Angew. Chem. Int. Ed.*, **41**, 1790–1793; (b) List, B. (2002) Direct catalytic asymmetric α-amination of aldehydes. *J. Am. Chem. Soc.*, **124**, 5656–5657; (c) Suri, J.T., Steiner, D.D., and Barbas, C.F. III (2005) Organocatalytic enantioselective synthesis of metabotropic glutamate receptor ligands. *Org. Lett.*, **7**, 3885–3888.

18. (a) Sabitha, G., Fatima, N., Reddy, E.V., and Yadav, J.S. (2005) First examples of proline-catalyzed domino Knoevenagel/hetero-Diels-Alder/elimination reactions. *Adv. Synth. Catal.*, **347**, 1353–1355; (b) Thayumanavan, R., Dhevalapally, B., Sakthivel, K., Tanaka, F., and Barbas, C.F. III (2002) Amine-catalyzed direct Diels-Alder reactions of α′β-unsaturated ketones with nitro olefins. *Tetrahedron Lett.*, **43**, 3817–3820; (c) Ramachary, D.B., Chowdari, N.S., and Barbas, C.F. III (2002) Amine-catalyzed direct self Diels-Alder reactions of α′β-unsaturated ketones in water: synthesis of pro-chiral cyclohexanes. *Tetrahedron Lett.*, **43**, 6743–6746.

19. (a) Shi, M., Jiang, J.K., and Li, C.Q. (2002) Lewis base and L-proline co-catalyzed Baylis– Hillman reaction of arylaldehydes with methyl vinyl ketone. *Tetrahedron Lett.*, **43**, 127–130; (b) Chen, S.H., Hong, B.C., Su, C.F., and Sarshar, S. (2005) An unexpected inversion of enantioselectivity in the proline catalyzed intramolecular Baylis–Hillman reaction. *Tetrahedron Lett.*, **46**, 8899–8903; (c) Imbriglio, J.E., Vasbinder, M.M., and Miller, S.J. (2003) Dual catalyst control in the amino acid-peptide-catalyzed enantioselective Baylis-Hillman reaction. *Org. Lett.*, **5**, 3741–3743.

20. (a) Utsumi, N., Zhang, H., Tanaka, F., and Barbas, C.F. III (2007) A way to highly enantiomerically enriched aza-Morita-Baylis-Hillman-type products. *Angew. Chem. Int. Ed.*, **46**, 1878–1880; (b) Vesely, J., Dziedzic, P., and Córdova, A. (2007) Aza-Morita-Baylis-Hillman-type reactions : highly enantioselective organocatalytic addition of unmodified α,β-unsaturated aldehydes to N-Boc protected imines. *Tetrahedron Lett.*, **48**, 6900–6904.

21. Wang, J., Li, H., Mei, Y., Lou, B., Xu, D., Xie, D., Guo, H., and Wang, W. (2005) Direct, facile aldehyde and ketone α-selenenylation reactions promoted by L-prolinamide and pyrrolidine sulfonamide organocatalysts. *J. Org. Chem.*, **70**, 5678–5687.

22. (a) Zhong, G. (2003) A facile and rapid route to highly enantiopure 1,2-diols by novel catalytic asymmetric -aminoxylation of aldehydes. *Angew. Chem. Int. Ed.*, **42**, 4247–4250; (b) Brown, S.P., Brochu, M.P., Sinz, C.J., and MacMillan, D.W.C. (2003) The direct and enantioselective organocatalytic α-oxidation of aldehydes. *J. Am. Chem. Soc.*, **125**, 10808–10809; (c) Hayashi, Y., Yamaguchi, J., Hibino, K., and Shoji, M. (2003) Direct proline catalyzed asymmetric α-aminooxylation of aldehydes. *Tetrahedron Lett.*, **44**, 8293–8296; (d) Bøgevig, A., Sundén, H., and Córdova, A. (2004) Direct catalytic enantioselective -aminoxylation of ketones: a stereoselective synthesis of α-hydroxy and α′α′-dihydroxy ketones. *Angew. Chem. Int. Ed.*, **43**, 1109–1112; (e) Hayashi, Y., Yamaguchi, J., Hibino, K., and Shoji, M. (2004) Direct proline-catalyzed asymmetric α-aminoxylation of ketones. *Angew. Chem. Int. Ed.*, **43**, 1112–1115; (f) Hayashi, Y., Yamaguchi, J., Sumiya, T., Hibino, K., and Shoji, M. (2004) Direct proline-catalyzed asymmetric α-aminoxylation of aldehydes and

ketones. *J. Org. Chem.*, **69**, 5966–5973; (g) Córdova, A., Sundén, H., Bøgevig, A., Johansson, M., and Himo, F. (2004) The direct catalytic asymmetric α-aminooxylation reaction: development of stereoselective routes to 1,2-diols and 1,2-amino alcohols and density functional calculations. *Chem. Eur. J.*, **10**, 3673–3684.

23. Brochu, M.P., Brown, S.P., and MacMillan, D.W.C. (2004) Direct and enantioselective organocatalytic α-chlorination of aldehydes. *J. Am. Chem. Soc.*, **126**, 4108–4109.

24. (a) Wallbaum, S. and Martens, J. (1992) Asymmetric syntheses with chiral oxazaborolidines. *Tetrahedron: Asymmetry*, **3**, 1475–1504; (b) Rispens, M.T., Zondervan, C., and Feringa, B.L. (1995) Catalytic enantioselective allylic oxidation. *Tetrahedron: Asymmetry*, **6**, 661–664.

25. Para, N.A. and MacMillan, D.W.C. (2001) New strategies in organic catalysis: the first enantioselective organocatalytic Friedel-Crafts alkylation. *J. Am. Chem. Soc.*, **123**, 4370–4371.

26. (a) Beeson, T.D., Mastracchio, A., Hong, J.-B., Ashton, K., and MacMillan, D.W.C. (2007) Enantioselective organocatalysis using SOMO activation. *Science*, **316**, 582–585; (b) Kim, H. and MacMillan, D.W.C. (2007) Enantioselective organo-SOMO catalysis: the α-vinylation of aldehydes. *J. Am. Chem. Soc.*, **130**, 398–399.

27. Gruttadauria, M., Giacalone, F., and Noto, R. (2008) Supported proline and proline-derivatives as recyclable organocatalysts. *Chem. Soc. Rev.*, **37**, 1666–1688.

28. Gruttadauria, M., Giacalone, F., and Noto, R. (2009) Water in stereoselective organocatalytic reactions. *Adv. Synth. Catal.*, **351**, 33–57.

29. (a) Mukaiyama, T. (1999) The unexpected and the unpredictable in organic synthesis. *Tetrahedron*, **55**, 8609–8670; (b) Nicolaou, K.C., Vourloumis, D., Wissinger, N., and Baran, P.S. (2000) The art and science of total synthesis at the dawn of the twenty-first century. *Angew. Chem. Int. Ed.*, **39**, 44–122; (c) Evans, D.A., Ratz, A.M.B., Huff, E., and Sheppard, G.S. (1995) Total synthesis of the polyether antibiotic Lonomycin A (Emericid). *J. Am. Chem. Soc.*, **117**, 3448–3467; (d) Narasaka, K. and Pai, F.-C. (1984) Stereoselective reduction of β hydroxyketones to 1,3-diols highly selective 1,3-asymmetric induction via boron chelates. *Tetrahedron*, **40**, 2233–2238; (e) Keck, G.E., Wager, C.A., Sell, T., and Wager, T.T. (1999) An especially convenient stereoselective reduction of β-hydroxy ketones to anti 1,3 diols using samarium diiodide. *J. Org. Chem.*, **64**, 2172–2173.

30. (a) Machajewski, T.D. and Wong, C.-H. (2000) The catalytic asymmetric aldol reaction. *Angew. Chem. Int. Ed.*, **39**, 1352; (b) Alcaide, B. and Almendros, P. (2003) The direct catalytic asymmetric cross-aldol reaction of aldehydes. *Angew. Chem. Int. Ed.*, **42**, 858–860; (c) Palomo, C., Oiarbide, M., and Garcia, J.M. (2002) The aldol addition reaction: an old transformation at constant rebirth. *Chem. Eur. J.*, **8**, 36; (d) Mahrwald, R. (ed.) (2004) *Modern Aldol Reactions*, vols. **1**, **2**, Wiley-VCH Verlag GmbH, Weinheim.

31. Wong, C.-H., Halcomb, R.L., Ichikawa, Y., and Kajimoto, T. (1995) Enzymes in organic synthesis: application to the problems of carbohydrate recognition. *Angew. Chem. Int. Ed. Engl.*, **34**, 412–432.

32. (a) Welton, T. (1999) Room-temperature ionic liquids. Solvents for synthesis and catalysis. *Chem. Rev.*, **99**, 2071–2083; (b) Zhao, D., Wu, M., Kou, Y., and Min, E. (2002) Ionic liquids: applications in catalysis. *Catal. Today*, **74**, 157–189.

33. (a) Kotrusz, P., Kmentová, I., Gotov, B., Toma, S., and Solčaniová, E. (2002) Proline-catalysed asymmetric aldol reaction in the room temperature ionic liquid [bmim]PF6. *Chem. Commun.*, 2510–2511; (b) Loh, T.-P., Feng, L.-C., Yang, H.Y., and Yang, J.-Y. (2002) L-Proline in an ionic liquid as an efficient and reusable catalyst for direct asymmetric aldol reactions. *Tetrahedron Lett.*, **43**, 8741–8743; (c) Córdova, A. (2004) Direct catalytic asymmetric cross-aldol reactions in ionic liquid media. *Tetrahedron Lett.*, **45**, 3949–3952.

34. (a) Miao, W. and Chan, T.H. (2006) Ionic-liquid-supported organocatalyst: efficient and recyclable ionic-liquid-anchored proline for asymmetric aldol reaction. *Adv. Synth. Catal.*, **348**, 1711–1718; (b) Luo, S., Mi, X., Zhang, L., Liu, S., Xu, H., and Cheng, J.-P. (2006) Functionalized chiral ionic liquids as highly efficient asymmetric organocatalysts for Michael addition to nitroolefins. *Angew. Chem. Int. Ed.*, **45**, 3093–3097; (c) Lombardo, M., Pasi, F., Easwar, S., and Trombini, C. (2007) An improved protocol for the direct asymmetric aldol reaction in ionic liquids, catalysed by onium ion-tagged prolines. *Adv. Synth. Catal.*, **349**, 2061–2065.
35. Siyutkin, D.E., Kucherenko, A.S., Struchkova, M.I., and Zlotin, S.G. (2008) A novel (S)-proline-modified task-specific chiral ionic liquid – an amphiphilic recoverable catalyst for direct asymmetric aldol reactions in water. *Tetrahedron Lett.*, **49**, 1212–1216.
36. (a) Mehnert, C.P., Cook, R.A., Dispenziere, N.C., and Afeworki, M. (2002) Supported ionic liquid catalysis – a new concept for homogeneous hydroformylation catalysis. *J. Am. Chem. Soc.*, **124**, 12932–12933; (b) Mehnert, C.P. (2005) Supported ionic liquid catalysis. *Chem. Eur. J.*, **11**, 50–56.
37. (a) Gruttadauria, M., Riela, S., Lo Meo, P., D'Anna, F., and Noto, R. (2004) Supported ionic liquid asymmetric catalysis. a new method for chiral catalysts recycling. the case of proline-catalyzed aldol reaction. *Tetrahedron Lett.*, **45**, 6113–6116; (b) Gruttadauria, M., Riela, S., Aprile, C., Lo Meo, P., D'Anna, F., and Noto, R. (2006) Supported ionic liquids. New recyclable materials for the L-proline-catalyzed aldol reaction. *Adv. Synth. Catal.*, **348**, 82–92.
38. Krattiger, P., Kovasy, R., Revell, J.D., Ivan, S., and Wennemers, H. (2005) Increased structural complexity leads to higher activity: peptides as efficient and versatile catalysts for asymmetric aldol reactions. *Org. Lett.*, **7**, 1101–1103.
39. Aprile, C., Giacalone, F., Gruttadauria, M., Mossuto Marculescu, A., Noto, R., Revell, J.D., and Wennemers, H. (2007) New ionic liquid-modified silica gels as recyclable materials for L-proline- or H-Pro- Pro-Asp-NH2-catalyzed aldol reaction. *Green Chem.*, **9**, 1328–1334.
40. (a) Benaglia, M., Celentano, G., and Cozzi, F. (2001) Enantioselective aldol condensations catalyzed by poly(ethylene glycol)-supported proline. *Adv. Synth. Catal.*, **343**, 171–173; (b) Benaglia, M., Cinquini, M., Cozzi, F., Puglisi, A., and Celentano, G. (2002) Poly(ethylene glycol)-supported proline: a versatile catalyst for the enantioselective aldol and iminoaldol reactions. *Adv. Synth. Catal.*, **344**, 533–542.
41. Benaglia, M., Cinquini, M., Cozzi, F., Puglisi, A., and Celentano, G. (2003) Poly(ethylene-glycol)-supported proline: a recyclable aminocatalyst for the enantioselective synthesis of γ-nitroketones by conjugate addition. *J. Mol. Catal. A: Chem.*, **204-205**, 157–163.
42. Gu, L., Wu, Y., Zhang, Y., and Zhao, G. (2007) A new class of efficient poly(ethylene-glycol)-supported catalyst based on proline for the asymmetric Michael addition of ketones to nitrostyrenes. *J. Mol. Catal. A: Chem.*, **263**, 186–194.
43. Font, D., Jimeno, C., and Pericàs, M.A. (2006) Polystyrene-supported hydroxyproline: an insoluble, recyclable organocatalyst for the asymmetric aldol reaction in water. *Org. Lett.*, **8**, 4653–4656.
44. Font, D., Sayalero, S., Bastero, A., and Jimeno, C. (2008) Toward an artificial aldolase. M. A. Pericàs. *Org. Lett.*, **10**, 337–340.
45. Font, D., Bastero, A., Sayalero, S., Jimeno, C., and Pericàs, M.A. (2007) Highly enantioselective α-aminoxylation of aldehydes and ketones with a polymer-supported organocatalyst. *Org. Lett.*, **9**, 1943–1946.
46. (a) Calderón, F., Fernández, R., Sánchez, F., and Fernández-Mayoralas, A. (2005) Asymmetric aldol reaction using immobilized proline on mesoporous support. *Adv. Synth. Catal.*, **347**, 1395–1403; (b) Doyagüez, E.G., Calderón, F., Fernández, R., Sánchez, F., and Fernández-Mayoralas, A. (2007) Asymmetric aldol reaction catalyzed by a

heterogenized proline on a mesoporous support. The role of the nature of solvents. *J. Org. Chem.*, **72**, 9353–9356.
47. Chouhan, G., Wang, D., and Alper, H. (2007) Magnetic nanoparticle-supported proline as a recyclable and recoverable ligand for the CuI catalyzed arylation of nitrogen nucleophiles. *Chem. Commun.*, 4809–4810.
48. Giacalone, F., Gruttadauria, M., Mossuto Marculescu, A., and Noto, R. (2007) Polystyrene-supported proline and prolinamide. Versatile heterogeneous organocatalysts both for asymmetric aldol reaction in water and α-selenenylation of aldehydes. *Tetrahedron Lett.*, **48**, 255–259.
49. (a) Back, T.G. (ed.) (1999) *Organoselenium Chemistry: A Practical Approach*, Oxford University Press, New York; (b) Wirth, T. (ed.) (2000) *Organoselenium Chemistry. Modern Developments in Organic Chemistry*, Topics in Current Chemistry, Vol. **208**, Springer-Verlag, Berlin; (c) Paulmier, C. (1986) *Selenium Reagents and Intermediates in Organic Synthesis*, Pergamon Press, Oxford; (d) Reich, H.J. (1979) Functional group manipulation using organoselenium reagents. *Acc. Chem. Res.*, **12**, 22–30.
50. Hillman, M.E.D. and Baylis, A.B. (1973) Reaction of acrylic type compounds with aldehydes and certain ketones. US Patent 3,743,669, filed Nov. 6, 1970 and issued Jul. 3, 1973.
51. (a) Basavaiah, D., Jaganmohan Rao, A., and Satyanarayana, T. (2003) Recent advances in the Baylis-Hillman reaction and applications. *Chem. Rev.*, **103**, 811–892; (b) Basavaiah, D., Venkateswara Rao, K., and Jannapu Reddy, R. (2007) The Baylis–Hillman reaction: a novel source of attraction, opportunities, and challenges in synthetic chemistry. *Chem. Soc. Rev.*, **36**, 1581–1588.
52. Shi, M., Jiang, J.-K., and Li, C.-Q. (2002) Lewis base and L-proline co-catalyzed Baylis– Hillman reaction of arylaldehydes with methyl vinyl ketone. *Tetrahedron Lett.*, **43**, 127–130.
53. Davies, A.H.J., Ruda, M., and Tomkinson, N.C.O. (2007) Aminocatalysis of the Baylis– Hillman reaction: an important solvent effect. *Tetrahedron Lett.*, **48**, 1461–1464.
54. (a) Corma, A., García, H., and Leyva, A. (2003) Heterogeneous Baylis– Hillman using a polystyrene-bound 4-(N-benzyl-N-methylamino)pyridine as reusable catalyst. *Chem. Commun.*, 2806–2807; (b) Chen, H.-T., Huh, S., Wiench, J.W., Pruski, M., and Lin, V.S.-Y. (2005) Dialkylaminopyridine-functionalized Mesoporous Silica nanosphere as an efficient and highly stable heterogeneous nucleophilic catalyst. *J. Am. Chem. Soc.*, **127**, 13305–13311; (c) Helms, B., Guillaudeu, S.J., Xie, Y., McMurdo, M., Hawker, C.J., and Fréchet, J.M.J. (2005) One-pot reaction cascades using star polymers with core-confined catalysts. *Angew. Chem. Int. Ed.*, **44**, 6384.
55. Giacalone, F., Gruttadauria, M., Mossuto Marculescu, A., D'Anna, F., and Noto, R. (2008) Polystyrene-supported proline as recyclable catalyst in the Baylis-Hillman reaction of arylaldehydes and methyl or ethyl vinyl ketone. *Catal. Commun.*, **9**, 1477–1481.
56. Cordova, A., Notz, W., and Barbas, C.F. III (2002) Direct organocatalytic aldol reactions in buffered aqueous media. *Chem. Commun.*, 3024.
57. (a) Mase, N., Nakai, Y., Ohara, N., Yoda, H., Takabe, K., Tanaka, F., and Barbas, C.F. III (2006) Organocatalytic direct asymmetric aldol reactions in water. *J. Am. Chem. Soc.*, **128**, 734–735; (b) Hayashi, Y., Sumiya, T., Takahashi, J., Gotoh, H., Urushima, T., and Shoji, M. (2006) Highly diastereo- and enantioselective direct aldol reactions in water. *Angew. Chem. Int. Ed.*, **45**, 958–961.
58. Gruttadauria, M., Giacalone, F., Mossuto Marculescu, A., Lo Meo, P., Riela, S., and Noto, R. (2007) Hydrophobically directed aldol reactions: polystyrene-supported L-proline as recyclable catalyst for direct asymmetric aldol reaction in the presence of water. *Eur. J. Org. Chem.*, 4688–4698.
59. (a) Raj, M., Maya, V., Ginotra, S.K., and Singh, V.K. (2006) Highly enantioselective direct aldol reaction catalyzed

by organic molecules. *Org. Lett.*, **8**, 4097–4099; (b) Tang, Z., Jiang, F., Yu, L.T., Cui, X., Gong, L.Z., Mi, A.Q., Jiang, Y.Z., and Wu, Y.D. (2003) Novel small organic molecules for a highly enantioselective direct aldol reaction. *J. Am. Chem. Soc.*, **125**, 5262–5263; (c) Tang, Z., Jiang, F., Cui, F., Gong, L.Z., Mi, A.Q., Jiang, Y.Z., and Wu, Y.D. (2004) Enantioselective direct aldol reactions catalyzed by l-prolinamide derivatives. *Proc. Natl. Acad. Sci. U.S.A.*, **101**, 5755–5760; (d) Tang, Z., Yang, Z.H., Chen, X.H., Cun, L.F., Mi, A.Q., Jiang, Y.Z., and Gong, L.Z. (2005) A highly efficient organocatalyst for direct aldol reactions of ketones with aldehydes. *J. Am. Chem. Soc.*, **127**, 9285–9289.

60. (a) Zhao, J.-F., He, L., Jiang, J., Tang, Z., Cun, L.-F., and Gong, L.-Z. (2008) Organo-catalyzed highly diastereo- and enantio-selective direct aldol reactions in water. *Tetrahedron Lett.*, **49**, 3372–3375; (b) Maya, V., Raj, M., and Singh, V.K. (2007) Highly enantioselective organocatalytic direct aldol reaction in an aqueous medium. *Org. Lett.*, **9**, 2593–2595; (c) Chen, X.H., Tang, Z., Luo, S.W., Cun, L.F., Mi, A.Q., Jiang, Y.Z., and Gong, L.Z. (2007) Organocatalyzed highly enantioselective direct aldol reactions of aldehydes with hydroxyacetone and fluoroacetone in aqueous media: the use of water to control regioselectivity. *Chem. Eur. J.*, **13**, 689–701; (d) Zu, L., Xie, H., Li, H., Wang, J., and Wang, W. (2008) Highly enantioselective aldol reactions catalyzed by a recyclable fluorous (S) pyrrolidine sulfonamide on water. *Org. Lett.*, **10**, 1211–1214; (e) Russo, A., Botta, G., and Lattanzi, A. (2007) Highly stereoselective direct aldol reactions catalyzed by (S)-NOBIN-l-prolinamide. *Tetrahedron*, **63**, 11886–11892; (f) Guizzetti, S., Benaglia, M., Raimondi, L., and Celentano, G. (2007) Enantioselective direct aldol reaction "on water" promoted by chiral organic catalysts. *Org. Lett.*, **9**, 1247–1250; (g) Wang, C., Jiang, Y., Zhang, X.-X., Huang, Y., Li, B.-G., and Zhang, G.-L. (2007) Rationally designed organocatalyst for direct asymmetric aldol reaction in the presence of water. *Tetrahedron Lett.*, **48**, 4281–4285; (h) Lei, M., Shi, L., Li, G., Chen, S., Fang, W., Ge, Z., Cheng, T., and Li, R. (2007) Dipeptide-catalyzed direct asymmetric aldol reactions in the presence of water. *Tetrahedron*, **63**, 7892–7898; (i) Wu, Y., Zhang, Y., Yu, M., Zhao, G., and Wang, S. (2006) Highly efficient and reusable dendritic catalysts derived from N-prolylsulfonamide for the asymmetric direct aldol reaction in water. *Org. Lett.*, **8**, 4417–4420; (j) Sathapornvajana, S. and Vilaivan, T. (2007) Prolinamides derived from aminophenols as organocatalysts for asymmetric direct aldol reactions. *Tetrahedron*, **63**, 10253–10259; (k) Fu, Y.-Q., Li, Z.-C., Ding, L.-N., Tao, J.-C., Zhang, S.-H., and Tang, M.-S. (2006) Direct asymmetric aldol reaction catalyzed by simple prolinamide phenols. *Tetrahedron: Asymmetry*, **17**, 3351–3357; (l) Gryko, D. and Saletra, W.J. (2007) Organocatalytic asymmetric aldol reaction in the presence of water. *Org. Biomol. Chem.*, **5**, 2148–2153.

61. (a) Kehat, T. and Portnoy, M. (2007) Polymer-supported proline-decorated dendrons: dendritic effect in asymmetric aldol reaction. *Chem. Commun.*, 2823–2825; (b) Dhar, D., Beadham, I., and Chandrasekaran, S. (2003) Proline and benzylpenicillin derivatives grafted into mesoporous MCM-41: Novel organic– inorganic hybrid catalysts for direct aldol reaction. *Proc. Indian Acad. Sci., Chem. Sci.*, **115**, 365–372; (c) Luo, S., Mi, X., Zhang, L., Liu, S., Xu, H., and Cheng, J.-P. (2007) Functionalized ionic liquids catalyzed direct aldol reactions. *Tetrahedron*, **63**, 1923–1930.

62. (a) Giacalone, F., Gruttadauria, M., Lo Meo, P., Riela, S., and Noto, R. (2008) New simple hydrophobic proline derivatives as highly active and stereoselective catalysts for the direct asymmetric aldol reaction in aqueous medium. *Adv. Synth. Catal.*, **350**, 2747–2760; (b) Gruttadauria, M., Giacalone, F., Mossuto Marculescu, A., Salvo, A.M.P., and Noto, R. (2009) Stereoselective aldol reaction catalyzed by

a highly recyclable polystyrene supported substituted prolinamide catalyst. *ARKIVOC*, (viii), 5–15.

63. (a) Gryko, D. and Lipiński, R. (2006) Asymmetric direct aldol reaction catalysed by L-prolinethioamides. *Eur. J. Org. Chem.*, 3864–3876; (b) Oh, C.-H., Dong, H.-G., Cho, H.-W., Park, S.J., Hong, J.H., Baek, D., and Cho, J.-H. (2002) Synthesis and antibacterial activity of 1 β-methylcarbapenems having a 2,2-disubstituted-1,3-diazabicyclo[3.3.0]octan-4-one moiety and related compounds. Part III. *Arch. Pharm. Pharm. Med. Chem.*, **335**, 200–206.

64. Bak, A., Fich, M., Larsen, B.D., Frokjaer, S., and Friis, G.J. (1999) N-terminal 4-imidazolidinone prodrugs of Leu-enkephalin: synthesis, chemical and enzymatic stability studies. *Eur. J. Pharm. Sci.*, **7**, 317–323.

65. (a) Kano, T., Takai, J., Tokuda, O., and Maruoka, K. (2005) Design of an axially chiral amino acid with a binaphthyl backbone as an organocatalyst for a direct asymmetric aldol reaction. *Angew. Chem. Int. Ed.*, **44**, 3055–3057; (b) Rodríguez, B. and Bolm, C. (2006) Thermal effects in the organocatalytic asymmetric Mannich reaction. *J. Org. Chem.*, **71**, 2888–2891; (c) Aratake, S., Itoh, T., Okano, T., Nagae, N., Sumiya, T., Shoji, M., and Hayashi, Y. (2007) Highly diastereo- and enantioselective direct aldol reactions of aldehydes and ketones catalyzed by siloxyproline in the presence of water. *Chem. Eur. J.*, **13**, 10246–10256; (d) Kotke, M. and Schreiner, P.R. (2006) Acid-free, organocatalytic acetalization. *Tetrahedron*, **62**, 434–439.

66. Lombardo, M., Easwar, S., Pasi, F., and Trombini, C. (2009) The ion tag strategy as a route to highly efficient organocatalysts for the direct asymmetric aldol reaction. *Adv. Synth. Catal.*, **351**, 276–282.

67. (a) Marigo, M., Wabnitz, T.C., Fielenbach, D., and Jørgensen, K.A. (2005) Enantioselective organocatalyzed α sulfenylation of aldehydes. *Angew. Chem. Int. Ed.*, **44**, 794–797; (b) Alemán, J., Cabrera, S., Maerten, E., Overgaard, J., and Jørgensen, K.A. (2007) Asymmetric organocatalytic α-arylation of aldehydes. *Angew. Chem. Int. Ed.*, **46**, 5520–5523; (c) Cabrera, S., Reyes, E., Alemán, J., Milelli, A., Kobbelgaard, S., and Jørgensen, K.A. (2008) Organocatalytic asymmetric synthesis of α,α-disubstituted α-amino acids and derivatives. *J. Am. Chem. Soc.*, **130**, 12031–12037.

68. Palomo, C. and Mielgo, A. (2006) Diarylprolinol ethers: expanding the potential of enamine/iminium-ion catalysis. *Angew. Chem. Int. Ed.*, **45**, 7876–7880.

69. Franzén, J., Marigo, M., Fielenbach, D., Wabnitz, T.C., Kjærsgaard, A., and Jørgensen, K.A. (2005) A general organocatalyst for direct α-functionalization of aldehydes: stereoselective C-C, C-N, C-F, C-Br, and C-S bond-forming reactions. Scope and mechanistic insights. *J. Am. Chem. Soc.*, **127**, 18296–18304.

70. Zhao, G.-L., Ibrahem, I., Dziedzic, P., Sun, J., Bonneau, C., and Córdova, A. (2008) One-pot catalytic enantioselective domino nitro-Michael/Michael synthesis of cyclopentanes with four stereocenters. *Chem. Eur. J.*, **14**, 10007–10011.

71. Fustero, S., Moscardó, J., Jiménez, D., Pérez-Carrión, M.D., Sánchez-Roselló, M., and del Pozo, C. (2008) Organocatalytic approach to benzofused nitrogen-containing heterocycles: enantioselective total synthesis of (+)-Angustureine. *Chem. Eur. J.*, **14**, 9868–9872.

72. Liu, K., Chougnet, A., and Woggon, W.-D. (2008) A short route to α-Tocopherol. *Angew. Chem. Int. Ed.*, **47**, 5827–5829.

73. Tasis, D., Tagmatarchis, N., Bianco, A., and Prato, M. (2006) Chemistry of carbon nanotubes. *Chem. Rev.*, **106**, 1105–1136.

74. (a) Eddaoudi, M., Moler, D.B., Li, H., Chen, B., Reineke, T.M., O'Keeffe, M., and Yaghi, O.M. (2001) Modular chemistry: secondary building units as a basis for the design of highly porous and robust metal-organic carboxylate frameworks. *Acc. Chem. Res.*, **34**, 319–330; (b) Li, H., Eddaoudi, M., O'Keeffe, M., and Yaghi, O.M. (1999) Design and synthesis of an exceptionally stable and

highly porous metal- organic framework. *Nature*, **402**, 276–279; (c) James, S.L. (2003) Metal-organic frameworks. *Chem. Soc. Rev.*, **32**, 276–288.

75. (a) Henschel, A., Gedrich, K., Kraehnert, R., and Kaskel, S. (2008) Catalytic properties of MIL-101. *Chem. Commun.*, 4192–4194; (b) Wu, C.-D., Hu, A., Zhang, L., and Lin, W. (2005) A homochiral porous metal-organic framework for highly enantioselective heterogeneous asymmetric catalysis. *J. Am Chem. Soc.*, **127**, 8940–8941.

76. (a) Gascon, J., Aktay, U., Hernandez-Alonso, M.D., van Klink, G.P.M., and Kapteijn, F. (2009) Amino-based metal-organic frameworks as stable, highly active basic catalysts. *J. Catal.*, **261**, 75–87; (b) Qiu, L.-G., Gu, L.-N., Hu, G., and Zhang, L.-D. (2009) Synthesis, structural characterization and selectively catalytic properties of metal– organic frameworks with nano-sized channels: a modular design strategy. *J. Sol. State Chem.* doi: 10.1016/j.jssc.2008.11.018.

5
The Complex-Induced Proximity Effect in Organolithium Chemistry and Its Importance in the Lithiation of Tertiary Amines

Viktoria H. Gessner

5.1
Introduction

Organolithium compounds are amongst the most-often applied reagents in synthetic chemistry. Since the first investigation by Schlenk and Holtz numerous developments have turned lithiumorganics into versatile compounds in various fields of research. Their applications range from simple deprotonation, addition, or substitution reactions to polymerization and asymmetric synthesis. Especially, combinations of organolithium compounds and Lewis bases, above all nitrogen ligands, have gained interest due to the possible tuning of their reactivity and the introduction of stereoselectivity. This control is based on the coordination of the ligand to the lithiumalkyl leading to the deaggregation of the oligomeric organolithium compounds to smaller aggregates like dimers or even monomers. Yet, recently, several examples have proven that even the Lewis bases themselves possess a distinct reactivity toward decomposition reactions with the lithium base. In the case of nitrogen ligands, this side reaction does not only result in the loss of base and in the formation of by-products but also in the synthesis of nitrogen-containing lithiated compounds that offer a variety of further applications. Especially, α-lithiated tertiary amines have become intriguing building blocks in many fields of sciences. These systems allow a facile introduction of an amino-substituted carbon atom, which facilitates the synthesis of many nitrogen-containing compounds such as chiral ligand systems, catalysts, or compounds of biological significance, including amongst others alkaloids and amino acids. This chapter gives an overview of nitrogen ligands in the coordination chemistry of organolithium compounds with special focus on the direct deprotonation of tertiary amines to α-lithiated species. A combination of structure elucidation and computational studies gives insight into mechanistic features and observed selectivities. The process and selectivities are explained by means of the spatial proximity of reactive groups in intermediate, precoordinated adducts of the amine, and the organolithium compounds according to the complex-induced proximity effect (CIPE).

Ideas in Chemistry and Molecular Sciences: Advances in Synthetic Chemistry. Edited by Bruno Pignataro
Copyright © 2010 WILEY-VCH Verlag GmbH & Co. KGaA, Weinheim
ISBN: 978-3-527-32539-9

5.2
State of the Art

5.2.1
Structure Formation Patterns of Organolithium Compounds

The structure formation of organolithium compounds is a multifarious and wide-ranging field of research, on which many groups have concentrated in recent years. Owing to the strong structure–reactivity relationship, these investigations are not only limited to the structure elucidation of interesting adducts and aggregates of most diverse compositions, but the understanding of reaction mechanisms and selectivities of these compounds is also enhanced. In spite of their high reactivity and sensitivity toward oxygen, carbon dioxide, and humidity, lithiumorganic compounds have become important reagents in synthetic chemistry [1]. The high reactivity is based on the highly polar lithium–carbon bond with dominantly ionic character; therefore, the structural properties can be explained by electrostatic interactions of counterions comparable to lithium salts, such as lithium fluoride. Accordingly, organolithium compounds possess a distinct tendency to form oligomeric aggregates in solution as well as in solid state to achieve optimal charge distribution [2]. Therefore, structure formation is not limited to symmetric adducts such as tetramers or hexamers, but a huge variety of multifarious aggregates of most different compositions and degrees of aggregation are also formed.

The reactivity of organolithium compounds is strongly connected with the structure and the degree of aggregation of the reactive lithium species. In general, oligomerization and charge distribution limit the reactivity, so that deaggregation to smaller, more reactive compounds is desired. Therefore, aggregation and deaggregation are influenced by the addition of Lewis bases, above all ether compounds and nitrogen ligands. Thus, in many reactions Lewis bases are added to deaggregate the oligomeric structures and enhance the reactivity. Generally, spatially demanding Lewis bases with a strong coordination ability as well as spatially demanding lithiumalkyls are required to stabilize small adducts, such as dimers or monomers. Therefore, almost all simple monomeric organolithium compounds that have been isolated so far are adducts of *tert*-butyllithium in combination with spatially demanding nitrogen ligands, such as (−)-sparteine [2].

Figure 5.1 depicts the deaggregation process of *tert*-butyllithium. The tetrameric parent structure $(tBuLi)_4$ is cleaved by the addition of diethylether to the dimeric adduct $(tBuLi \cdot Et_2O)_2$. With (−)-sparteine as Lewis base even monomeric $tBuLi \cdot$(−)-sparteine can be stabilized [3]. Generally, such monomeric alkyllithiums are the reactive and rate-determining species in reactions and are therefore the desired target of the deaggregation process. However, deaggregation to small adducts, above all monomers, is an energy-consuming process and therefore only few monomers have been isolated so far [4]. In reactions with higher aggregated lithium species this energy remains for the progress of the reaction itself. In recent years, monomers have been proven as reactive species in organolithium chemistry,

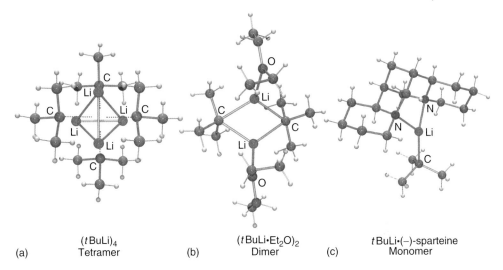

Figure 5.1 Deaggregation of *tert*-butyllithium: from tetrameric (*t*BuLi)$_4$ (a) to dimeric (*t*BuLi·Et$_2$O)$_2$ (b) and monomeric *t*BuLi·(−)-sparteine (c).

and dimer- or higher aggregate-based mechanisms or reactions via the so-called triple ions have been discussed [5]. Thus, for a more detailed understanding of the reactivity of these compounds, knowledge of the structure of the reactive species is necessary. Owing to the extremely high reactivity of these compounds, cryo-X-ray analysis and computational studies have turned into powerful techniques for the handling and the understanding of these systems.

5.2.2
The Complex-Induced Proximity Effect (CIPE)

Besides the use in deaggregation and activation of the oligomeric lithiumalkyls, Lewis bases are also widely applied for the introduction of stereoinformation. Selective deprotonation and addition reactions with chiral organolithium adducts are still one of the most emerging fields in organolithium chemistry [6]. Many investigations focus on the design of novel chiral ligand systems to gain highly selective reactions. The mechanism of such asymmetric deprotonation reactions generally proceeds via an intermediate precoordinated complex, in which reactive groups approximate each other. The spatial proximity of these groups results in an increased reactivity toward each other and thus in observed selectivities. This reaction mechanism via such an intermediate precoordinated complex is called *complex-induced proximity effect* [7]. Scheme 5.1 depicts this CIPE mechanism by means of the asymmetric deprotonation of carbamates, which was first introduced by Hoppe in 1990 and has found wide applications in synthesis [6a]. The overall application of organolithium compounds in asymmetric synthesis is beyond the scope of this review article. These types of reactions have extensively been discussed

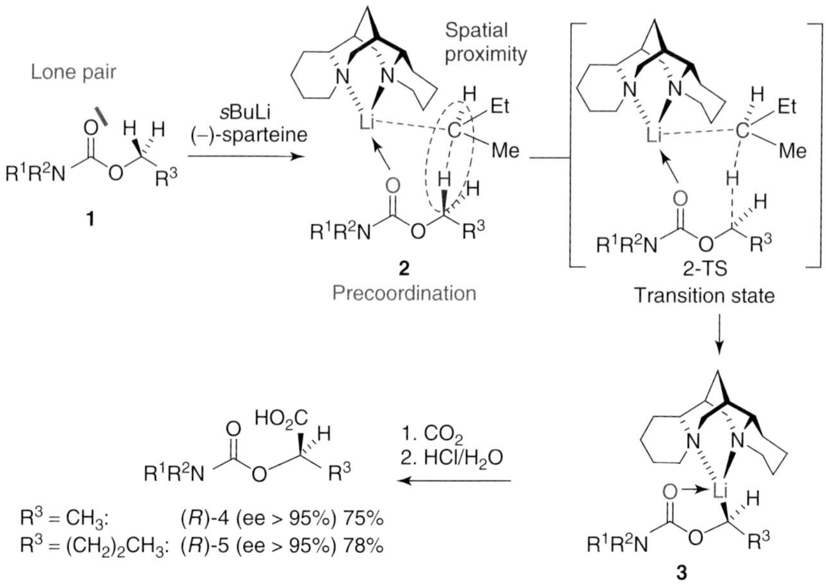

Scheme 5.1 Asymmetric deprotonation via precoordination according to the complex-induced proximity effect.

elsewhere [8]. However, this reaction is a vivid example for the description of the CIPE mechanism and its efficiency in the determination of selectivities, which is of importance for the following explanations. In the first step of the carbamate deprotonation, the lithiumalkyl-(−)-sparteine adduct is coordinated by the oxygen atom of the carbamate function forming the precoordinated complex **2**. In this intermediate, the carbanionic center comes into close proximity to one of the two methylene hydrogens, resulting in a decrease of the reaction barrier and thus in the kinetical favoritism of the abstraction of this hydrogen in the transition state **2-TS**. The configuration of the formed lithiated species (**3**) is stabilized by coordination of the oxygen and thus epimerization is prevented at low temperatures. Consequently, **3** can be trapped with electrophiles giving high enantioselectivities.

The CIPE concept is not only used for the description and explanation of observed stereoselectivities in asymmetric deprotonation or addition reactions, but can also be applied for further reactions such as the directed *ortho*-metallation (DOM) [7]. In principle, the CIPE is solely a special case of kinetically controlled reactions. The differences in the reaction barriers, which are – in the case of CIPE – caused by the spatial proximity of reactive groups, lead to the observed selectivities. In reactions without possible precoordination, for example, of substrates without functional groups, these differences in the reaction barriers can be the result of sterical hindrance or electrostatic interactions occurring during the approximation of the reactants. However, in each case the configurative stabilization of the intermediate lithiated compound is required to prevent epimerization and to obtain high selectivities. If no equilibrium is reached between the isomers, the selectivities are solely determined by the differences in the reaction barriers. The formation of precoordinated complexes according to CIPE enables the isolation and characterization of these relatively stable intermediates during the reaction process. This enables a direct insight into the on-going processes and thus facilitates the elucidation of the reaction mechanism and the understanding of the observed selectivities.

5.2.3
Synthesis of α-Lithiated Tertiary Amines

Lewis bases, above all nitrogen ligands, are widely used in organolithium chemistry for the deaggregation of their oligomeric structures and for the control of reactivity. Typically, these ligands are supposed to be inert under the reaction conditions. However, recent investigations have proven a distinct reactivity of the Lewis base toward direct deprotonation reactions with the lithiumalkyl. Prominent examples are the decomposition of ether compounds and tertiary methyl amines [9]. These decomposition reactions result in the loss of base and in the formation of side products. Yet, it has been shown that this reaction also provides access to α-lithiated tertiary amines, which represent interesting building blocks in many fields of sciences. They offer a facile introduction of nitrogen functions and thus the synthesis and variation of nitrogen-containing compounds, such as ligand systems, natural products, or catalysts.

Owing to the low acidity of the α-hydrogen atoms the preparation of such α-lithiated amines by direct deprotonation has long been considered as impossible [10]. Instead, the preparation was generally carried out by transmetallation (especially of the corresponding organotin compounds) (Scheme 5.2, A) [11], or reductive bond cleavage (B) of α-thio [12], α-cyano [13], and α-alkoxy amines [14] or α-imines [15] by lithium or potassium metals. Only few examples of the direct deprotonation (C) of tertiary amines have been reported, such as N-methylpiperidine or N,N-dimethylaniline [16]. This direct synthesis generally requires stronger bases, such as nBuLi/tBuOK, and elevated reaction temperatures. However, only low

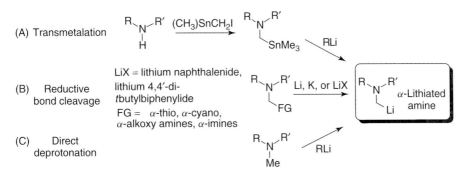

Scheme 5.2 Synthetic methodologies for the preparation of α-lithiated tertiary methyl amines.

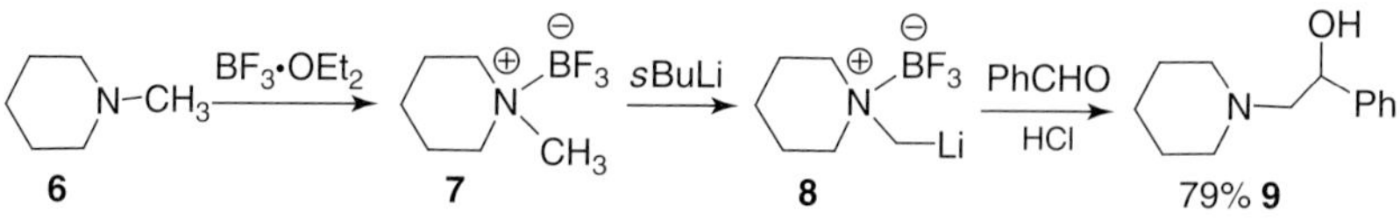

Scheme 5.3 Direct deprotonation of N-methylpiperidine via Lewis acid activation.

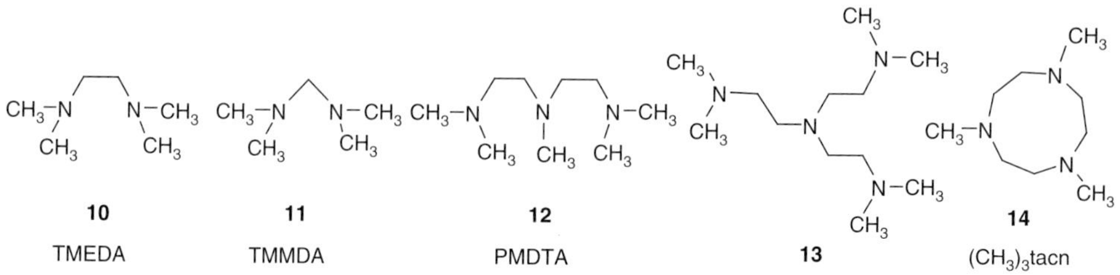

Figure 5.2 Tertiary methyl amines that undergo direct lithiation under mild reaction conditions.

yields of the lithiated products are accessible and thus this methodology is, in general, not suitable for preparative applications. One of the best results was obtained with N-methylpiperidine, yielding 73% of the corresponding alcohol after trapping with a carbonyl compound [16e]. An alternative activation of the amine can be achieved by the formation of a Lewis acid complex above all with boron trifluoride etherate. The resulting aminoborane such as **7** can be deprotonated readily even at low temperatures [17]. The first example of this activation has been reported by Kessar et al. with N-methylpiperidine **6** giving good yields of the corresponding alcohol **9** (Scheme 5.3).

A further successful methodology to direct deprotonation reactions revealed to be the intramolecular activation of the amine through a second donor center in the molecule. Thus, the deprotonation of a series of di- or triamines could be achieved under mild conditions with good yields of the trapping products, amongst the commonly used ligands are N,N,N',N'-tetramethylethylenediamine (TMEDA, **10**) and N,N,N',N'',N''-pentamethyldiethylenetriamine (PMDTA, **12**) as well as the cyclic ligand N,N',N''-trimethyl-1,4,7-triazacyclononane ((CH$_3$)$_3$tacn, **14**) (Figure 5.2) [4b, 18]. In the case of N,N,N',N'-tetramethylmethylenediamine (TMMDA, **11**) even dilithiation of one methyl group at both nitrogen atoms has been observed [19]. In the following sections, the mechanism of this direct lithiation of tertiary amines with the involved activation is discussed. Observed selectivities are explained by the isolation of intermediate, precoordinated adducts in combination with computational studies. Thus, mechanisms via both monomeric organolithium species and dimers and higher aggregated adducts are also presented.

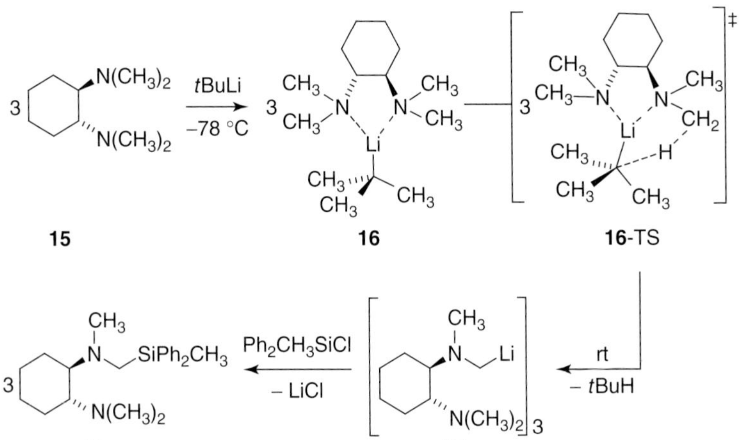

Scheme 5.4 Direct deprotonation of (R,R)-TMCDA (**15**) to the α-lithiated amine **17** via monomeric *t*BuLi·(R,R)-TMCDA (**16**).

5.3
Latest Developments

5.3.1
Precoordination as Key to Direct Deprotonation of Tertiary Amines

The activation of tertiary methyl amines for direct α-lithiation by further nitrogen atoms in the molecule gave hint to a reaction mechanism via a precoordinated intermediate, in which the lithiumalkyl is coordinated by the donor atoms of the amine. This assumption could be proven by the isolation of such an intermediate during the deprotonation of the chiral nitrogen ligand (1R,2R)-N,N,N′,N′-tetramethylcyclohexane-1,2-diamine [(R,R)-TMCDA, **15**] [20]. (R,R)-TMCDA is selectively deprotonated at one of its methyl groups by simple warming of a solution of the amine with *tert*-butyllithium to room temperature (Scheme 5.4). This reaction has also been observed with further alkyllithiums, such as *sec*- and *n*-butyllithium as well as *iso*-propyllithium. Furthermore, the preparative use of the formed α-lithiated amine was shown exemplarily by means of the synthesis of the aminomethyl-functionalized silane **18**. The lithiated product **17** could be characterized as a trimeric compound.

During the reaction process, monomeric *t*BuLi·(R,R)-TMCDA (**16**) is formed as precoordinated intermediate complex of the lithiation reaction. Analogous complexes are built with further alkyllithium compounds, such as monomeric *s*BuLi·(R,R)-TMCDA and dimeric *n*BuLi, *i*PrLi, and MeLi adducts, which have been isolated and characterized by X-ray diffraction analysis [21]. In the monomeric intermediate complexes (**16**), the α-hydrogen atom and the carbanionic group come

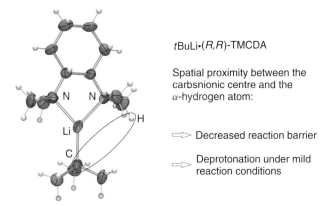

Figure 5.3 Molecular structure of tBuLi·(R,R)-TMCDA (**16**); proximity of reactive groups in **16** enables direct deprotonation.

into close proximity to each other (Figure 5.3). This proximity is clearly visible in the crystal structure of **16**, showing a distance of only 3.377 Å between these groups. Finally, the precoordination – according to the CIPE (Section 5.2.2) – results in the decrease of the reaction barrier and thus in the facile deprotonation of the methyl group under mild reaction conditions. Density functional theory (DFT) calculations confirm these results showing a barrier of 98 kJ mol^{-1} for the deprotonation with *tert*-butyllithium as lithium base, which can be afforded at room temperature [20]. An analogous spatial proximity could be detected in the (R,R)-TMCDA adducts with different organolithium bases [21].

The reaction barrier of such deprotonation reactions is typically calculated as energetic difference between the transition state and the precoordinated complex. Therefore, the energy, which is required for the cleavage of the oligomeric lithium species to the precoordinated complex, is not considered separately. However, this deaggregation process can crucially influence the reaction barrier, when – contrary to *t*BuLi·(R,R)-TMCDA (**16**) – the stabilization of small aggregates is energetically disfavored [17]. This is, for example, often the case for methyllithium, which is most difficult to stabilize in small degrees of aggregation. For instance, while the cleavage of tetrameric (MeLi(CH$_3$)$_2$O)$_4$ to monomeric MeLi(R,R)-TMCDA affords 42 kJ mol^{-1}, cleavage of tetrameric (*t*BuLi)$_4$ to **16** is favored by 14 kJ mol^{-1} [20, 21]. Thus, contrary to *tert*-butyllithium, methyllithium forms a dimeric adduct with (R,R)-TMCDA, so that calculating a monomer-based mechanism for MeLi might not reflect the situation in experiment. Therefore, calculations are more reliable when also taking the deaggregation process into account or starting from precoordinated species such as *t*BuLi·(R,R)-TMCDA (**16**), which have been detected in the experiment.

The feasibility of the direct lithiation of (R,R)-TMCDA proves that precoordination is an essential step for enabling deprotonation reactions of tertiary methyl amines under mild reaction conditions. However, not only precoordination but also spatial proximity of reactive groups is required for the sufficient decrease of the reaction

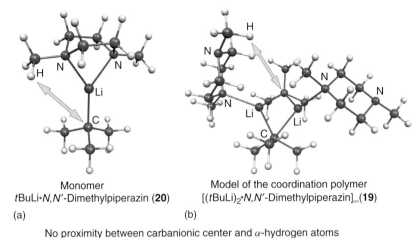

Monomer	Model of the coordination polymer
tBuLi·N,N'-Dimethylpiperazin (20)	[(tBuLi)$_2$·N,N'-Dimethylpiperazin]$_\infty$(19)
(a)	(b)

No proximity between carbanionic center and α-hydrogen atoms

Figure 5.4 No proximity of reactive groups in polymeric [(tBuLi)$_2$·N,N'-dimethylpiperazine]$_\infty$ (**19**) and monomeric tBuLi·N,N'-dimethylpiperazine (**20**) as starting complexes for the deprotonation of N,N'-dimethylpiperazine.

barrier. The missing approximation of reactive groups is, for example, the reason why N,N'-dimethylpiperazine does not readily undergo deprotonation under mild conditions. For an equimolar combination of this amine with *tert*-butyllithium, a coordination polymer [(tBuLi)$_2$·N,N'-dimethylpiperazine]$_\infty$ (**19**) could be isolated, in which tBuLi dimers are linked with each other by the diamine ligands [20]. In this adduct, the methyl groups do not point in direction to the carbanionic center. Thus, the distances between the carbanionic center and the α-hydrogen atoms are considerably longer (at least 4.169 Å) than that in monomeric tBuLi·(R,R)-TMCDA (**16**), which showed a distance of only 3.377 Å (Figure 5.3). This increased contact between the reactive groups results in an increased reaction barrier of 132 kJ mol^{-1} for a dimeric intermediate as model system for the coordination polymer, which was found in the crystal (Figure 5.4b). The progress via a hypothetic monomeric species, (tBuLi·N,N'-dimethylpiperazine) (**20**), also showed an elevated reaction barrier of 131 kJ mol^{-1} (Figure 5.4a). Both reaction barriers are too high to be readily overcome at room temperature, thus explaining the absence of the direct deprotonation under mild reaction conditions. Nevertheless, lithiation of N,N'-dimethylpiperazine is possible via transmetallation reaction [22].

Former studies on the correlation between the proximity of reactive groups and the reaction barriers of lithiation reactions, such as DOM or asymmetric deprotonation, have also proven precoordination as a crucial step for the feasibility of the direct deprotonation. In these examples, C···H distances between the carbanionic center and the hydrogen atom in the range of 3.1 and 3.7 Å indicated highly preorganized systems, resulting in decreased reaction barriers [23]. The proximity features in these complexes allowed predicting and understanding direct lithiations and selectivities.

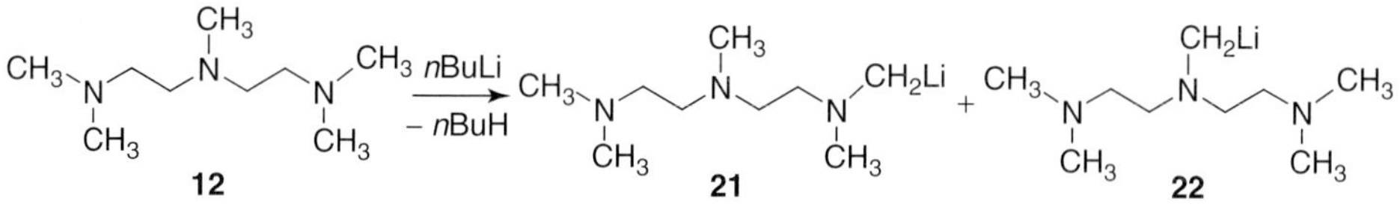

Scheme 5.5 Regioselectivity of the deprotonation of PMDTA (**12**) with n-butyllithium.

a) 1 eq. nBuLi, ratio: **21**:**22** = 80:20
b) 2 eq. nBuLi, ratio: **21**:**22** = 63:37

5.3.2
Regioselective α-Lithiation

The spatial proximity between reactive groups is necessary for enabling direct deprotonation reactions, and is the reason for the observation of regioselective lithiation reactions of tertiary methyl amines with different α-hydrogen atoms. One example is the tridentate ligand PMDTA (**12**) that belongs to the most frequently used Lewis bases in organolithium chemistry due to its strong coordination ability. PMDTA can be lithiated at its terminal and central methyl groups to the α-lithiated amines **21** and **22**. Klumpp and coworkers have shown that the regioselectivity of this deprotonation depends on the amount of n-butyllithium used [18c,d]. While 1 equiv. of the lithiumalkyl gives an 80 : 20 mixture in favor of the lithiation of the terminal methyl group, 2 equiv. results in an increase of the deprotonation of the central methyl group (Scheme 5.5). These varying selectivities hint that reaction mechanisms of the deprotonation occur via two different intermediate complexes.

The dependency on the amount of n-butyllithium used could be explained by isolation and structural characterization of the n-butyllithium adduct [(nBuLi)$_2$·PMDTA]$_2$ (**23**) (Figure 5.5). Compound **23** represents an interesting organolithium aggregate formally consisting of a dimeric n-butyllithium, in which a monomeric nBuLi·PMDTA inserts (**24**) are present. While this complex can be regarded as the precoordinated intermediate of the deprotonation with the 2 : 1 ratio of lithiumalkyl and amine, a monomeric adduct (1 : 1 ratio) can be assumed for the lithiation with 1 equiv. of the lithium base. Such monomeric PMDTA adducts have, for example, been isolated with phenyllithium and can thus be assumed as potential reactive intermediate during the reaction process [4a]. In such a hypothetic monomer, nBuLi·PMDTA (**24**), the deprotonation of the terminal methyl group possesses a barrier of only 99 kJ mol^{-1}. In contrast, lithiation of the central methyl is energetically disfavored by 23 kJ mol^{-1}, which is due to the highly deformed geometry of the adduct in the transition state (Figure 5.6a). These energetic differences in the reaction barriers of the monomer-based mechanism cause the favored deprotonation of the terminal methyl of PMDTA (**12**) group observed in the experiment.

Regarding the deprotonation via the isolated 2 : 1 adduct (**23**), the energetic difference between the barriers for the terminal and central lithiation becomes smaller (16 kJ mol^{-1}), thus resulting in the observed decrease of the regioselectivity.

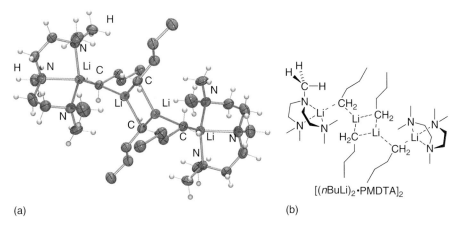

Figure 5.5 (a) Molecular structure (hydrogen atoms of the butyl groups omitted for clarity). (b) Schematic display of [(nBuLi)$_2$·PMDTA]$_2$ (23), intermediate of the deprotonation of PMDTA (12).

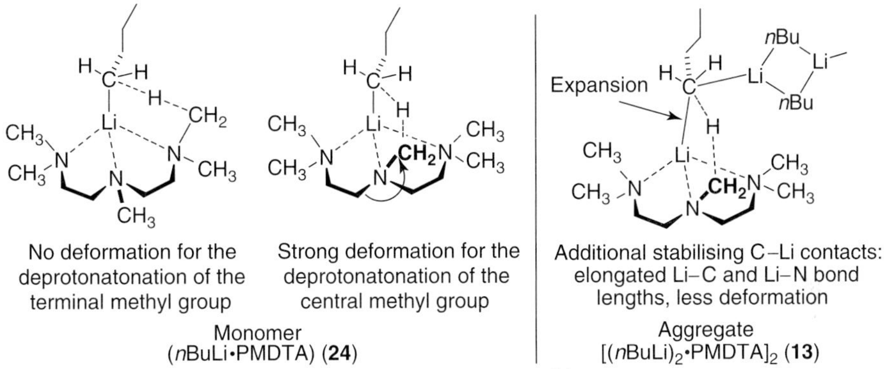

Figure 5.6 Transition states for the deprotonation of the terminal and central methyl groups of PMDTA via monomeric nBuLi·PMDTA (23) (a) and aggregate [(nBuLi)$_2$·PMDTA]$_2$ (24) (b).

This energetic approximation between both transition states is due to additional stabilizing effects in the aggregate [(nBuLi)$_2$·PMDTA]$_2$ (23) (Figure 5.6b). The second carbon lithium contact in 23 does not only result in a further stabilization of the aggregate, but also in the expansion of the contacts between the lithiumalkyl and the PMDTA ligand. These elongated Li−C and Li−N contacts finally lead to a reduced tension and deformation in the transition state of the deprotonation of the

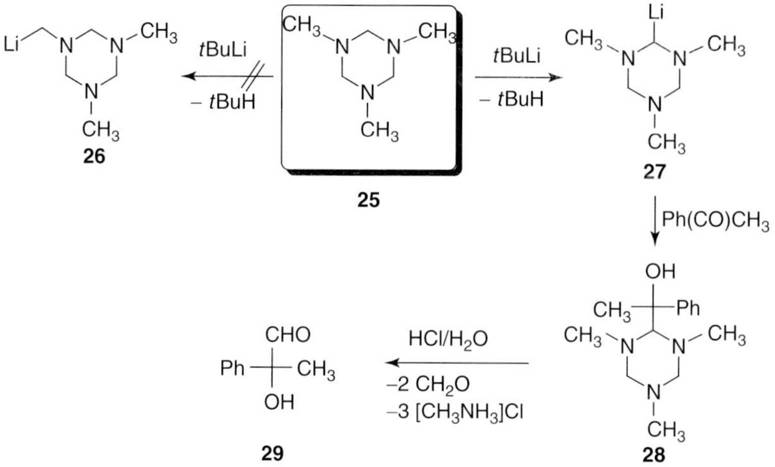

Scheme 5.6 Regioselective deprotonation of the methylene group of triazacyclohexane **25**.

central methyl group [24]. Consequently, the dependency of the deprotonation of PMDTA (**12**) on the amount of n-butyllithium can be referred to different reaction mechanisms via a monomeric species **24** and via aggregate [(nBuLi)$_2$·PMDTA]$_2$ (**23**).

A further example for regioselective, direct deprotonation of a tertiary amine with different α-hydrogen atoms is the cyclic triamine 1,3,5-trimethyl-hexahydro-1,3,5-triazine (**25**). This tridentate ligand can potentially be lithiated at its methyl group and methylene bridge. Mitzel and coworkers showed that in the experiment selective deprotonation of the methylene bridge occurs at room temperature with tert-butyllithium without lithation of the methyl group to compound **26** (Scheme 5.6) [25]. The lithiated triazacyclohexane **27** was used for the preparation of alcohols such as **28** in good isolated yields up to 70%. Hydrolytic workup under acidic conditions gave the acylation product **29** as a result of degradation of the polyaminal-type system [25a]. Thus, the overall reaction represents a mercury- and thallium-free alternative to the classical Corey–Seebach method [26].

As intermediate complex of this reaction, [(tBuLi)$_3$·C$_6$H$_{15}$N$_3$] (**30**) could be isolated and characterized by X-ray diffraction analysis (Figure 5.7) [25b]. In this adduct, three tert-butyllithium units are coordinated on top of the tridentate cyclic ligand forming a six-membered Li−C−Li−C−Li−C ring. The carbanionic centers of the lithiumalkyl are directly located above the methylene units. This results in a distinct proximity between the hydrogen atoms of the methylene bridge and the carbanionic groups, and thus in an enhanced reactivity toward methylene deprotonation. DFT studies of both possible transition states (Figure 5.8a) show a barrier of 107 kJ mol^{-1} for the abstraction of the hydrogen of the methylene bridge and significantly higher barrier of 136 kJ mol^{-1} for the deprotonation of the methyl group. This confirms the process of the reaction at room temperature and the observed selectivities.

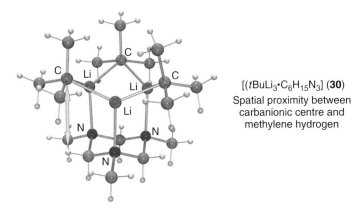

Figure 5.7 Molecular structure of the intermediate complex, [(tBuLi)$_3$·C$_6$H$_{15}$N$_3$] (30).

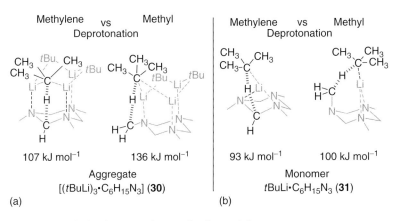

Figure 5.8 Calculated reaction barriers for the methylene and methyl deprotonation of triazacyclohexane (25) via (a) aggregate 30 and (b) a hypothetic monomer tBuLi·C$_6$H$_{15}$N$_3$ (31).

Interestingly, this closer contact between the methylene unit and the carbanionic center is also pronounced in a hypothetic monomer, tBuLi·C$_6$H$_{15}$N$_3$ (31), or dimeric tert-butyllithium adduct, confirming the differences in the reaction barriers and the observed selective deprotonation (Figure 5.8b). Such pathways via different aggregates than via 30 are important for the deprotonation of 25 with further alkyllithium bases than tert-butyllithium. In the experiment, also n-butyllithium and sec-butyllithium were found to readily undergo such a lithiation reaction with the same preference for the abstraction of the methylene hydrogen atom [25a]. However, due to different spatial demand, these lithiumalkyls should prefer the formation of other adducts with the triazacyclohexane than with aggregate 30 during the reaction process. Accordingly, a dimeric structure has been observed with phenyllithium [25b]. DFT calculations confirm that the

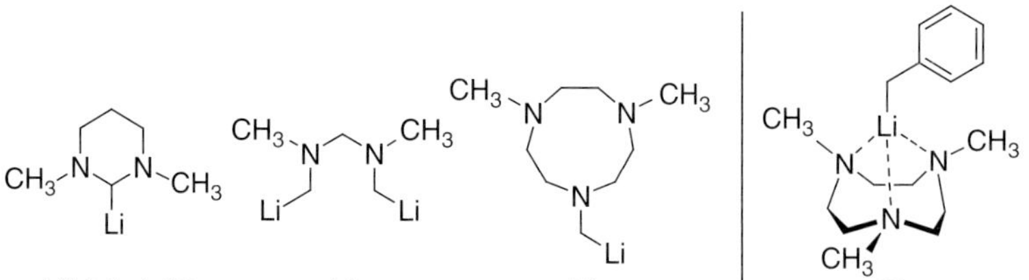

Figure 5.9 (a) α-lithiated diazacyclohexane **32**, TMMDA (**11**), and (CH$_3$)$_3$tacn (**14**); (b) schematic of the molecular structure of benzyllithium·(CH$_3$)$_3$tacn (**33**): proximity between the methyl groups and carbanionic center.

deprotonation of the methylene group is also preferred in such dimeric adduct by 37 kJ mol^{-1} (dimer-based mechanism with methyllithium as deprotonation agent instead of *tert*-butyllithium). Regarding the monomer-based mechanism depicted in Figure 5.8, a preference of 7 kJ mol^{-1} has been observed. For this process via *t*BuLi·$_6$H$_{15}$N$_3$ (**31**), even a conformational change of the methyl group to the axial position is required for the methyl deprotonation.

Altogether, the regioselectivity of direct deprotonation reactions of tertiary amines with different α-hydrogen atoms can be explained by the precoordination according to the CIPE (Section 5.2.3) [7]. Analogous to **25**, the bidentate ligand, 1,3-dimethyl-1,3-diazacyclohexane (**32**), also undergoes selective deprotonation of the methylene bridge (Figure 5.9) and not of its methyl groups [25a]. This selectivity suggests the formation of intermediate species comparable to complexes **30** and **31** during the triazacyclohexane lithiation, which possess a close contact between the carbanionic center and the methylene unit, thus preferring methylene deprotonation. Yet, there is no experimental proof of such an intermediate adduct of this amine so far.

Contrary to **25** and **32**, selective deprotonation even of both methyl groups was observed for TMMDA (**11**), although this ligand possesses the same arrangement of methylene and methyl groups, CH$_3$NCH$_2$NCH$_3$ [19]. Selective lithiation of the methyl group was also observed for the tridentate ligand (CH$_3$)$_3$tacn (**14**). In the case of this amine, a monomeric benzyllithium adduct (**33**) could be isolated and characterized by X-ray diffraction analysis. In benzyllithium·(CH$_3$)$_3$tacn (**33**), the methyl groups of the triazacyclononane **14** are arranged in the direction of the carbanionic center (Figure 5.9b) [4b]. This arrangement in **33** is comparable to adducts of the tridentate ligand PMDTA (**12**) (Figure 5.6) and thus explains the selective deprotonation of the methyl group.

5.3.3
α-Lithiation versus β-Lithiation

TMEDA, **10** is one of the most-often applied Lewis bases for the activation of organolithium compounds in synthetic chemistry. However, investigations by

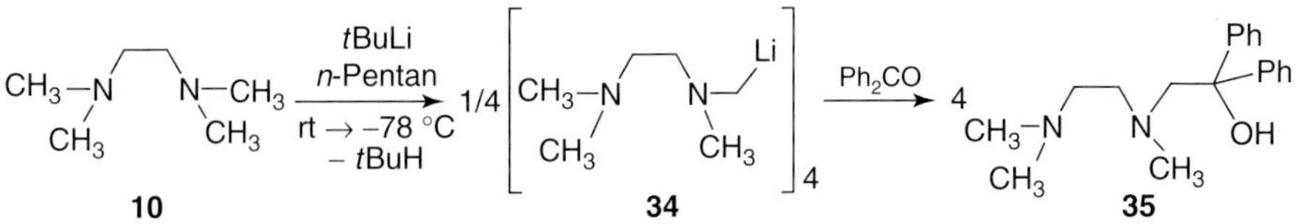

Scheme 5.7 Direct α-deprotonation of TMEDA **10** to the lithiated amine **34** and alcohol **35** after trapping.

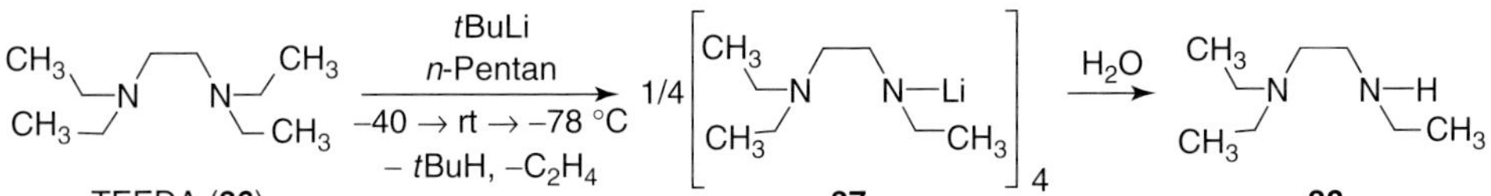

Scheme 5.8 Direct β-deprotonation of TEEDA **36** to lithium amide **37** after elimination of ethene and trapping product **38** after hydrolysis.

Köhler and coworkers have shown that TMEDA also undergoes a decomposition reaction with strong bases. In this case, the deprotonation reaction occurs at the ethylene bridge and/or at the methyl group depending on the deprotonation agent used [18a]. With *tert*-butyllithium, lithiation mainly occurs at the methyl group (Scheme 5.7); however, with lower yields compared to (*R,R*)-TMCDA (**15**) or 1,3,5-trimethylhexahydro-1,3,5-triazine (**25**), due to the competing deprotonation of the ethylene bridge. The α-lithiated amine (**34**) could be isolated and structurally characterized as a tetrameric compound and trapped with benzophenone to the tridentate ligand (**35**) in yields up to 51% [27]. For synthetic applications, however, lithiated TMEDA and the dilithiated analog have been prepared by transmetallation of the corresponding organotin compound or reductive C–S bond cleavage giving higher yields than the direct deprotonation [10b, 11d].

Changing from TMEDA to the ethyl substituted analog *N,N,N′,N′*-tetraethylethylenediamine (TEEDA, **36**), deprotonation can potentially occur in α- as well as in β-position to the nitrogen atom. In the experiment, selective deprotonation in β-position was observed even below room temperature. The lithiated amine instantly reacts under elimination of ethene to the corresponding lithium amide (**37**) (Scheme 5.8), which was isolated and characterized by X-ray analysis and hydrolyzed to the unsymmetrical diamine (**38**). As an intermediate of this selective β-lithiation reaction, the monomeric adduct of **36** with *tert*-butyllithium, *t*BuLi·TEEDA (**39**), could be isolated. This highly reactive precoordinated complex (Figure 5.10a) results in the approximation of the carbanionic center and the β-hydrogen atom. In the crystal, *t*BuLi·TEEDA (**39**) possesses a considerably closer contact between the carbanionic center and the β-hydrogen atom (3.15(2) Å) compared to the α-hydrogen atoms (at least 3.95(2) Å) [23]. This proximity finally results in a decreased reaction barrier for the β-lithiation and thus in the selective deprotonation observed in the experiment. Computational studies of both possible transition states of the monomer-based mechanism confirm this observation showing a reaction barrier

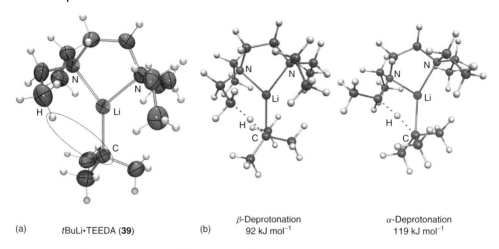

(a) *t*BuLi·TEEDA (**39**)   (b) β-Deprotonation 92 kJ mol^{-1} α-Deprotonation 119 kJ mol^{-1}

Figure 5.10 (a) Molecular structure of *t*BuLi·TEEDA (**39**): close contact between the carbanionic center and β-hydrogen atom; (b) transition states of the β- and α-deprotonation of TEEDA (**36**).

of 119 kJ mol^{-1} for the α-lithiation and a significantly lower barrier of 92 kJ mol^{-1} for the β-lithiation (Figure 5.10b). This energetic difference is sufficiently high to explain the regioselectivity of this reaction [21a].

Besides *tert*-butyllithium, *n*BuLi and *i*PrLi too undergo such a selective β-lithiation of TEEDA (**36**). With *iso*-propyllithium and this diamine, an unsymmetrical aggregate, [(*i*PrLi)$_3$·(TEEDA)$_2$], could be isolated, indicating a potential aggregate-based mechanism for the deprotonation [4e]. The selective β-lithiation via monomeric *t*BuLi·TEEDA (**39**) suggests an analogous mechanism of the decomposition of diethylether via a precoordinated adduct according to the CIPE. This decomposition of ether compounds is of fundamental importance in organolithium chemistry as many alkyllithiums are commercially available in solutions of ether, and often reactions with these reagents are carried out in these solvents [9].

5.4
Conclusions and Outlook

Access to α-lithiated amines via direct deprotonation reactions offers an easy introduction of nitrogen functions by simple substitution reactions. The enhanced mechanistic knowledge allows a more profound understanding of the progress of the reaction and of the observed selectivities. In general, the mechanism of this lithiation occurs – according to the CIPE – via an intermediate complex, in which the organolithium compound is precoordinated by the amine ligand. In this complex, reactive groups approximate each other, thus resulting in a decreased reaction barrier and finally enabling the direct deprotonation. The distinct proximity of the carbanionic center and different CH acidic groups enables the differentiation between these hydrogen atoms and the selective abstraction of only one specific hydrogen atom. As a consequence, the precomplexation enables both regioselective α-lithiation and β-deprotonation reactions. While the mechanistic features of these direct lithiation reactions of tertiary methyl amines and its selectivities have been elucidated and investigated in detail, the application of this knowledge is still in its infancy. However, due to the great demand of nitrogen-containing compounds, the direct deprotonation undoubtedly offers huge potential in synthetic chemistry as has already been shown exemplarily in the case of TMCDA (**15**) or TMMDA (**11**) [28]. Therefore, interesting building blocks become accessible enabling the preparation of new ligand systems, catalysts, natural products, or organometallic compounds. With the deprotonation of chiral amines, such as (R,R)-TMCDA, even systems with stereoinformation in the molecule become easily accessible.

The example of direct lithiation of tertiary amines demonstrates the huge potential of the combination of X-ray diffraction analysis and quantumchemical studies for the elucidation of reaction mechanisms. With these methods the isolation of reactive intermediate species becomes possible giving a direct insight into the reaction mechanism and thus information about reactive species, the progress of the reaction, and observed selectivities. This approach to mechanistic knowledge is not only limited to the α-lithiation of tertiary methylamines and the decomposition of Lewis bases in organolithium chemistry, but can also be applied on further reactions via stable intermediates. Therefore in future, a further gain of knowledge about reaction mechanisms can be expected by the application of this method.

Acknowledgments

I specially thank my supervisor Prof. Dr Carsten Strohmann for his inimitable support and advices. I also thank the *Fond der Chemischen Industrie* for a doctoral scholarship and the *Deutsche Forschungsgemeinschaft* for financial support as well as the universities of *Dortmund* and *Würzburg* for the facilities. I also thank Christian Däschlein for his permanent support and help during my PhD work.

References

1. Schlenk, W. and Holtz, J. (1917) *Ber. Dtsch. Chem. Ges.*, **50**, 262–274.
2. (a) Stey, T. and Stalke, D. (2004) in *The Chemistry of Organolithium Compounds* (eds Z.Rappoport and I. Marek), John Wiley & Sons, Ltd, Chichester, pp. 47–120; (b) Gessner, V.H., Däschlein, C., and Strohmann, C. (2009) *Chem. Eur. J.*, **15** (14), 3320–3334, and reference therein.
3. (a) Kottke, T. and Stalke, D. (1993) *Angew. Chem. Int. Ed. Engl.*, **32** (4), 580–582; (b) Strohmann, C., Seibel, T., and Strohfeldt, K. (2003) *Angew. Chem. Int. Ed.*, **42** (37), 4531–4533.
4. Examples of monomeric organolithiums: (a) Schümann, U., Kopf, J., and Weiss, E. (1985) *Angew. Chem. Int. Ed. Engl.*, **24** (3), 215–216; (b) Arnold, J., Knapp, V., Schmidt, J.A.R., and Shafir, A. (2002) *J.*

Chem. Soc., Dalton Trans., 3273–3274; (c) Zarges, W., Marsch, M., Harms, K., and Boche, G. (1989) Chem. Ber., **122** (12), 2303–2309.; (d) Strohmann, C., Seidel, T., and Schildbach, D. (2004) J. Am. Chem. Soc., **126** (32), 9876–9877; (e) Strohmann, C., Gessner, V.H., and Damme, A. (2008) Chem. Commun., (29), 3381–3383.

5. (a) Bernstein, M.P. and Collum, D.B. (1993) J. Am. Chem. Soc., **115** (2), 789–790; (b) Liao, S. and Collum, D.B. (2003) J. Am. Chem. Soc., **125** (15), 114–15127; (c) Sun, X. and Collum, D.B. (2000) J. Am. Chem. Soc., **122** (11), 2452–2458; (d) Sun, X., Kenkre, S.L., Remenar, J.F., Gilchrist, J.H., and Collum, D.B. (1997) J. Am. Chem. Soc., **119** (20), 4765–4766; (e) Remenar, J.F. and Collum, D.B. (1997) J. Am. Chem. Soc., **119** (24), 5573–5582; (f) Jones, A.C., Sanders, A.W., Sikorski, W.H., Jansen, K.L., and Reich, H.J. (2008) J. Am. Chem. Soc., **130** (19), 6060–6061; (g) Reich, H.J., Sikorski, W.H., Thompson, J.L., Sanders, A.W., and Jones, A.C. (2006) Org. Lett., **8** (18), 4003–4006.

6. For examples, see: (a) Hoppe, D., Hintze, F., and Tebben, P. (1990) Angew. Chem. Int. Ed. Engl., **29** (12), 1422–1424; (b) Kerrick, S.T. and Beak, P. (1991) J. Am. Chem. Soc., **113** (25), 9708–9710; (c) Whisler, M.C. and Beak, P. (2003) J. Org. Chem., **68** (4), 1207–1215; (d) Coldham, I., Copley, R.C.B., Haxell, T.F.N., and Howard, S. (2003) Org. Biomol. Chem., **1** (9), 1532–1544; (e) Metallinos, C., Szillat, H., Taylor, N.J., and Snieckus, V. (2003) Adv. Synth. Catal., **345** (3), 370–382; (f) Würthwein, E.-U., Behrens, K., and Hoppe, D. (1999) Chem. Eur. J., **5** (12), 3459–3463, and references therein.

7. (a) Whisler, C.M., MacNeil, S., Snieckus, V., and Beak, P. (2004) Angew. Chem. Int. Ed., **43** (17), 2206–2225; (b) Hartung, C.G. and Snieckus, V. (2002) in Modern Arene Chemistry (ed. D.Astruc), Wiley-VCH Verlag GmbH, Weinheim, pp. 330–367; (c) Beak, P. and Meyers, A.I. (1986) Acc. Chem. Res., **19** (11), 356–363.

8. For review, see (a) Kizirian, J.-C. (2008) Chem. Rev., **108** (1), 140–205; (b) Yamataka, H., Yamada, K., and Tomioka, K. (2004) in The Chemistry of Organolithium Compounds (eds Z. Rappoport and I. Marek), John Wiley & Sons, Inc., New York, pp. 901–939; (c) Hoppe, D. and Christoph, G. (2004) in The Chemistry of Organolithium Compounds (eds Z. Rappoport and I. Marek), John Wiley & Sons, Inc., New York, pp. 1055–1164.

9. Maercker, A. and Demuth, W. (1973) Angew. Chem. Int. Ed. Engl., **12** (1), 75–76.

10. Rondan, N.G., Houk, K.N., Beak, P., Zajdel, W.J., Chandrasekhar, J., and von Schleyer, P.R. (1981) J. Org. Chem., **46** (20), 4108–4110.

11. (a) Seyferth, D. and Weiner, M.A. (1959) J. Org. Chem., **24** (9), 1395–1396; (b) Tian, X., Fröhlich, R., Pape, T., and Mitzel, N.W. (2005) Organometallics, **24** (22), 5294–5298; (c) Peterson, D.J. (1967) J. Organomet. Chem., **9** (2), 373–374; (d) Peterson, D.J. (1971) J. Am. Chem. Soc., **93** (16), 4027–4031; (e) Bruhn, C., Becke, F., and Steinborn, D. (1998) Organometallics, **17** (10), 2124–1426; (f) Becke, F., Heinemann, F.W., Rüffer, T., Wiegeleben, P., Boese, R., Bläser, D., and Steinborn, D. (1997) J. Organomet. Chem., **548** (2), 205–210; (g) Gawley, R.E. and Zhang, Q. (1995) J. Org. Chem., **60** (18), 5763–5769, and references therein.

12. (a) Strohmann, C. and Abele, B.C. (1996) Angew. Chem. Int. Ed. Engl., **35** (20), 2378–2380; (b) Broka, C.A. and Shen, T. (1989) J. Am. Chem. Soc., **111** (8), 2981–2984; (c) Florio, S., Capriati, V., Gallo, A., and Cohen, T. (1995) Tetrahedron Lett., **36** (25), 4463–4466; (d) Katritzky, A.R., Qi, M., and Feng, D. (1998) J. Org. Chem., **63** (19), 6712–6714; (e) Alonso, D.A., Alonso, E., Najera, C., Ramon, D.J., and Yus, M. (1997) Tetrahedron, **53** (13), 4835–4856; (f) Katritzky, A.R., Feng, D., and Qi, M. (1997) J. Org. Chem., **62** (18), 6222–6225.

13. (a) Zeller, E., Sajus, H., and Grierson, D.S. (1991) Synlett, (1), 44–46; (b) Bonin, M., Romem, J.R., Grierson, D.S.,

and Husson, H.-P. (1982) *Tetrahedron Lett.*, **23** (33), 3369–3372.
14. Azzena, U., Melloni, G., and Nigra, C. (1993) *J. Org. Chem.*, **58** (24), 6707–6711.
15. Guijarro, D. and Yus, M. (1993) *Tetrahedron*, **49** (35), 7761–7768.
16. (a) Peterson, D.J. and Hays, H.R.J. (1965) *J. Org. Chem.*, **30** (6), 1939–1942; (b) Lepley, A.R. and Giumanini, A.G. (1966) *J. Org. Chem.*, **31** (7), 2055–2060; (c) Lepley, A.R. and Khan, W.A. (1966) *J. Org. Chem.*, **31** (7), 2061–2064; (d) Smith, W.N. (1974) *Adv. Chem. Ser.*, **130**, 23–55; (e) Ahlbrecht, H. and Dollinger, H. (1984) *Tetrahedron Lett.*, **25** (13), 1353–1356; (f) Lepley, A.R., Khan, W.A., Giumanini, A.B., and Giumanini, A.G. (1966) *J. Org. Chem.*, **31** (7), 2047–2051.
17. (a) Kessar, S.V., Singh, P., Vohra, R., Kaur, N.P., and Singh, K.N.J. (1991) *Chem. Soc., Chem. Commun.*, (8), 568–570; (b) Kessar, S.V. and Singh, P. (1997) *Chem. Rev.*, **97** (3), 721–737; (c) Ebden, M.R., Simpkins, N.S., and Fox, D.N.A. (1995) *Tetrahedron Lett.*, **36** (47), 8697–8700; (d) Vedejs, E. and Kendall, J.T. (1997) *J. Am. Chem. Soc.*, **119** (29), 6941–6942, and references therein.
18. (a) Köhler, F.H., Hertkorn, N., and Blümel, J. (1987) *Chem. Ber.*, **120** (12), 2081–2082; (b) Harder, S. and Lutz, M. (1994) *Organometallics*, **13** (12), 5173–5176; (c) Schakel, M., Aarnts, M.P., and Klumpp, G.W. (1990) *Recl. Trav. Chim. Pays-Bas*, **109** (4), 305–306; (d) Klumpp, G.W., Luitjes, H., Schakel, M., de Kanter, E.J.J., Schmitz, R.F., and van Eikema Hommes, N.J.R. (1992) *Angew. Chem. Int. Ed. Engl.*, **31** (5), 633–635; (e) Luitjes, H., Schakel, M., Aamts, M.P., Schmitz, R.F., de Kanter, F.J.J., and Klumpp, G.W. (1997) *Tetrahedron*, **53** (29), 9977–9988; (f) Hildebrand, A., Lönnecke, P., Silaghi-Dumitrescu, L., Silaghi-Dumitrescu, I., and Hey-Hawkins, E. (2006) *J. Chem. Soc., Dalton Trans.*, (7), 967–974.
19. Karsch, H.H. (1996) *Chem. Ber.*, **129** (5), 483–484.
20. Strohmann, C. and Gessner, V.H. (2007) *Angew. Chem. Int. Ed.*, **46** (43), 8281–8283.
21. (a) Strohmann, C. and Gessner, V.H. (2007) *J. Am. Chem. Soc.*, **129** (29), 8952–8953; (b) Strohmann, C. and Gessner, V.H. (2008) *J. Am. Chem. Soc.*, **130** (35), 11719–11725.
22. Tian, X., Fröhlich, R., and Mitzel, N.W. (2006) *Z. Anorg. Allg. Chem.*, **632** (2), 307–312.
23. (a) Saá, J.M., Morey, J., Frontera, A., and Deyá, P.M. (1995) *J. Am. Chem. Soc.*, **117** (3), 1105–1116; (b) Wiberg, K.B. and Bailey, W.F. (2001) *J. Am. Chem. Soc.*, **123** (34), 8231–8238; (c) Wiberg, K.B. and Bailey, W.F. (2000) *Angew. Chem. Int. Ed.*, **39** (12), 2127–2129.
24. Strohmann, C. and Gessner, V.H. (2007) *Angew. Chem. Int. Ed.*, **46** (24), 4566–4569.
25. (a) Bojer, D., Kamps, I., Tian, X., Hepp, A., Pape, T., Fröhlich, R., and Mitzel, N.W. (2007) *Angew. Chem. Int. Ed.*, **46** (22), 4176–4179; (b) Strohmann, C. and Gessner, V.H. (2008) *Chem. Asian J.*, **3** (11), 1929–1934; (c) Köhn, R.D., Seifert, G., and Kociok-Köhn, G. (1996) *Chem. Ber.*, **129** (11), 1327–1333.
26. (a) Seebach, D. and Corey, E.J. (1975) *J. Org. Chem.*, **40** (2), 231–237; (b) Seebach, D. (1979) *Angew. Chem. Int. Ed. Engl.*, **18** (4), 239–258.
27. (a) Gessner, V.H. and Strohmann, C. (2008) *J. Am. Chem. Soc.*, **130** (44), 14412–14413; (b) Gessner, V.H., Däschlein, C., and Strohmann, C. (2009) *Acta Cryst. E*, **E65** (2), o383.
28. Examples of the synthetic application of α-lithiated amines: (a) Murakami, M., Hayashi, M., and Ito, Y. (1992) *J. Org. Chem.*, **57** (3), 793–794; (b) Murata, Y. and Nakai, T. (1990) *Chem. Lett.*, **19** (11), 2069–2072; (c) Gawley, R.E., Zhang, Q., and Campagna, S. (1995) *J. Am. Chem. Soc.*, **117** (47), 11817–11818; (d) Coldham, I., Hutton, R., and Snowden, D.J. (1996) *J. Am. Chem. Soc.*, **118** (22), 5322–5323; (e) Hammerschmidt, F. and Hanbauer, M. (2000) *J. Org. Chem.*, **65** (19), 6121–6131.

Part II
Predictive Tools in Organic Chemical Reactions

6
Double Hydrogen Bonding in Asymmetric Organocatalysis: A Mechanistic Perspective

Tommaso Marcelli

6.1
Introduction

The use of small organic molecules as catalysts, widely known as *organocatalysis*, hardly requires any introduction. In the first decade of its existence as a distinct research field, it has repeatedly thrilled the synthetic community by providing synthetic protocols for an amazing array of chemical transformations, sometimes with performances matching or even exceeding those obtained with transition metal catalysts [1–4].

Nevertheless, organocatalysis still suffers from a number of major limitations precluding its widespread use in industrial production. Such problems can be revealed by evaluating the discrepancies between the supposed benefits of organocatalysis and the real state of things. For instance, it is a commonplace that organocatalysis is inherently environmentally friendly due to the absence of toxic transition metals. In most cases, organocatalytic processes require high catalyst loadings, considerable amounts of solvent, low temperatures, and long reaction times. All these factors negatively affect the "greenness" of organocatalytic reactions. Next to its presumed environmentally friendliness, organocatalysis is often praised for being "biomimetic." Although some striking similarities in the mode of action of organocatalysts and enzymes support this notion, the claim of biomimicry is often out of line. One can think at the substrate specificity typical of enzymes and, more importantly, at their turnover rates, routinely exceeding those of the best organocatalysts by several orders of magnitude. Although many of the organocatalytic protocols are valuable tools for the small-scale synthesis of chiral compounds in a chemical laboratory, their use in industrial applications is limited by all these factors.

Besides few remarkable examples, the development of organocatalysts is strongly experiment driven, as can be easily deduced by the number of structurally related compounds ordinarily screened for catalytic activity at the beginning of nearly every literature report. This is understandable, considering the mechanistic complexity of many organocatalytic reactions. Although the mode of action of the main classes of metal-free catalysts is schematically understood, detailed mechanistic

Ideas in Chemistry and Molecular Sciences: Advances in Synthetic Chemistry. Edited by Bruno Pignataro
Copyright © 2010 WILEY-VCH Verlag GmbH & Co. KGaA, Weinheim
ISBN: 978-3-527-32539-9

studies often reveal that successful organocatalytic protocols are the result of a subtle interplay of several factors. In particular, enantioselective reactions are very difficult to design considering that small variations in the energy difference between diastereomeric transition states have a profound impact on the optical purity of the products. Nevertheless, the understanding of organocatalytic reactions proceeds at a steady pace and, although rare, examples of mechanism-based *in silico* design of enantioselective catalysts can already be found in the recent literature [5].

Hydrogen bonding is surely the most important noncovalent interaction in asymmetric organocatalysis. A general strategy to increase the electrophilicity of a substrate involves either explicit protonation or hydrogen bonding with one or two acidic protons of the catalyst. In particular, many successful organocatalysts contain moieties capable of donating two hydrogen bonds to a Lewis basic site of a substrate. Such organocatalysts have often been compared to the "oxyanion hole" of serine proteases, where two amide protons of one glycine and one serine residues interact with the amide carbonyl of the substrate to stabilize the incipient negative charge on the oxygen resulting from the attack of a serine hydroxyl group [6]. Other catalysts contain two spatially close acidic protons, although only one of the hydrogen bonds present in the key transition state of the catalyzed reaction takes place with the substrate while the second hydrogen bond is intramolecular and connects two functionalities of the catalyst. A third class of hydrogen bonding organocatalysts features a Brønsted acid (BA) site and a Lewis basic atom. Such catalysts can also engage in two hydrogen bonds with either one or two substrates. The following pages discuss the mode of action of organocatalysts sharing the presence of two hydrogen bonds in the catalyst/substrate(s) reactive complexes (Figure 6.1).

Besides being clearly behind the scope of this contribution, a comprehensive overview of the literature on this topic would overlap with a number of excellent reviews [7–10]. Instead, this chapter features a gallery of recent examples unraveling important mechanistic aspects of double hydrogen bonding organocatalysis in synthetically relevant transformations. In addition, some striking recent results in the development of novel types of double hydrogen bonding organocatalysts are briefly discussed.

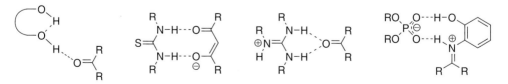

Figure 6.1 Representative examples of double hydrogen bonding in organocatalysis.

6.2
Diols and Amidoalcohols

Although chiral diols have been extensively used as ligands for transition metals in asymmetric transformations, their potentialities as catalysts have long been overlooked. In general, catalysts featuring two spatially close alcohol functions can efficiently activate carbonyl compounds, nitroso compounds, and boronic esters. In this respect, especially (axially chiral) C_2-symmetric diols have found several applications as catalysts or cocatalysts. Alternatively, diols containing an additional Lewis basic moiety (such as a tertiary amine or phosphine) have been shown to be efficient asymmetric catalysts for the simultaneous activation of a nucleophile and an electrophile [9]. The examples described in the following pages highlight two aspects of "pure" double hydrogen bonding catalysis for the same transformation, namely, a hetero-Diels–Alder (hDA) reaction. For this reason, this section includes both diols and amidoalcohols – double hydrogen bonding catalysts – that could find many applications in the future due to their ready availability.

6.2.1
Single-Point versus Two-Point Activation

The $\alpha,\alpha,\alpha',\alpha'$-tetraaryl-1,3-dioxolan-4,5-dimethanol (TADDOL)-catalyzed asymmetric hDA reaction, developed by Rawal and coworkers, surely stands among the most significant achievements in organocatalysis [11]. This work clearly established that simple chiral hydrogen bond donors can be used to promote organic reactions in synthetically useful yields and enantioselectivities. Successively, this protocol was extended to all-carbon Diels–Alder (DA) reactions, considerably broadening its scope and usefulness [12]. Furthermore, the same group showed that also axially chiral diols (such as 1,1'-biaryl-2,2'-dimethanol (BAMOL)) are able to promote hDA reactions with an excellent degree of asymmetric induction (Scheme 6.1) [13].

On the basis of crystallographic data, it was proposed that the chiral diol activates the dienophile by establishing only one intermolecular hydrogen bond with the aldehyde oxygen, while the other hydroxy group of the catalyst engages in an intramolecular hydrogen bond. This type of catalyst/substrate interaction is referred to as *single-point activation* (or cooperative hydrogen bonding), in contraposition with the possibly more intuitive two-point activation, in which catalyst and substrate are held together by two intermolecular hydrogen bonds (Figure 6.2).

The elucidation of both reaction mechanism and origin of enantioselectivity in the (h)DA reaction has been the subject of a number of computational studies. Houk and coworkers performed density functional theory (DFT) calculations on both DA and hDA reactions using 1,4-butanediol as the catalyst [14]. For both reactions, it was found that single-point activation of the dienophile resulted in a more efficient lowering of the reaction barrier, thereby confirming Rawal's hypothesis. Shortly afterwards, the same group reported a combined molecular mechanics (MMs) and quantum mechanics/molecular mechanics (QMs/MMs) study addressing the reasons for enantioselection in the hDA reaction [15]. Besides

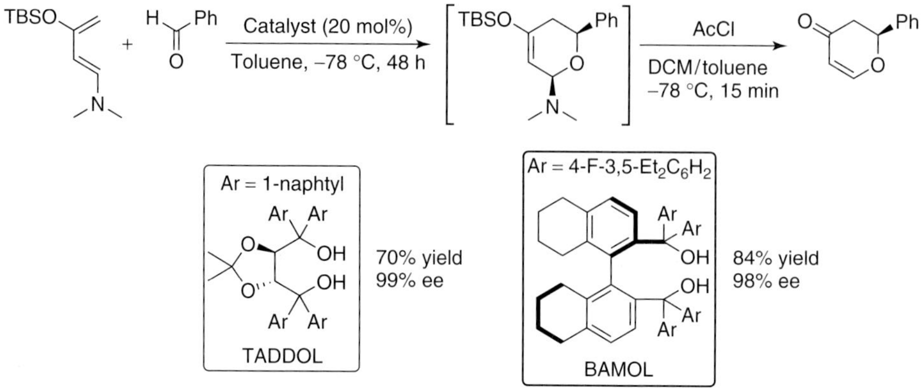

6 Double Hydrogen Bonding in Asymmetric Organocatalysis

Scheme 6.1 Asymmetric hDA reaction catalyzed by chiral diols. (TBS = *tert*-butyldimethylsilyl; TADDOL = $\alpha,\alpha,\alpha',\alpha'$-tetraaryl-1,3-dioxolan-4,5-dimethanol; BAMOL = 1,1′-biaryl-2,2′-dimethanol.)

the already-mentioned catalyst/dienophile key hydrogen bond, the lowest-energy transition state featured a stabilizing CH/π interaction between the aldehyde hydrogen and one of the naphthyl rings of TADDOL (Figure 6.3). This result slightly revised the models for enantioselection originally proposed by Rawal *et al.* [12] and, subsequently, by other authors based on computational studies at lower levels of theory, all of them featuring a π/π interaction between the phenyl ring of benzaldehyde and one naphthyl ring of TADDOL [16, 17].

6.2.2
The Impact of Acidity on Enantioselectivity

Sigman and coworkers developed a family of organocatalysts based on an oxazoline-containing scaffold featuring two nonidentical hydrogen bond donors

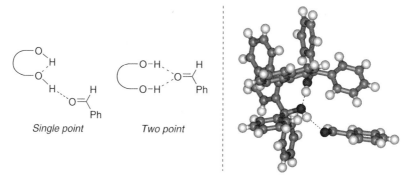

Figure 6.2 Alternative catalyst/dienophile hydrogen bonding patterns and crystal structure of a BAMOL/benzaldehyde complex.

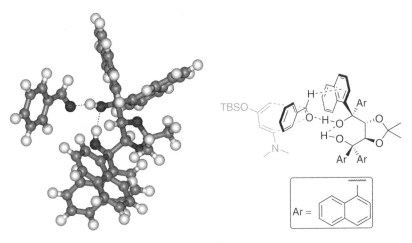

Figure 6.3 Lowest-energy transition state for the TADDOL-catalyzed hDA reaction (diene not shown for clarity in the 3D representation).

[18]. Although there are no computational or crystallographic studies supporting this hypothesis, given the structural features of this class of catalysts, two-point activation seems to be the only mechanistic option for the activation of the dienophile. The highly modular nature of the catalyst design allowed independent tuning of the acidity of both hydrogen bond donors. This feature was exploited to carry out an experimental mechanistic study on the impact that catalyst acidity has on rate and selectivity in the previously described hDA reaction (Scheme 6.2) [19].

These experiments highlighted a remarkable correlation between the acidity of the substituted acetamides (estimated from the pK_a of the parent acid) and both rates of formation and enantiomeric excesses of the hDA products. Analysis of the initial rates of formation of the two enantiomers revealed that while product formation becomes faster with more acidic catalysts for both enantiomers, the increase in rate for the major enantiomer is definitely more pronounced, resulting in higher

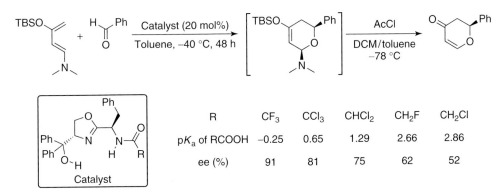

Scheme 6.2 Impact of catalyst acidity on the enantiomeric excess of the product.

asymmetric inductions with more acidic catalysts. This study provided systematic experimental evidence for one specific reaction to support the assumption that increasing the acidity of a (double) hydrogen bonding donor results in improved catalytic performances.

6.3 (Thio)ureas

Ureas and thioureas have been popular in crystal engineering since the late 1980s [20, 21]. Their capability of donating two relatively strong and directional hydrogen bonds makes them attractive functionalities for the achievement of tight complexation with Lewis basic sites. Following some early observations on their potential in catalysis [22, 23], the (thio)urea moiety has been introduced in a steadily increasing number of organocatalysts. Together with phosphoric acids (see below), thioureas have now become the most popular moieties for the activation of hydrogen bond acceptors toward nucleophilic attack [24]. Although the mode of action of thiourea-containing organocatalysts might often seem straightforward, there are some mechanistic aspects playing an important role on the catalytic performances that have been addressed only recently.

6.3.1
Mechanistic Duality in Aminothiourea Catalysis

Many thiourea catalysts also contain a tertiary amine. Nearly the entirety of the early work on such organocatalysts was based on the mechanistic assumption that the thiourea would activate the electrophile by double hydrogen bonding, while the amino group would deprotonate the nucleophile, resulting in the formation of an ion pair (pathway A, Scheme 6.3). However, an alternative mechanism featuring a reversal of the binding geometry of electrophile and nucleophile would also lead to product formation (pathway B).

The main hurdle in the identification of which of these two pathways is active in a certain reaction is their kinetic indistinguishability. In such cases, computational techniques can provide significant help in understanding both the mechanism and the reasons for the stereochemical outcome of a reaction. This mechanistic duality was first identified by Soós, Papai *et al.* who performed a DFT study of the catalytic enantioselective addition of acetylacetone to nitroolefins [25], previously developed by Takemoto and coworkers (Scheme 6.4) [26, 27].

These calculations clearly showed that pathway B is energetically less demanding, therefore contradicting the mechanistic hypothesis formulated by Takemoto *et al.* (Figure 6.4) [26, 27]. Interestingly, for both pathways A and B, a preference for the formation of the experimentally observed major enantiomer was found.

This mechanistic duality has been addressed also for 1,2-additions. Hiemstra *et al.* developed the first highly enantioselective organocatalyst for the Henry reaction of aromatic aldehydes, a C6ʹ-thiourea *Cinchona* derivative (Scheme 6.5). This catalyst

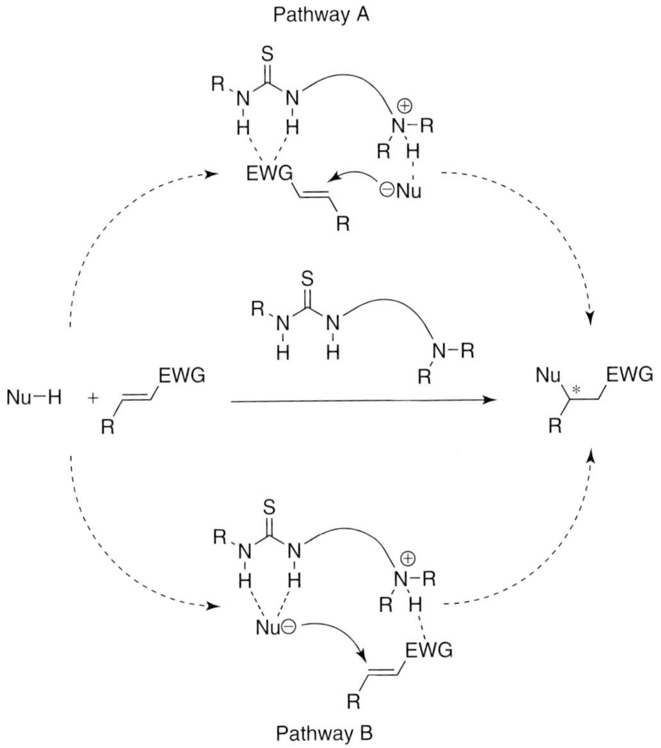

Scheme 6.3 Mechanistic duality in aminothiourea catalysis, exemplified for a conjugate addition (EWG = electron-withdrawing group).

was designed starting from the observation that cupreidines (C6′–OH quinidine derivatives) promoted the Henry reaction of nitromethane with activated aromatic aldehydes in moderate enantiomeric excesses [28], whereas the use of quinidine as catalyst yielded substantially racemic product [29]. Replacement of the phenol with an activated thiourea significantly improved both rate and enantioselectivity [30].

DFT calculations on a model system showed that in this case both pathways have very similar reaction barriers and, therefore, they might both be operative under the reaction conditions. However, only for pathway A, which is energetically

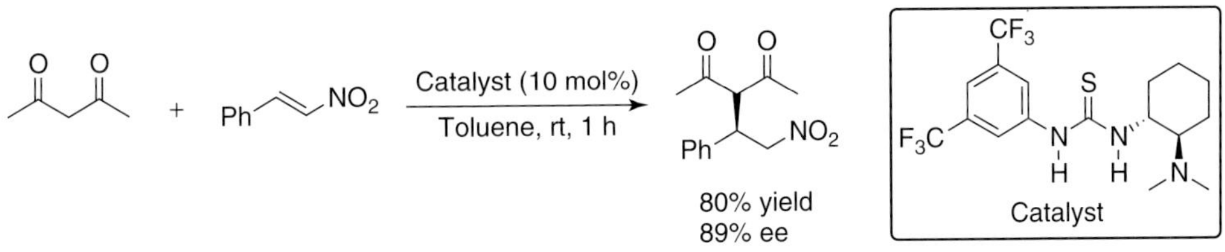

Scheme 6.4 Enantioselective addition of acetylacetone to nitroolefins.

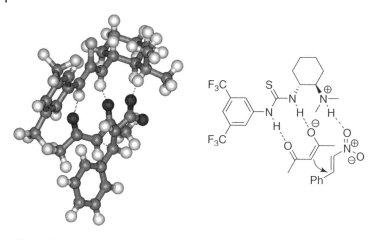

Figure 6.4 Lowest-energy transition state in pathway B for the addition of acetylacetone to nitrostyrene.

slightly favored, the observed enantioselectivity could be successfully reproduced (Figure 6.5) [31].

On the other hand, a DFT study by Zuend and Jacobsen showed that in the organocatalytic cyanation of ketones, previously developed in the same laboratory [32], pathway A is significantly more accessible [33]. Comparison of calculated and experimental isotope effects further corroborated this conclusion. In this case, it is worth noting that the cyanide ion is activated by two hydrogen bonds with an ammonium ion and with a secondary amide (Figure 6.6). Moreover, investigation of the reasons for enantioselectivity provided information that was used to design an improved catalyst, considerably broadening the scope of the reaction.

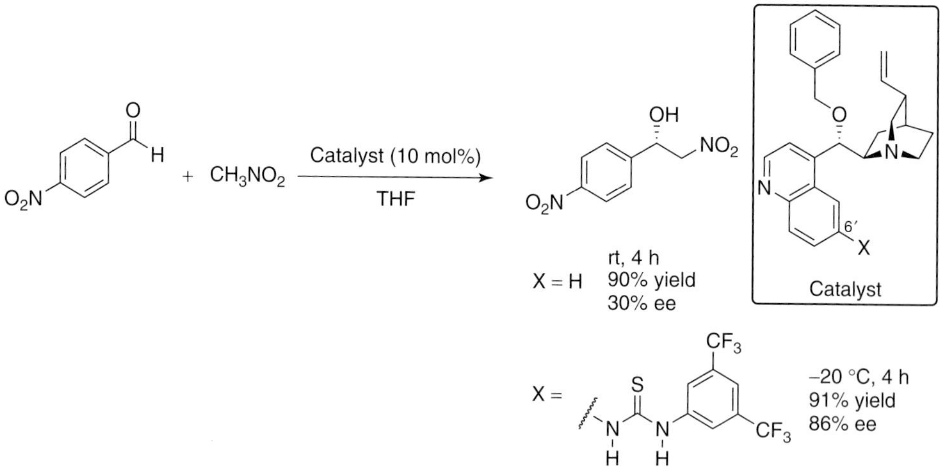

Scheme 6.5 Enantioselective Henry reaction using *Cinchona*-derived organocatalysts.

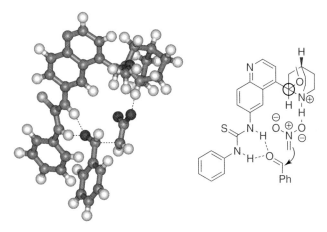

Figure 6.5 Lowest-energy transition state in pathway A for the Henry reaction of benzaldehyde.

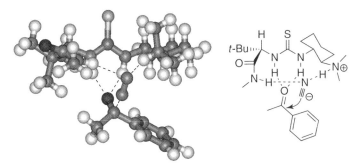

Figure 6.6 Transition state for the enantioselective addition of HCN to acetophenone.

6.3.2
Mono- versus Bidentate Coordination

In catalysts containing a thiourea and a primary amine, operating with an enamine mechanism, the above mentioned mechanistic duality is obviously not possible. Yet, there are other nontrivial aspects playing an important role on the stereochemical outcome of the reactions catalyzed by these compounds. For instance, in the most intuitive complexation geometry between a nitro group and a thiourea, each N–H engages in a hydrogen bond with one of the oxygens (bidentate coordination). Alternatively, both thiourea hydrogens could coordinate to the same oxygen of the nitro group (monodentate coordination, Figure 6.7).

Tsogoeva *et al.* developed an aminothiourea-catalyzed addition of acetone to nitroolefins [34]. DFT calculations showed that, although the two hydrogen bonding patterns lead to very similar energies, only the monodentate complex is productive in the C–C bond forming step. This observation contradicted previous working hypotheses on thiourea-catalyzed additions to nitroolefins [26, 27]. Also, from

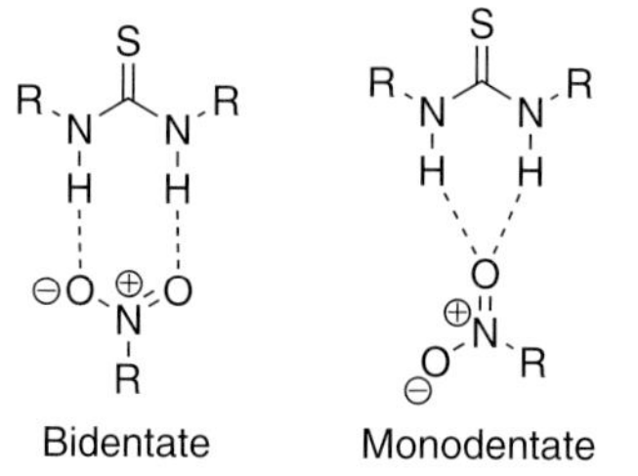

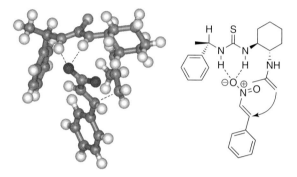

Figure 6.7 Complexation geometries of thioureas with nitro groups.

Figure 6.8 Lowest-energy transition state in the enantioselective addition of acetone to nitrostyrene.

an intuitive point of view, the computational results pointed to a six-membered transition state instead of the eight-membered structure required for bidentate coordination of the nitroolefin to the catalyst. This observation might prove useful in the rationalization of organocatalytic reactions involving other electrophiles, such as α,β-unsaturated imides (Figure 6.8) [35].

6.3.3
Catalyst Self-Association

Besides being excellent hydrogen bond donating moieties, ureas possess a fairly Lewis basic carbonyl oxygen that can in turn accept hydrogen bonds from other species, such as another urea molecule. This partly explains the success of thioureas in asymmetric catalysis since, in this latter class of compounds, the formation of supramolecular-stacked architectures is prevented by the poor hydrogen bond accepting properties of the thiocarbonyl sulfur atom. As a consequence, thioureas are generally much more soluble than the corresponding ureas. Nevertheless, thiourea organocatalysts can still form other types of noncovalently bound dimers (or possibly oligomers) and this is not necessarily a drawback for catalytic performances.

Tárkányi, Soós and coworkers carried out an extensive NMR study [36] to unravel the solution behavior of a C9 *epi-Cinchona* thiourea catalyst, previously developed in the same group [37]. NOESY data obtained at low temperature in apolar solvents

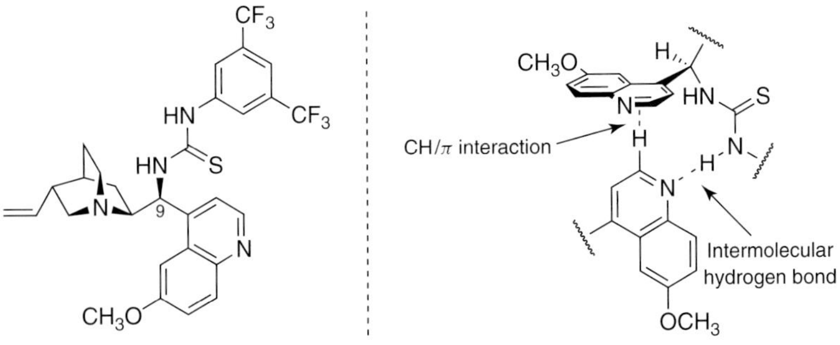

Figure 6.9 C9-*epi*-*Cinchona* thiourea catalyst and schematic representation of the dimer.

clearly indicated that this catalyst undergoes a noncovalent self-dimerization to yield an asymmetric complex with a well-defined geometry. In more detail, a T-shaped CH/π interaction between the two 6-methoxyquinoline rings and a hydrogen bond between a thiourea proton and one quinoline nitrogen were identified as the most significant interactions holding the dimer together (Figure 6.9). The authors pointed out that, since the best catalytic performances were obtained in apolar solvents, it is likely that the noncovalent dimer might actually be responsible for the high activities and enantioselectivities obtained in the addition of nitromethane to chalcones [37].

Chin, Song *et al.* reported the use of the same catalyst in the asymmetric methanolysis of *meso*-anhydrides. In this case, however, the best results were obtained with aprotic, hydrogen bond accepting solvents and at low catalyst concentrations, indicating that the monomer might be the species responsible for efficient asymmetric induction [38]. The authors obtained further support for this hypothesis by replacing the thiourea moiety of the catalyst with a sulfonamide. Use of the resulting *Cinchona* derivative in the same reaction substantially decreased the influence of temperature, solvent, and concentration on the enantiomeric excess of the products (Scheme 6.6) [39].

At the light of these results, the ability of oxygenated solvents to disrupt noncovalent intermolecular catalyst assemblies might also be at the basis of the unique solvent dependence of the enantiomeric excess found by Hiemstra *et al.* for the previously described asymmetric Henry reaction [30]. In this case, while no correlation was found between enantiomeric excess of the product and dielectric constant of the solvent, a qualitative trend indicating better performances with solvents of higher Lewis basicity was clearly established, possibly due to inhibition of catalyst self-association [31].

6.3.4
Halide Binding

As previously mentioned, one of the most interesting features of thioureas, exploited in supramolecular chemistry long before their advent in organocatalysis, is their ability to coordinate a variety of anions, including halides [40]. Until very

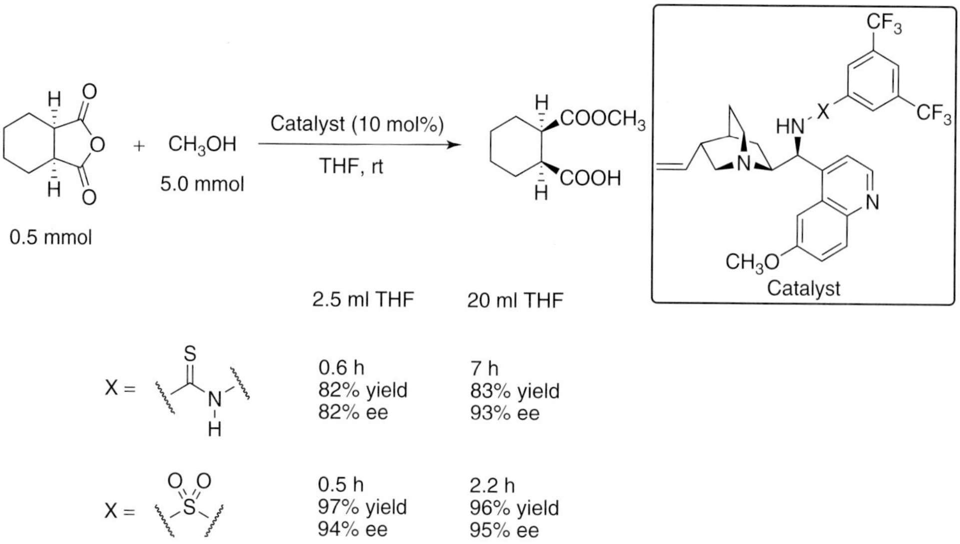

Scheme 6.6 Enantioselective opening of *meso*-anhydrides and concentration dependence of the catalytic performances.

recently, this well-known phenomenon has not been believed to play an important role in organocatalytic reactions. However, in some cases, complexation of a thiourea organocatalyst with a halide anion rather than with the electrophile might actually be responsible for the observed asymmetric inductions in nucleophilic additions. Jacobsen and coworkers developed a thiourea-catalyzed highly enantioselective Pictet–Spengler-type cyclization of indoles on *N*-acyliminium ions [41] (Scheme 6.7).

The reaction, carried out in the presence of a chloride source, begins with the irreversible formation of a chlorolactam. From here, there are two alternative mechanistic possibilities for product formation: (i) an S_N2 process, with the indole C2 attacking the lactam and displacing the chloride and (ii) an S_N1 mechanism, involving loss of chloride to generate an *N*-acyliminium ion that is subsequently trapped by the indole. A combination of spectroscopic studies and systematic variation of the reaction conditions strongly supported the second mechanistic scenario. DFT calculations on a model system as well as pronounced counterion and solvent effects prompted the authors to propose a thiourea–chloride complexation

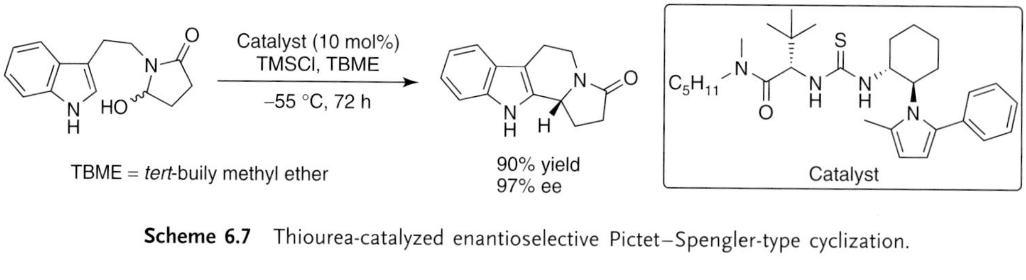

Scheme 6.7 Thiourea-catalyzed enantioselective Pictet–Spengler-type cyclization.

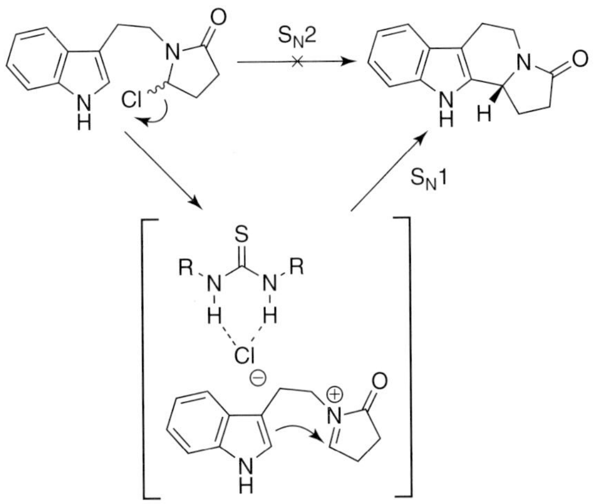

Scheme 6.8 Mechanism of the Pictet–Spengler-type cyclization, involving the formation of a thiourea–chloride complex.

in the C–C bond forming step. In other words, the thiourea moiety does not interact directly with the substrate in this step but rather forms a complex with the chloride anion. Therefore, the enantioselectivity arises from the interaction of the cationic intermediate (the N-acyliminium ion) with the chiral counterion (the thiourea–chloride complex) (Scheme 6.8).

The same group successfully extended this strategy also to the addition of silyl ketene acetals to oxocarbenium ions [42]. These results suggest that chiral thiourea–chloride complexes might be the catalytic species responsible for the high enantiomeric excesses observed in other previously developed transformations [43, 44]. The use of chiral counterions as stereocontrollers is an emerging strategy in both transition metal [45] and iminium catalysis [46] and might find other important applications in thiourea catalysis.

6.4
Phosphoric Acids

Shortly after their first applications as organocatalysts were reported, phosphoric acids became extremely popular for the activation of various electrophiles, especially imines [47–49]. In particular, axially chiral C_2-symmetric catalysts with very bulky substituents in proximity of the acid moiety can promote a remarkable array of reactions with excellent enantioselectivities at occasionally low catalyst loadings. Although phosphoric acids are monoprotic BAs, they engage in double hydrogen bonding in most of the transformations they promote. The following pages focus on this aspect as well as recent examples of intriguing substrate combinations that can be activated by phosphoric acids.

6.4.1
Bifunctionality

The hypothesis that phosphoric acids act as Lewis base–Brønsted acid (LBBA) bifunctional catalysts was already formulated by Uraguchi and Terada in their seminal report describing the enantioselective addition of acetylacetone to N-carbamoyl imines [50]. In the same year, Akiyama and coworkers developed an addition of ketene silyl acetals to N-aryl imines [51]. Interestingly, the enantioselectivity was found to be strongly dependent on the nature of the N-aryl substituent (Scheme 6.9).

The presence of an *ortho*-hydroxy group was found to be crucial, suggesting its participation in the complexation with the catalyst. In a later DFT study on a model system, two alternative imine/phosphoric acid coordination modes were considered, in analogy to what is described in Section 6.3.2 (Figure 6.10). For all reaction intermediates and transition states, it was found that bidentate coordination is energetically favored [52].

These calculations also indicated that, in the transition states for C–C bond formation, the rigidity of the dicoordinated iminium/phosphate complex forces the N-aryl imine substituent to stack with one p-nitrophenyl ring of the catalyst. While attack on the *re*-face of the imine does not experience unfavorable steric interactions (Figure 6.11), addition to the *si*-face of the electrophile (not shown) leads to considerable repulsions between the substrates and one nitrophenyl group of the catalyst.

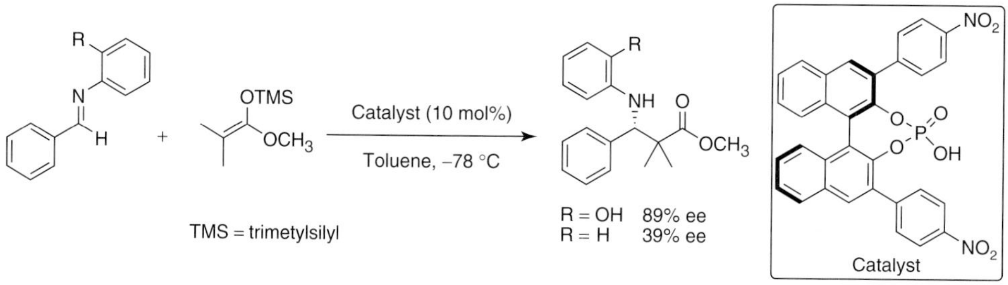

Scheme 6.9 Phosphoric acid–catalyzed addition of ketene silyl acetals to N-aryl imines.

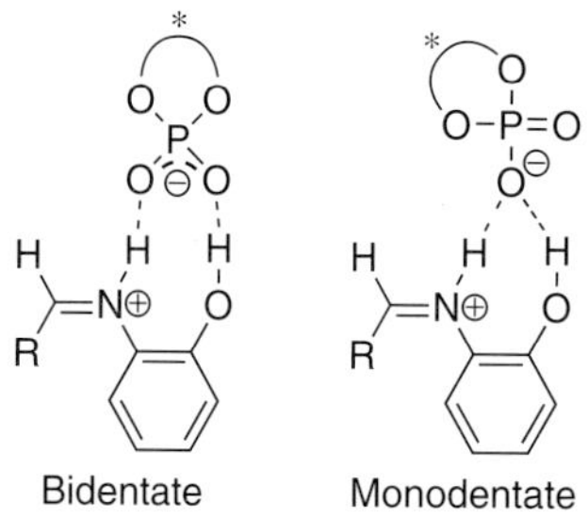

Figure 6.10 Alternative coordination modes of a phosphate with an N-(*ortho*-hydroxy)aryl iminium.

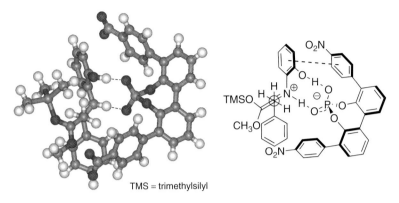

Figure 6.11 Lowest-energy transition state in the enantioselective Mannich reaction of ketene silyl acetals.

In the last years, phosphoric acids have been extensively shown to be efficient promoters for the enantioselective reduction of various electrophiles using dihydropyridines (mainly Hantzsch esters) as hydrogen donors [53, 54]. Curiously, although no explicit proposals were formulated on the mechanism of this type of transformations, the possibility of specific interactions between the dihydropyridine and the phosphoric acid catalyst was generally overlooked. Himo et al. reported a DFT study of both reaction mechanism and origin of enantioselectivity for the phosphoric acid–catalyzed transfer hydrogenation of N-aryl imines and quinolines [55]. Three different mechanistic hypotheses were considered for the hydride transfer: simultaneous activation of both reaction partners via either a bidentate (di-LBBA) or a monodentate complex (mono-LBBA), or, alternatively, activation of the phosphate with the iminium but no interactions between catalyst and dihydropyridine (BA, Scheme 6.10).

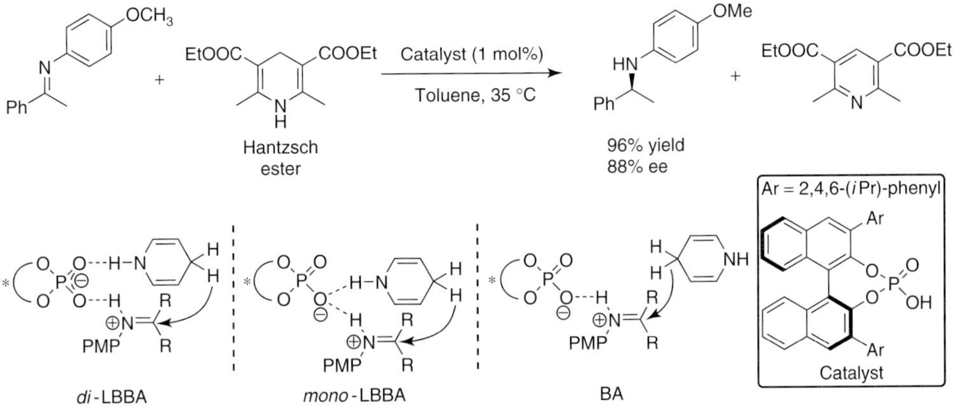

Scheme 6.10 Phosphoric acid–catalyzed transfer hydrogenation of ketimines using Hantzsch esters and mechanistic hypotheses for the hydride transfer (PMP = p-methoxyphenyl).

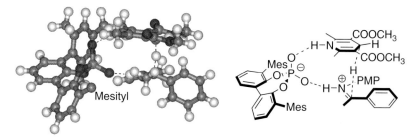

Figure 6.12 Lowest-energy transition state in the enantioselective hydride transfer to an N-aryl ketimine (for clarity, a mesityl group has been removed from the 3D representation).

Calculations on a model system revealed that the *di*-LBBA pathway is by far favored over the other two hypotheses. Interestingly, it was found that although the *E*-ketimine is considerably more stable than its *Z*-isomer, upon protonation and complexation with the phosphate, *E*- and *Z*-iminium ions have very similar energies and the barrier for hydride transfer is accessible for both of them. This consideration turned out to play an important role on the stereochemical outcome of the reaction. Calculations with a chiral phosphoric acid successfully reproduced the sense and magnitude of the experimentally observed enantioselectivity, the lowest-energy transition state being hydride transfer to the *re*-face of the *Z*-iminium (Figure 6.12). The enantioselectivity was rationalized considering the steric interactions between the aryl rings of the ketimine and the bulky mesityl groups of phosphoric acid used for modeling the reaction.

Further calculations successfully reproduced the reversal of enantioselectivity observed in the transfer hydrogenation of 2-arylquinolines to the corresponding tetrahydroquinolines derivative. In this case, the cyclic iminium ion can only have an *E*-geometry and hydride transfer to the *si*-face is energetically favored [55]. A successive study by the same group addressed the enantioselective reduction of racemic α-branched aldimines; also in this case the calculations correctly reproduced the sense of enantioselectivity [56]. Simón and Goodman independently reported a QM/MM study of the enantioselective transfer hydrogenation of N-aryl ketimines, reaching substantially identical conclusions with respect to mechanism and enantioselectivity [57]. The same authors also investigated the phosphoric acid–catalyzed enantioselective hydrocyanation of imines [58]. Once again, the *E*/*Z* isomerism of the iminium ion was found to have a major impact on the stereochemical outcome of the reaction. In more detail, the switch of enantioselectivity observed on changing the imine N-substituent from phenyl to benzyl was successfully rationalized in terms of the relative stabilities of the two geometrical isomers of the iminium ion.

6.4.2
Competent Substrates

Although most of the literature on chiral phosphoric acid catalysis describes nucleophilic additions to imines, these catalysts can efficiently activate also other types of electrophiles, as shown by some recent examples. Akiyama and coworkers reported an enantioselective protocol for the addition of indoles to nitroalkanes (Scheme 6.11) [59].

The resulting adducts could be obtained in good yields and excellent enantiomeric excesses when a significant amount of powdered molecular sieves was added to the reaction mixture. This indicated that even small amounts of moisture considerably inhibit product formation (while substantially maintaining the asymmetric induction), a rather unusual phenomenon in asymmetric organocatalysis. The authors proposed that phosphoric acid promotes the reaction by forming a bidentate hydrogen-bound complex with one oxygen of the nitro group and the indole N−H (the reaction of N-methyl indole is sluggish and leads to the formation of racemic product). Although further studies are needed to rationalize these results, it is possible that, given the relative weakness of catalyst/substrate interactions, even a small amount of water can disrupt the hydrogen bonding network, thereby considerably lowering the catalytic efficiency. Zhou, He *et al.* investigated the phosphoric acid−catalyzed addition of indoles to chalcones [60]. In this case, however, the obtained enantioselectivities were considerably lower (below 60% ee). The authors evaluated the influence of a variety of additives, including acid, bases, and oxidants (but not drying agents) and all of them had a detrimental effect on the asymmetric induction. Antilla and coworkers developed a highly enantioselective protocol for the opening of *meso*-aziridines with trimethylsilyl azide using a vaulted phosphoric acid catalyst [61]. Screening of various reaction parameters revealed a strong solvent dependence of the enantioselectivity. Moreover, the presence of an electron poor N-benzoyl substituent on the aziridine was crucial to obtain satisfactory results (Scheme 6.12).

The use of other azide donors (such as sodium azide or tetrabutylammonium azide) resulted in no reaction. Analysis of the reaction mixture by ^{1}H-NMR spectroscopy prompted the authors to propose a mechanism involving an initial reaction of phosphoric acid with the trimethylsilyl azide to form a chiral silane, which would then activate the aziridine for ring opening. In other words, the role

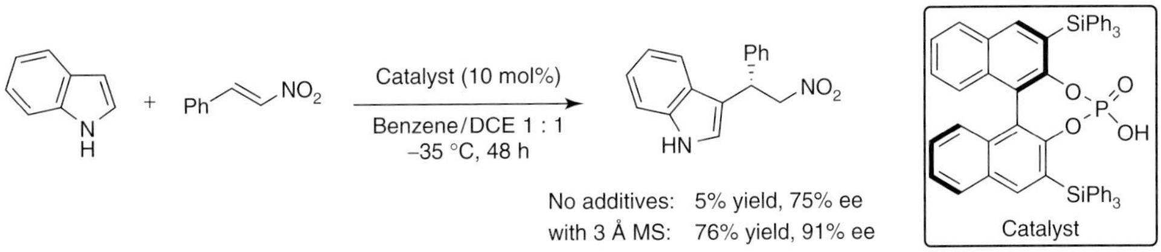

Scheme 6.11 Phosphoric acid−catalyzed enantioselective addition of indoles to nitroalkenes.

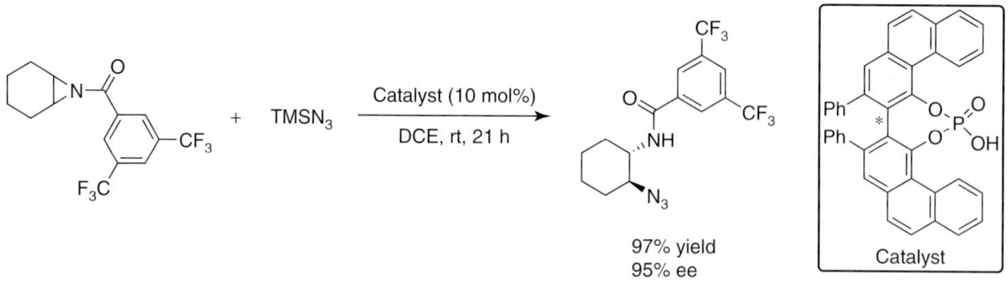

Scheme 6.12 Phosphoric acid–catalyzed enantioselective opening of *meso*-aziridines.

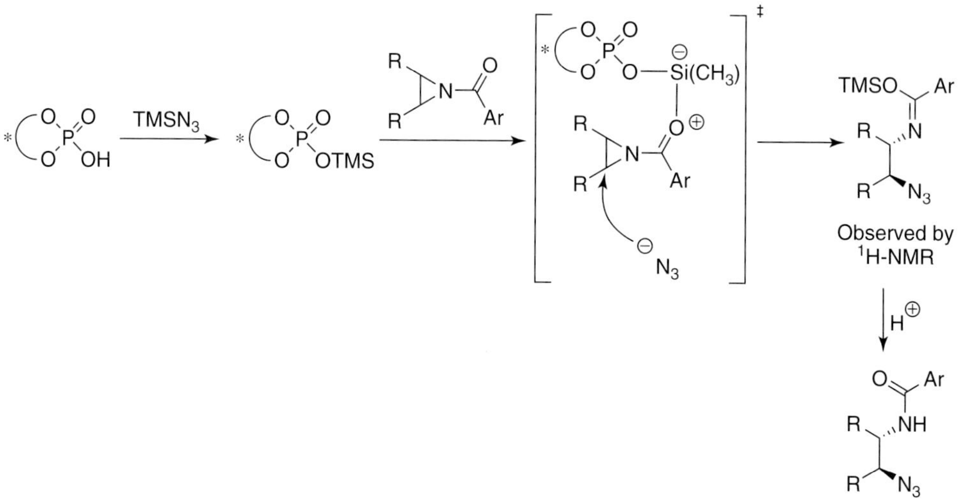

Scheme 6.13 Proposed mechanism for the enantioselective opening of *meso*-aziridines.

of phosphoric acid in this reaction would be to generate *in situ* a chiral Lewis acid from the pronucleophile (Scheme 6.13). Considering the ready availability of a wide variety of silane reagents, this concept might find other important applications in asymmetric catalysis.

6.5
Other Catalysts

6.5.1
Guanidiniums

Many biocatalytic processes involve interactions of guanidinium ions with negatively charged species. At physiological pH, guanidines are generally protonated and can interact with several anionic species (such as carboxylates and phosphates) through double hydrogen bonding. In the last few years, several chiral

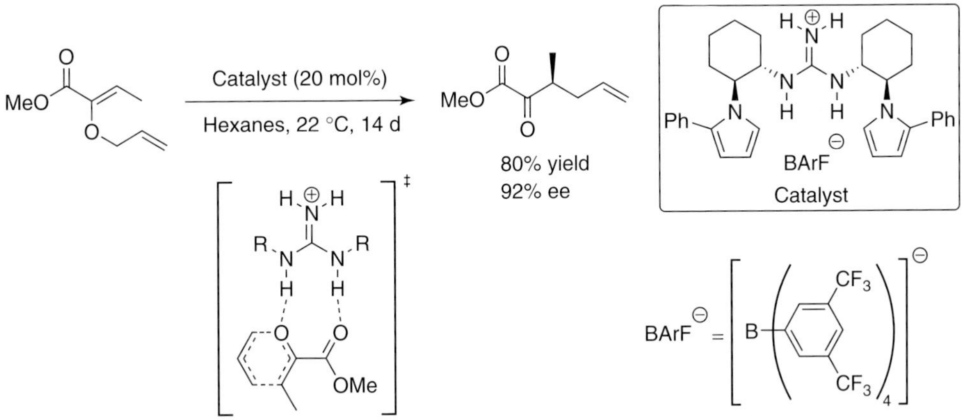

Scheme 6.14 Guanidinium-catalyzed enantioselective Claisen rearrangement and proposed transition state.

guanidine-catalyzed reactions have been developed [62]. In organocatalytic reactions, guanidines generally fulfill two functions: deprotonation of a pronucleophile and interaction of the resulting guanidinium with an anionic species through double hydrogen bonding. The first example of asymmetric "pure" hydrogen bonding catalysis by guanidinium ions was recently reported by Uyeda and Jacobsen who developed a highly enantioselective Claisen rearrangement catalyzed by a chiral C_2-symmetric catalyst [63]. While the use of a catalytic amount of ureas and thioureas did not lead to synthetically useful rate accelerations at room temperature [22], simple achiral guanidiniums promoted the reaction of a variety of allyl vinyl ethers with more acceptable reaction times. However, good enantioselectivities could be obtained only with substrates bearing an additional ester group on the vinylic α-carbon (Scheme 6.14). It is worth mentioning that the best asymmetric inductions were obtained in hexanes, although the catalyst is nearly insoluble in this solvent.

DFT calculations on a model system showed that in this reaction two different oxygens coordinate to the guanidinium moiety, resulting in a maximization of the charge separation. This might explain the better results obtained with these substrates compared to other allyl vinyl ethers lacking an additional Lewis basic moiety. Considering that, in aprotic media, the guanidinium catalyst used in this study has very similar acidity to N,N'-diarylthioureas, the reason for its superior catalytic activity is unclear.

6.5.2
Squaramides

Rawal and coworkers designed a *Cinchona*-derived bifunctional squaramide-amine organocatalyst and tested it in the addition of β-dicarbonyl compounds to nitroolefins (Scheme 6.15) [64]. This catalyst gave excellent results in terms of rate

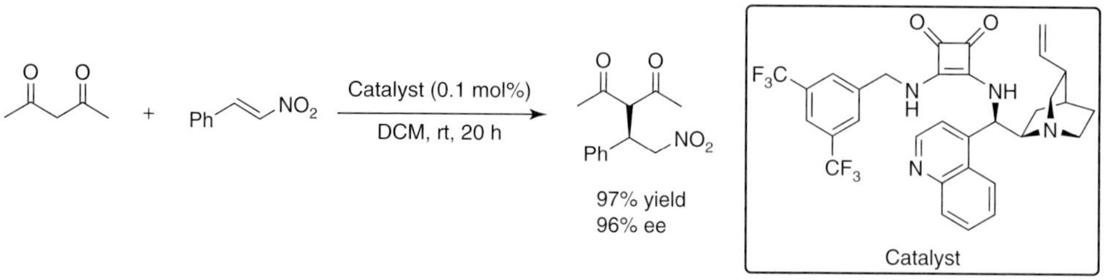

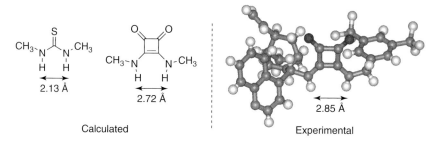

Scheme 6.15 Squaramide-amine catalyzed addition of acetylacetone to nitroolefins.

Figure 6.13 H–H distances in double hydrogen bonding organocatalysts.

and enantioselectivity even at loadings as low as 0.1 mol%, clearly outperforming thiourea-based catalysts [26].

Simple MMs calculations showed that the distance between the two hydrogens involved in substrate coordination is considerably longer in N,N′-dimethylsquaramide than that in N,N′-dimethylthiourea; this observation was further confirmed by a crystal structure of the *Cinchona* squaramide derivative (Figure 6.13). The impact of this structural feature on the catalytic properties as well as on the geometry of substrate coordination has not been established yet. However, it is clear that the squaramide moiety might find many applications in organocatalysis [65], also in light of its straightforward and modular installment on a chiral scaffold.

6.5.3
Quinolinium Thioamides

Ganesh and Seidel explored the use of several protonated hydrogen bonding catalysts in the enantioselective addition of indoles to nitroalkenes [66]. Their design principle was based on systematic structural variations starting from the hydroxythiourea catalyst developed by Ricci *et al.* for the same transformation [67]. It was envisioned that a properly located pyridinium could engage in a hydrogen bond with the sulfur atom of the thiourea, rigidifying its structure and enhancing its acidity [23]. The N–H–S interaction was indeed detected by ^{1}H-NMR for several of the considered structures. This led to considerable rate acceleration compared to a simple thiourea but no improvements in the asymmetric induction. On the other

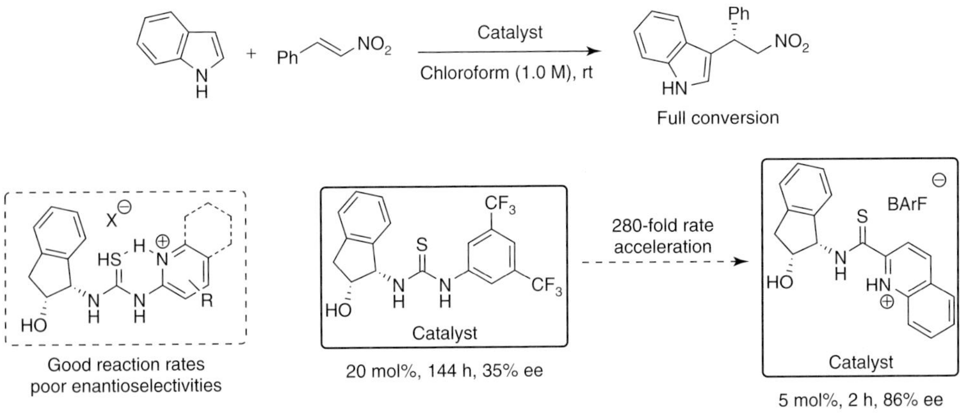

Scheme 6.16 Quinolinium thioamide–catalyzed enantioselective addition of indoles to nitroalkanes.

hand, the use of a quinolinium thioamide catalyst gave excellent results, both in terms of rate and enantioselectivity (Scheme 6.16).

Both NMR and X-ray data indicated that the quinolinium proton does not interact with the thiocarbonyl sulfur while it could participate in substrate binding. Unfortunately, unambiguous experimental determination of this postulated interaction by ^{1}H-NMR titration of the catalyst with nitrostyrene was hampered by the rapid exchange of the quinolinium and the alcoholic proton. It is envisioned that, given the impressive performances obtained in the studied conjugate addition, the use of 2-(thio)acyl pyridinium catalysts might find useful applications also in other types of transformations.

6.6
Conclusions and Outlook

In these last years, asymmetric organocatalysis has gone a long way, winning the attention of many synthetic and computational chemists. The development of highly stereoselective catalytic systems for an impressive array of both "classic" organic reactions and new complex transformations has abundantly demonstrated that efficient enantio- and diastereoselective man-made catalysts are not confined to the realm of transition metal complexes. In order to realize the full potential of asymmetric organocatalysis, future research will have to focus on the development of considerably more active catalytic systems. Hydrogen bonding plays a major role in enzymatic processes and it will likely be further exploited in upcoming organocatalytic systems, together with other noncovalent interactions such as ion pairing, π stacking, and CH/π interactions. Advancements in the understanding of organocatalytic reactions will allow the rational introduction of hydrogen bond donors/acceptors in the catalyst design to improve stabilization of the desired transition states, resulting in improved reaction rates and selectivities. In this respect, especially chemists working in asymmetric organocatalysis will likely continue to take advantage of the ability of double hydrogen bonding patterns to rigidify the complexation geometry between catalyst and substrate(s). Although relatively unexplored, the use of intramolecular hydrogen bonds to control the solution structure of the catalyst might prove a powerful strategy for the future [68]. Computational techniques have abundantly proven to be a highly valuable tool for the rationalization of experimental results. Although high-level quantum mechanical studies still require dedicated computer resources, simple calculations on model systems can be easily performed on a normal personal computer and provide synthetic chemists with very useful insights on organocatalytic reactions. It seems therefore reasonable to assume that, in the coming years, the importance of computational tools in the development of new catalytic systems is destined to grow.

The results described in the previous pages significantly improved our understanding of organocatalytic reactions, often by pointing out important aspects that had been overlooked in the initial catalyst design. Together with the inexhaustible source of inspiration represented by enzymatic processes, the use of these mechanistic insights in catalyst design will likely result in significant progress toward the development of truly biomimetic organocatalytic systems.

References

1. Dondoni, A. and Massi, A. (2008) Asymmetric organocatalysis: from infancy to adolescence. *Angew. Chem. Int. Ed.*, **47**, 4638–4660.
2. Dalko, P.I. (ed.) (2007) *Enantioselective Organocatalysis: Reactions and Experimental Procedures*, Wiley-VCH Verlag GmbH, Weinheim.
3. Berkessel, A. and Gröger, H. (2005) *Asymmetric Organocatalysis – From Biomimetic Concepts to Applications in Asymmetric Synthesis*, Wiley-VCH Verlag GmbH, Weinheim.
4. Seayad, J. and List, B. (2005) Asymmetric organocatalysis. *Org. Biomol. Chem.*, **3**, 719–724.
5. Houk, K.N. and Cheong, P.H. (2008) Computational prediction of small-molecule catalysts. *Nature*, **455**, 309–313.
6. Ema, T., Tanida, D., Matsukawa, T., and Sakai, T. (2008) Biomimetic trifunctional organocatalyst showing a great acceleration for the transesterification between vinyl ester and alcohol. *Chem. Commun.*, 957–959.
7. Taylor, M.S. and Jacobsen, E.N. (2006) Asymmetric catalysis by chiral hydrogen-bond donors. *Angew. Chem. Int. Ed.*, **45**, 1520–1543.
8. Doyle, A.G. and Jacobsen, E.N. (2007) Small-molecule H-bond donors in asymmetric catalysis. *Chem. Rev.*, **107**, 5713–5743.
9. Yu, X. and Wang, W. (2008) Hydrogen-bond-mediated asymmetric catalysis. *Chem. Asian J.*, **3**, 516–532.
10. Pihko, P.M. (2004) Activation of carbonyl compounds by double hydrogen bonding: an emerging tool in asymmetric catalysis. *Angew. Chem. Int. Ed.*, **43**, 2062–2064.
11. Huang, Y., Unni, A.K., Thadani, A.N., and Rawal, V.H. (2003) Hydrogen bonding: single enantiomers from a chiral-alcohol catalyst. *Nature*, **424**, 146.
12. Thadani, A.N., Stankovic, A.R., and Rawal, V.H. (2004) Enantioselective Diels–Alder reactions catalyzed by hydrogen bonding. *Proc. Natl. Acad. Sci. U.S.A.*, **101**, 5846–5850.
13. Unni, A.K., Takenaka, N., Yamamoto, H., and Rawal, V.H. (2005)

Axially chiral biaryl diols catalyze highly enantioselective hetero-Diels-Alder reactions through hydrogen bonding. *J. Am. Chem. Soc.*, **127**, 1336–1337.
14. Gordillo, R., Dudding, T., Anderson, C.D. and Houk, K.N. (2007) Hydrogen bonding catalysis operates by charge stabilization in highly polar Diels-Alder reactions. *Org. Lett.*, **9**, 501–503.
15. Anderson, C.D., Dudding, T., Gordillo, R., and Houk, K.N. (2008) Origin of enantioselection in hetero-Diels-Alder reactions catalyzed by naphthyl-TADDOL. *Org. Lett.*, **10**, 2749–2752.
16. Zhang, X., Du, H., Wang, Z., Wu, Y., and Ding, K. (2006) Experimental and theoretical studies on the hydrogen-bond-promoted enantioselective hetero-Diels-Alder reaction of Danishefsky's diene with benzaldehyde. *J. Org. Chem.*, **71**, 2862–2869.
17. Harriman, D.J., Lambropoulos, A., and Deslongchamps, G. (2007) In silico correlation of enantioselectivity for the TADDOL catalyzed asymmetric hetero-Diels-Alder reaction. *Tetrahedron Lett.*, **48**, 689–692.
18. Rajaram, S. and Sigman, M.S. (2005) Design of hydrogen bond catalysts based on a modular oxazoline template: application to an enantioselective hetero Diels-Alder reaction. *Org. Lett.*, **7**, 5473–5475.
19. Jensen, K. and Sigman, M. (2007) Systematically probing the effect of catalyst acidity in a hydrogen-bond-catalyzed enantioselective reaction. *Angew. Chem. Int. Ed.*, **46**, 4748–4750.
20. Etter, M.C. and Panunto, T.W. (1988) 1,3-Bis(m-nitrophenyl)urea: an exceptionally good complexing agent for proton acceptors. *J. Am. Chem. Soc.*, **110**, 5896–5897.
21. Etter, M.C., Urbanczyk-Lipkowska, Z., Zia-Ebrahimi, M., and Panunto, T.W. (1990) Hydrogen bond-directed cocrystallization and molecular recognition properties of diarylureas. *J. Am. Chem. Soc.*, **112**, 8415–8426.
22. Curran, D.P. and Lung, H.K. (1995) Acceleration of a dipolar Claisen rearrangement by hydrogen bonding to a soluble diaryl urea. *Tetrahedron Lett.*, **36**, 6647–6650.
23. Wittkopp, A. and Schreiner, P.R. (2003) Metal-free, noncovalent catalysis of Diels-Alder reactions by neutral hydrogen bond donors in organic solvents and in water. *Chem. Eur. J.*, **9**, 407–414.
24. Zhang, Z. and Schreiner, P.R. (2009) (Thio)urea organocatalysis – What can be learnt from anion recognition? *Chem. Soc. Rev.*, **38**. doi: 10.1039/b801793j.
25. Hamza, A., Schubert, G., Soós, T., and Papai, I. (2006) Theoretical studies on the bifunctionality of chiral thiourea-based organocatalysts: competing routes to C-C bond formation. *J. Am. Chem. Soc.*, **128**, 13151–13160.
26. Okino, T., Hoashi, Y., and Takemoto, Y. (2003) Enantioselective Michael reaction of malonates to nitroolefins catalyzed by bifunctional organocatalysts. *J. Am. Chem. Soc.*, **125**, 12672–12673.
27. Okino, T., Hoashi, Y., Furukawa, T., Xu, X., and Takemoto, Y. (2005) Enantio- and diastereoselective Michael reaction of 1,3-dicarbonyl compounds to nitroolefins catalyzed by a bifunctional thiourea. *J. Am. Chem. Soc.*, **127**, 119–125.
28. Marcelli, T., van Maarseveen, J.H., and Hiemstra, H. (2006) Cupreines and cupreidines: an emerging class of bifunctional cinchona organocatalysts. *Angew. Chem. Int. Ed.*, **45**, 7496–7504.
29. Marcelli, T., van der Haas, R.N.S., van Maarseveen, J.H., and Hiemstra, H. (2005) Cinchona derivatives as bifunctional organocatalysts for the direct asymmetric nitroaldol (Henry) reaction. *Synlett*, 2817–2819.
30. Marcelli, T., van der Haas, R.N.S., van Maarseveen, J.H., and Hiemstra, H. (2006) Asymmetric organocatalytic Henry reaction. *Angew. Chem. Int. Ed.*, **45**, 929–931.
31. Hammar, P., Marcelli, T., Hiemstra, H., and Himo, F. (2007) Density functional theory study of the Cinchona thiourea-catalyzed henry reaction: mechanism and enantioselectivity. *Adv. Synth. Catal.*, **349**, 2537–2548.
32. Fuerst, D.E. and Jacobsen, E.N. (2005) Thiourea-catalyzed enantioselective cyanosilylation of ketones. *J. Am. Chem. Soc.*, **127**, 8964–8965.

33. Zuend, S.J. and Jacobsen, E.N. (2007) Cooperative catalysis by tertiary amino-thioureas: mechanism and basis for enantioselectivity of ketone cyanosilylation. *J. Am. Chem. Soc.*, **129**, 15872–15883.
34. Yalalov, D., Tsogoeva, S., and Schmatz, S. (2006) Chiral thiourea-based bifunctional organocatalysts in the asymmetric Nitro-Michael addition: a joint experimental-theoretical study. *Adv. Synth. Catal.*, **348**, 826–832.
35. Hoashi, Y., Okino, T., and Takemoto, Y. (2005) Enantioselective Michael addition to α, β-unsaturated imides catalyzed by a bifunctional organocatalyst. *Angew. Chem. Int. Ed.*, **44**, 4032–4035.
36. Tárkányi, G., Király, P., Varga, S., Vakulya, B., and Soós, T. (2008) Edge-to-face CH/π aromatic interaction and molecular self-recognition in epi-Cinchona-based bifunctional thiourea organocatalysis. *Chem. Eur. J.*, **14**, 6078–6086.
37. Vakulya, B., Varga, S., Csámpai, A., and Soós, T. (2005) Highly enantioselective conjugate addition of nitromethane to chalcones using bifunctional cinchona organocatalysts. *Org. Lett.*, **7**, 1967–1969.
38. Rho, H.S., Oh, S.H., Lee, J.W., Lee, J.Y., Chin, J., and Song, C.E. (2008) Bifunctional organocatalyst for methanolytic desymmetrization of cyclic anhydrides: increasing enantioselectivity by catalyst dilution. *Chem. Commun.*, 1208–1210.
39. Oh, S., Rho, H., Lee, J., Lee, J., Youk, S., Chin, J., and Song, C. (2008) A highly reactive and enantioselective bifunctional organocatalyst for the methanolytic desymmetrization of cyclic anhydrides: prevention of catalyst aggregation. *Angew. Chem. Int. Ed.*, **47**, 7872–7875.
40. Schmidtchen, F.P. and Berger, M. (1997) Artificial organic host molecules for anions. *Chem. Rev.*, **97**, 1609–1646.
41. Raheem, I.T., Thiara, P.S., Peterson, E.A., and Jacobsen, E.N. (2007) Enantioselective Pictet-Spengler-type cyclizations of hydroxylactams: H-bond donor catalysis by anion binding. *J. Am. Chem. Soc.*, **129**, 13404–13405.
42. Reisman, S.E., Doyle, A.G., and Jacobsen, E.N. (2008) Enantioselective thiourea-catalyzed additions to oxocarbenium ions. *J. Am. Chem. Soc.*, **130**, 7198–7199.
43. Taylor, M.S. and Jacobsen, E.N. (2004) Highly enantioselective catalytic acyl-Pictet-Spengler reactions. *J. Am. Chem. Soc.*, **126**, 10558–10559.
44. Yamaoka, Y., Miyabe, H., and Takemoto, Y. (2007) Catalytic enantioselective petasis-type reaction of quinolines catalyzed by a newly designed thiourea catalyst. *J. Am. Chem. Soc.*, **129**, 6686–6687.
45. Hamilton, G.L., Kang, E.J., Mba, M., and Toste, F.D. (2007) A powerful chiral counterion strategy for asymmetric transition metal catalysis. *Science*, **317**, 496–499.
46. Mayer, S. and List, B. (2006) Asymmetric counteranion-directed catalysis. *Angew. Chem. Int. Ed.*, **45**, 4193–4195.
47. Connon, S.J. (2006) Chiral phosphoric acids: powerful organocatalysts for asymmetric addition reactions to imines. *Angew. Chem. Int. Ed.*, **45**, 3909–3912.
48. Akiyama, T., Itoh, J., and Fuchibe, K. (2006) Recent progress in chiral Brønsted acid catalysis. *Adv. Synth. Catal.*, **348**, 999–1010.
49. Akiyama, T. (2007) Stronger Brønsted acids. *Chem. Rev.*, **107**, 5744–5758.
50. Uraguchi, D. and Terada, M. (2004) Chiral Brønsted acid-catalyzed direct Mannich reactions via electrophilic activation. *J. Am. Chem. Soc.*, **126**, 5356–5357.
51. Akiyama, T., Itoh, J., Yokota, K., and Fuchibe, K. (2004) Enantioselective Mannich-type reaction catalyzed by a chiral Brønsted acid. *Angew. Chem. Int. Ed.*, **43**, 1566–1568.
52. Yamanaka, M., Itoh, J., Fuchibe, K., and Akiyama, T. (2007) Chiral Brønsted acid catalyzed enantioselective Mannich-type reaction. *J. Am. Chem. Soc.*, **129**, 6756–6764.
53. You, S. (2007) Recent developments in asymmetric transfer hydrogenation with Hantzsch esters: a biomimetic approach. *Chem. Asian J.*, **2**, 820–827.
54. Ouellet, S.G., Walji, A.M., and Macmillan, D.W.C. (2007) Enantioselective organocatalytic transfer

hydrogenation reactions using Hantzsch esters. *Acc. Chem. Res.*, **40**, 1327–1339.
55. Marcelli, T., Hammar, P., and Himo, F. (2008) Phosphoric acid catalyzed enantioselective transfer hydrogenation of imines: a density functional theory study of reaction mechanism and the origins of enantioselectivity. *Chem. Eur. J.*, **14**, 8562–8571.
56. Marcelli, T., Hammar, P., and Himo, F. (2009) Origin of enantioselectivity in the organocatalytic reductive amination of α-branched aldehydes. *Adv. Synth. Catal.*, **351**, 525–529.
57. Simo'n, L. and Goodman, J.M. (2008) Theoretical study of the mechanism of Hantzsch ester hydrogenation of imines catalyzed by chiral BINOL-phosphoric acids. *J. Am. Chem. Soc.*, **130**, 8741–8747.
58. Simo'n, L. and Goodman, J.M. (2009) Mechanism of BINOL-phosphoric acid-catalyzed Strecker reaction of benzyl imines. *J. Am. Chem. Soc.*, **131**. doi: 10.1021/ja808715j.
59. Itoh, J., Fuchibe, K., and Akiyama, T. (2008) Chiral phosphoric acid catalyzed enantioselective Friedel-Crafts alkylation of indoles with nitroalkenes: cooperative effect of 3 Å molecular sieves. *Angew. Chem. Int. Ed.*, **47**, 4016–4018.
60. Tang, H., Lu, A., Zhou, Z., Zhao, G., He, L., and Tang, C. (2008) Chiral phosphoric acid catalyzed asymmetric Friedel-Crafts alkylation of indoles with aimple α, β-unsaturated aromatic ketones. *Eur. J. Org. Chem.*, **2008**, 1406–1410.
61. Rowland, E.B., Rowland, G.B., Rivera-Otero, E., and Antilla, J.C. (2007) Brønsted acid-catalyzed desymmetrization of meso-aziridines. *J. Am. Chem. Soc.*, **129**, 12084–12085.
62. Leow, D. and Tan, C. (2009) Chiral guanidine catalyzed enantioselective reactions. *Chem. Asian J.*, **4**. doi: 10.1002/asia.200800361.
63. Uyeda, C. and Jacobsen, E.N. (2008) Enantioselective Claisen rearrangements with a hydrogen-bond donor catalyst. *J. Am. Chem. Soc.*, **130**, 9228–9229.
64. Malerich, J.P., Hagihara, K., and Rawal, V.H. (2008) Chiral squaramide derivatives are excellent hydrogen bond donor catalysts. *J. Am. Chem. Soc.*, **130**, 14416–14417.
65. Cheon, C.H. and Yamamoto, H. (2009) A new Brønsted acid derived from squaric acid and its application to Mukaiyama aldol and Michael reactions. *Tetrahedron Lett.*, **50**. doi: 10.1016/j.tetlet.2009.03.060.
66. Ganesh, M. and Seidel, D. (2008) Catalytic enantioselective additions of indoles to nitroalkenes. *J. Am. Chem. Soc.*, **130**, 16464–16465.
67. Herrera, R.P., Sgarzani, V., Bernardi, L., and Ricci, A. (2005) Catalytic enantioselective Friedel-Crafts alkylation of indoles with nitroalkenes by using a simple thiourea organocatalyst. *Angew. Chem. Int. Ed.*, **44**, 6576–6579.
68. Miller, S.J. (2004) In search of peptide-based catalysts for asymmetric organic synthesis. *Acc. Chem. Res.*, **37**, 601–610.

7
Dynamic Covalent Capture: A Sensitive Tool for Detecting Molecular Interactions

Leonard Prins

7.1
Introduction

Molecular recognition is at the center of many areas of chemistry. Examples are analytical chemistry (analyte sensor), catalysis (transition state-catalyst), medicinal chemistry (drug-biotarget), and advanced materials chemistry (building block A–building block B). Methodology that allows the rapid and precise detection of molecular recognition events is essential in all these fields. Traditionally, molecular recognition has been studied based on a rational design approach involving many iterative optimization loops, which makes it an energy- and time-consuming process. Additionally, it requires detailed knowledge about the target and the recognition process itself, information which is not always available. Currently, combinatorial methods are increasingly being used for detecting molecular recognition events, allowing the simultaneous screening of a vast amount of chemical compounds enabling a much larger part of chemical space to be explored. Dynamic covalent capture extends to the combinatorial approach for detecting molecular recognition events, but at a higher sensitivity level compared to conventional methodologies and with the novelty of self-selection by the target. The essential point of dynamic covalent capture is that a molecular recognition event is followed by the formation of a reversible covalent bond between the two molecules (Figure 7.1).

The term *covalent capture* emphasizes that the covalent bond serves to "capture" the noncovalent interaction [1]. Covalent capture using irreversible covalent bonds is not a new concept in chemistry. The first examples date back to the 1960s when the first suicide inhibitors for enzymes were reported [2]. These molecules bind to the active site of enzymes, but then form an irreversible covalent bond with the enzyme, which results in complete deactivation. The booming field of activity-based protein profiling (ABPP), in which the functional state of enzymes in entire proteomes is probed, is based on this concept [3]. Covalent capture is also a very useful tool for measuring noncovalent interactions between biomacromolecules (protein/DNA and protein/protein) [4], for discovering new chemical reactions using DNA-templated synthesis [5] and for the stabilization of self-assembled structures and foldamers [6], amongst others.

Ideas in Chemistry and Molecular Sciences: Advances in Synthetic Chemistry. Edited by Bruno Pignataro
Copyright © 2010 WILEY-VCH Verlag GmbH & Co. KGaA, Weinheim
ISBN: 978-3-527-32539-9

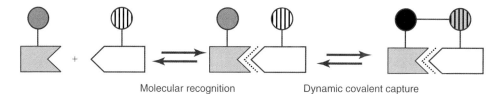

Figure 7.1 Dynamic covalent capture: a molecular recognition event is followed by the formation of a reversible covalent bond between two molecules.

The conceptual novelty of dynamic covalent capture originates from the use of *reversible* covalent bonds for capturing. Over the past decade, reversible covalent bond formation of imines for example and disulfides [7] has attracted an enormous interest as a result of their utility in the fields of dynamic combinatorial chemistry [8, 9] and templated synthesis under thermodynamic control [10]. Importantly, the application of reversible covalent bonds implies a paradigm shift in the application of covalent capture. Being under thermodynamic control, the product distribution is now determined by the relative thermodynamic ground state stabilities of the free components, the complex, and the reversibly captured complex. From a practical point of view, this is highly attractive since it does not require any knowledge about transition state structures, which determine product formation in irreversible covalent capture. A second major advantage is that the molecular recognition event has become intramolecular in the captured product. This allows the detection and quantification of weak noncovalent interactions between molecules, which would go unnoticed using conventional methods. This will be illustrated in this chapter by discussing a series of examples in which weak noncovalent interactions play a dominant role. In the final part, our own contributions in this area are discussed in more detail, illustrating the potential of dynamic covalent capture for catalyst discovery and illustrating novel analytical protocols specifically adapted for monitoring complex chemical systems.

7.2
Hydrogen Bond–Driven Self-Assembly in Aqueous Solution

The hydrogen bond (H bond)-driven self-assembly of molecules into well-defined aggregates in water remains one of the big challenges in the field of noncovalent synthesis. Considering the fact that biological self-assembly processes often heavily involve H bond formation, with the DNA helix being the most clamorous example, this illustrates that designed self-assembly processes have not yet reached the sophistication of those found in nature. The main success so far in inducing H bond formation in water has been obtained by creating hydrophobic microenvironments, for instance micelles [11, 12] or rodlike polymeric stacks [13, 14]. Very recently, it was shown by Gong *et al.* that the combined use of H bonds and dynamic covalent bonds results in the spontaneous formation of molecularly well-defined

7.2 Hydrogen Bond–Driven Self-Assembly in Aqueous Solution

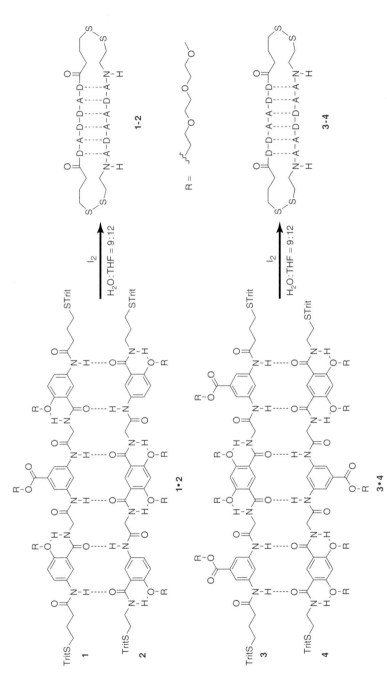

Figure 7.2 Sequence-specific association of oligoamides in aqueous media. Reproduced with permission from [1].

structures in water, in which the H bonds induce sequence selectivity [15, 16]. In nonpolar media, oligoamides **1** and **2** associate to give a highly stable duplex **1•2** held together by six H bonds (Figure 7.2). The complementary arrangement of H bond donors and acceptors in **1** (DADDAD) and **2** (ADAADA) ensures that the heteroduplex **1•2** is the only product. Similarly, oligoamides **3** (DDAADD) and **4** (AADDAA) give exclusively the heteroduplex **3•4**. In aqueous media, H bond formation alone is not enough to provide sufficient thermodynamic stability to the dimers. However, in combination with reversible disulfide formation, the exclusive formation of duplex **1-2** was observed in a 9 : 1 H_2O/THF mixture when **1** and **2** were mixed in an equimolar ratio and treated with iodine. The obvious question is to what extent duplex formation is driven by H bond formation. Control experiments indicate that in the absence of the complementary strands both **1** and **2** form a homocyclic structure via intramolecular disulfide formation. The presence of oligoamide **3** having a noncomplementary array of H bond donors and acceptors did not interfere in the product formation; duplex **1-2** was the only product observed. On the other hand, combining **3** with its complementary partner **4** resulted in the exclusive formation of duplex **3-4**. The importance of the recognition event was even more evident from a follow-up experiment in which six oligoamides (present as three couples of complementary strands with the ability to form either two, four, or six H bonds) were mixed in an equimolar ratio [16]. Although statistically a large number of duplexes could be formed, at thermodynamic equilibrium each strand had exclusively dimerized with its complementary partner. This phenomenon is called *self-sorting* and refers to the situation in which a system under thermodynamic control displays a nonstatistical composition [17, 18]. Rather unique, in this system self-sorting is based on a combination of noncovalent and dynamic covalent bonds.

7.3
Measuring Stability and Order in Biological Structures

7.3.1
Peptides

Knowledge of the forces that drive the folding of peptide strands into proteins is crucial for understanding the mechanisms behind diseases related to protein misfolding and for the engineering of artificial peptide constructs [19–21]. Protein folding processes are generally measured by monitoring changes in a spectroscopic indicator of the higher order structure, such as a circular dichroism (CD) signal, upon imposing conditions that induce denaturing of the protein (concentration, temperature, and ionic strength). Driven by the concern that the intrinsic stabilities of the limiting conformational states, native and denatured, are affected by the experimental conditions, Gellman *et al.* developed a new strategy for measuring higher order stability in polypeptides [22]. The so-called backbone thioester exchange (BTE) approach relies on the replacement of a backbone amide bond in

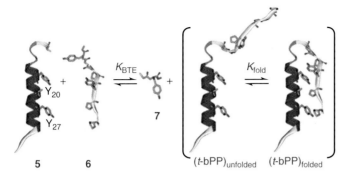

Figure 7.3 Thermodynamic equilibria involved in backbone thioester exchange. The parameter of interest is K_{fold}, which is related to the free binding energy between the two peptide fragments **5** and **6**. Reproduced with permission from [1].

the polypeptide with a thioester, which can undergo thioester/thiol exchange in aqueous solution at neutral pH (Figure 7.3).

The validity of this model was initially tested on bovine pancreatic polypeptide (bPP), a 36-residue protein that adopts a well-defined tertiary structure in which the N-terminal polyproline II segment **6** is folded back onto the C-terminal α-helix segment **5**. The amide bond between residues 9 and 10 was selected for replacement with the thioester, as this bond is located in the loop and does not participate in intramolecular hydrogen bonding. Addition of small thiol **7** (thioglycoyltyrosine N-methyl amide) results in the installment of an equilibrium in which this reference thiol competes with the C-terminal segment for thioester formation with the N-terminal segment. Assuming that the reference thiol has no tertiary interaction with the N-terminal segment and assuming that the two thioester bonds are isoenergetic, any shift in the equilibrium in favor of t-bPP can be directly correlated to a favorable tertiary interaction between the N- and C-terminal segments. The elegancy of this approach is that it allows for a rapid study of the effect of single mutations in the peptide chains on the stability of the folded structure. For bPP, such an analysis was performed by examining the K_{BTE} values upon an Ala substitution of the Tyr_{20} and Tyr_{27} residues, which are known to play a critical role in tertiary structure formation from NMR studies. In fact, the positive $\Delta G_{fold/BTE}$ values suggest that removal of these Tyr residues causes a loss of favorable contacts. In follow-up studies the same group exploited the BTE approach for studying the effect of Phe to F_5-Phe substitution on the conformational stability of the chicken villin headpiece subdomain (cVHP) [23] and for identifying preferred side chain constellations at antiparallel coiled coil interfaces [24]. With respect to the latter, it is interesting to notice that in his classical paper on the chiroselective replication of peptides, Ghadiri *et al.* had already used the same dynamic approach to demonstrate the stereospecificity of coiled coil formation [25]. The higher stability of a homochiral coiled coil over its heterochiral analog was studied either by a slow air oxidation of a racemic mixture

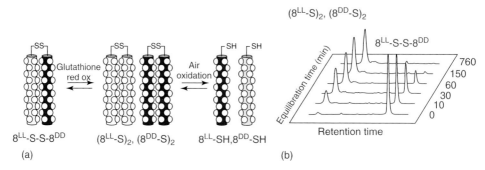

Figure 7.4 (a) Both slow air oxidation of a racemic mixture of coiled peptides 8^{DD}-SH and 8^{LL}-SH and disulfide exchange starting with heterochiral dimer 8^{DD}-S-S-8^{LL} in a glutathione redox buffer result in the exclusive presence of homochiral dimers $(8^{LL}$-S$)_2$ and $(8^{DD}$-S$)_2$ as (b) evidenced by HPLC analysis. Reproduced with permission from [1].

of coiled peptides 8^{DD}-SH and 8^{LL}-SH or by promoting disulfide exchange starting with heterochiral dimer 8^{DD}-S-S-8^{LL} in a glutathione redox buffer (Figure 7.4a). On statistical grounds, a 1 : 1 ratio of homochiral (8^{DD}-S-S-8^{DD}, 8^{LL}-S-S-8^{LL}) and heterochiral (8^{DD}-S-S-8^{LL}) coiled coils would be expected, but HPLC analysis clearly showed that in both experiments the heterochiral 8^{DD}-S-S-8^{LL} dimer was completely absent at thermodynamic equilibrium (Figure 7.4b).

Recently, the Gellman group [26] extended the BTE approach toward *secondary* peptide structures, in particular β-sheet formation, conceptually related to a prior study of Balasubramanian *et al.* [27]. In the Balasubramanian proof-of-principle study, the thermodynamic equilibrium was examined between two peptides **9** and **10** containing 1 and 4 Leu-Lys repeats, respectively, known to predispose peptides for β-sheet formation (Figure 7.5). Relying on conventional disulfide exchange they observed a strong amplification of dimeric peptide **10**-S-S-**10** with respect to the other dimeric peptides possible (**9**-S-S-**10**, **9**-S-S-**9**, **9**-S-S-**G**, and **10**-S-S-**G**, in which **G** denotes the reduced form of the glutathione redox buffer). 2D NMR studies (NOESY) indeed revealed contacts between the two peptide strands in support of the claim that the higher thermodynamic stability of the **10**-S-S-**10** dimer results from β-sheet formation. Later, the same approach was used for probing secondary structure in peptide nucleic acids [28].

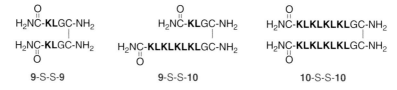

Figure 7.5 Possible products for the dimerization of a mixture of **9**-SH and **10**-SH. Reproduced with permission from [1].

7.3.2
Bilayer Membranes

The hypothesis of formation of clusters (or domains) of lipids within fluid biological membranes, in which the lipid distribution is not random, has been the subject of much controversial debate. This is not a trivial matter as a number of studies indicate that these clusters, also referred to as *lipid rafts*, may play key roles in the control of a variety of cellular processes, including signal transduction and membrane trafficking [29]. Furthermore, they have been associated with the production and cellular entry of viral particles [30]. Cholesterol is a major component of mammalian cell membranes. Despite numerous investigations involving membranes comprising cholesterol and phospholipids, the structural role of this sterol in producing condensed, fluid membranes has remained a mystery. Regen et al. have established a procedure, termed *nearest-neighbor recognition* (NNR), to quantitatively determine the tendency of lipids to cluster on the basis of the ability of a single lipid to react with its nearest neighbor. The technique relies on the reversible covalent capture of two adjacent lipids (or lipid and cholesterol) via the reversible formation of a disulfide bond [31, 32]. Regens's reasoning is very straightforward: if lipids L_A and L_B are connected via a reversible bond, equilibration will result in the formation of a mixture of homo- ($L_A L_A$ and $L_B L_B$) and heterodimers ($L_A L_B$). In the absence of any thermodynamic preference, the equilibrium distribution is such that $L_A L_A : L_B L_B : L_A L_B$ is 1 : 1 : 2, or, said in a different way, that the equilibrium constant K equals 4 (Figure 7.6). Any bias of the equilibrium in favor of the hetero- or homodimers is indicative of a nonstatistical distribution in the fluid bilayer.

Using NNR, Regen was able to prove the preference of cholesterol for lipids with long hydrocarbon chains (C16 and C18) when the sterol concentration reaches biologically relevant levels. This has provided experimental support to the theory of formation of "complexes" between cholesterol and lipids within biological membranes [33, 34]. Furthermore, evidence was obtained for the condensing effect of cholesterol on lipid bilayers. This implies that the flexible acyl chains of phospholipids are able to complement the shape of neighboring cholesterol molecules, resulting in a high number of hydrophobic contacts and tight packing.

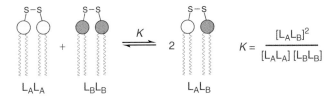

Figure 7.6 Schematic illustration of nearest-neighbor recognition (NNR) in which two exchangeable homodimers ($L_A L_A$ and $L_B L_B$) are equilibrated to form a mixture of homo- and heterodimers ($L_A L_B$). Reproduced with permission from [1].

Thus, cholesterol appears to act as a rigid hydrophobic template for the membrane lipids to maximize their interactions.

7.4
Drug Discovery

In all previous examples of reversible covalent capture, a small number of species and equilibria are studied. However, as will be shown next, dynamic covalent capture is also an attractive strategy for combinatorial drug discovery. Wells, Braisted and Erlanson and coworkers at Sunesis Pharmaceuticals have laid the conceptual foundation for a combinatorial screening by reporting on a tethering strategy for ligand discovery (Figure 7.7a) [35–37]. Their approach consists of using disulfide exchange between a Cys residue located near the target site and a small library of potential disulfide-containing ligands. Most of the library members will show no intrinsic affinity for the protein and the disulfide bond between ligand and protein will be easily reduced. However, binding interactions between a ligand and the target site of a protein stabilize the disulfide bond and shift the thermodynamic equilibrium toward the modified protein. Determination of the composition at thermodynamic equilibrium then reveals the library member with highest affinity for the target. The validity of the tethering concept was first tested on thymidylate synthase (TS), which has a Cys residue located in the active site [35]. This enzyme is involved in the synthesis of thymidine monophosphate (dTMP), and therefore a prominent anticancer drug target. Potential ligands were screened in sets of 10 with sufficient mass differences to allow identification of captured components by mass spectrometry. Tethering experiments were performed in the presence of an excess of 2-mercaptoethanol in order to impose reducing conditions and thus reversibility. Screening of a total of 1200 disulfide-containing fragments resulted in a strong selection of N-arylsulfonamideproline derivatives, represented by N-tosyl-D-proline. Importantly, as before, it is the noncovalent interaction between the ligand and the protein that determines the thermodynamic stability. This is evidenced by the fact that the observed distribution of kinetic products (in the absence of the reducing agent) is entirely different from the composition at thermodynamic equilibrium, showing only a moderate selection of N-tosyl-D-proline. The sensitivity of the tethering strategy in detecting weak interactions is illustrated by the low value obtained for the dissociation constant of 1.1 mM, which is a level that would be hard to detect with conventional high-throughput screening.

The scope of the tethering strategy was further increased by two subsequent developments: fragment assembly [38] and extended tethering [39]. Fragment assembly implies that the final drug is obtained via a merger of fragments selected individually from different screening experiments [40]. Binding of interleukin-2 (IL-2) to its receptor induces T-cell proliferation and is an important target for immune disorders. A series of interleukin-2 (IL-2) mutants were prepared in which Cys residues were introduced around the "hot spot" of the protein involved

7.4 Drug Discovery | 151

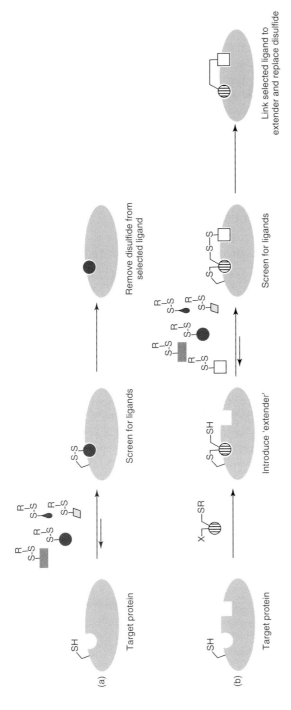

Figure 7.7 Schematic illustration of (a) the "tethering" and (b) the extended "tethering" strategy for drug discovery. Reproduced with permission from [1].

in binding to its receptor. Screening of all mutants against a library of 7000 disulfide-containing fragments resulted in the selection of a series of structurally related fragments with affinity for the area around the "hot spot." Next, a focused set of 20 compounds was prepared in which the selected fragments were merged with a known inhibitor for IL-20 obtained from a traditional drug discovery approach. Inhibition experiments revealed a 5–50-fold increase in affinity for 8 out of 20 components compared to the original inhibitor, bringing the IC_{50} value down to the nanomolar region. The extended tethering strategy implies the introduction of a Cys residue near the active site which is then covalently modified with an "extender," a small molecule with affinity for the protein that contains also a protected thiol group (Figure 7.7b). Deprotection liberates a thiol that can be used to perform a dynamic screening as described before. The selected fragments are then merged covalently with the extender to yield the final drug. Using the extended tethering strategy, a novel inhibitor was discovered for caspase-3, a protein involved in apoptosis. The results reported by Erlanson and coworkers, followed by others [41, 42], illustrate the added value of using reversible covalent capture in a drug discovery program. The main advantage of the tethering strategy is the ability to detect weak interactions between the target protein and library members. By itself this can be useful for discovering new lead compounds, but very promising is the improvement of binding affinity of known inhibitors by adding self-selected fragments or the unification of multiple weakly binding fragments into a single high-affinity binder.

7.5
Catalyst Discovery

Recently, we have shown that reversible covalent capture has potential within a catalyst discovery project [43]. Catalyst discovery is very challenging as it requires the identification of species able to stabilize elusive and transient transition states along the reaction pathway. Analogous to the approach used for raising catalytic antibodies or catalytically active molecularly imprinted polymers, a transition state analog was used as a target. Specifically, a nine-component hydrazide library was screened for components able to interact with benzaldehyde **11** functionalized with a phosphonate target, which is a transition state analog of the basic hydrolysis of an ester moiety (Figure 7.8). Functional groups able to develop stabilizing interactions with the phosphonate group should also accelerate the hydrolysis of an ester localized at the same position, because of transition state stabilization. Screening is straightforward, since the presence of stabilizing interactions enhances the thermodynamic stability of the corresponding hydrazone and leads to an increased concentration with respect to a control scaffold (2-methoxybenzaldehyde) that lacks the target [44]. Screening experiments (Section 7.6) revealed that hydrazones **11B**, **11C**, and **11I** are stabilized with respect to reference **11A** (in the order **11B** > **11C** ≈ **11I**) as a consequence of an intramolecular interaction between the phosphonate and the functional group present in the hydrazide unit. In order to

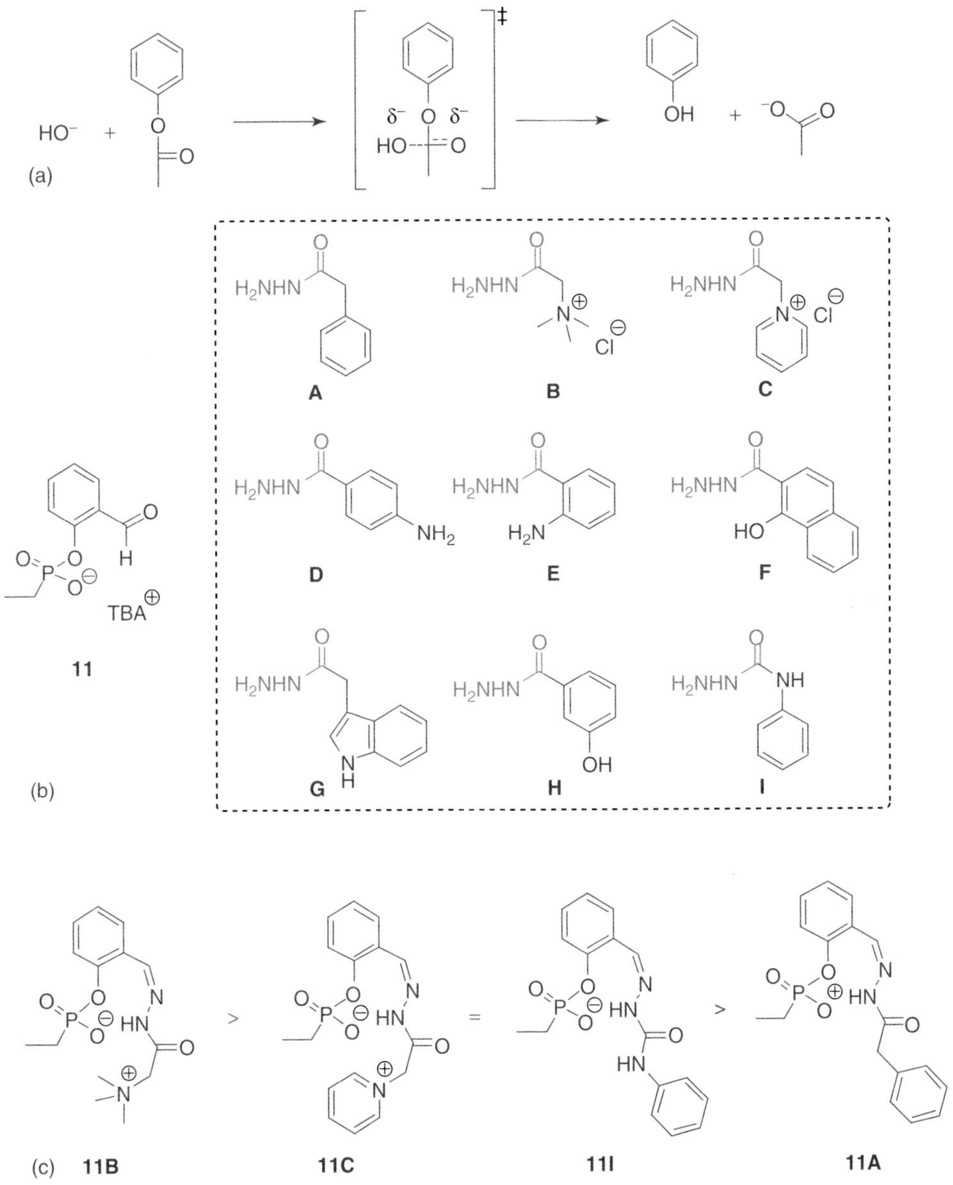

Figure 7.8 (a) Basic hydrolysis of phenylacetate including transition state. (b) Phosphonate target **11** and the nine-component hydrazide library used for screening. (c) Hydrazones for which the screening experiments have revealed the occurrence of stabilizing interactions between the phosphonate group and the functional group present in the hydrazide. The hydrazones are ordered in terms of strength of the interaction (**11B** > **11C** ≈ **11I**) with respect to the reference hydrazone **11A**.

establish a correlation between the observed amplification and catalytic efficiency, hydrazides **B** and **I** were studied in detail as they should express a different type and strength of interaction with the transition state (electrostatic and H bonding, respectively). Therefore, compounds **12B**, **12I**, and **12A** (which served as a reference) were prepared in which the structural elements of hydrazides **A**, **B**, and **I** were positioned in close proximity to a neighboring carboxylate ester (Figure 7.9a). Compared to the parent hydrazone structures, two small structural changes had to be introduced. The C=N double bond was reduced in order to render the structure stable under the basic conditions required for ester cleavage. Such a covalent postmodification is very common in imine-based dynamic combinatorial chemistry [45–47]. Secondly, the resulting secondary amine was methylated in order to prevent an intramolecular attack of the amine on the neighboring ester.

The effect of the presence of the ammonium and urea groups in **12B** and **12I**, respectively, on the ester cleavage was initially studied by measuring the methanolysis rates of **12A**, **12B**, and **12I**, because of the similarity to the conditions under which the amplification studies were performed (Figure 7.9b) The resulting pseudo-first-order rate constants ($k_{obs,12B} = 3.96 \times 10^{-2} s^{-1}$, $k_{obs,12I} = 1.82 \times 10^{-2} s^{-1}$, and $k_{obs,12A} = 0.82 \times 10^{-2} s^{-1}$) are in perfect agreement with the results of the amplification studies, both in terms of the order of reactivity ($k_{obs,12B} > k_{obs,12I} > k_{obs,12A}$) and the relative acceleration (4.8 : 1.6 : 1 for **12B**, **12I**, and **12A**, respectively). Several control experiments gave strong support for the hypothesis that the increased ester cleavage rate indeed resulted from transition state stabilization. First, the presence of 1 equiv. of tetramethylammonium chloride did not affect the methanolysis rate of compound **12A** at all. This excludes that the higher rate observed for **12B** is simply due to a change in ionic strength in the mixture. In other words, it shows that the ammonium ion needs to be present in close proximity to the carboxylic ester in order to induce a catalytic effect. Secondly, measuring the methanolysis rate of compound **12B** at decreasing substrate concentrations yielded the same rate constants, which is in strong support of intramolecular catalysis. Finally, we observed an enhanced catalytic effect upon decreasing the polarity of the medium, because of a lower solvation ability of the solvent. In a 1 : 9 mixture of $H_2O:CH_3CN$ compounds **12B** and **12I** were hydrolyzed 60 and 16 times faster, respectively (Figure 7.9c). This indicates that the ester moieties in **12B** and **12I** are cleaved at a higher rate due to a stabilization of the negative charges in the transition state by the neighboring groups.

These initial results are important for two reasons. First, they confirm the conceptual validity of dynamic covalent capture as a tool in catalyst discovery. A correlation is established between the thermodynamic amplification in the dynamic system and the efficiency of the selected chemical function in assisting in intramolecular catalysis. It should be noticed, though, that the obtained species is not yet a catalyst. Secondly, these results demonstrate the high sensitivity of the covalent capture strategy in detecting noncovalent interactions, which originates from the fact that these now occur intra- rather than intermolecularly.

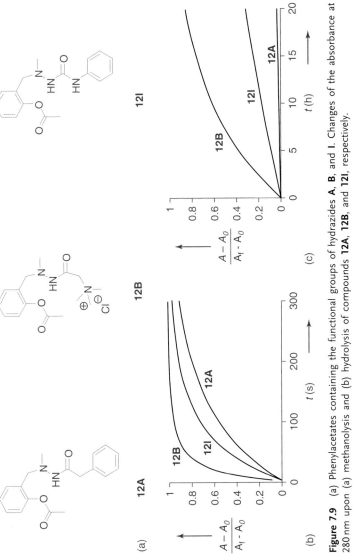

Figure 7.9 (a) Phenylacetates containing the functional groups of hydrazides **A**, **B**, and **I**. Changes of the absorbance at 280 nm upon (a) methanolysis and (b) hydrolysis of compounds **12A**, **12B**, and **12I**, respectively.

7.6
The Analysis of Complex Chemical Systems

7.6.1
^{1}H–^{13}C HSQC NMR Spectroscopy

The task of chemists applying dynamic covalent capture is to create libraries of molecules that have the potential to interact with the target. Essentially, the selection process is performed by the target itself. A crucial step is then to identify which library member was selected. Consequently, analytical methods that permit a rapid and continuous monitoring of the library composition are of eminent importance. Recently, we have developed protocols that permit a straightforward analysis of the composition of a dynamic system by NMR [48] and UV/Vis spectroscopy [49]. NMR spectroscopy is probably the most informative analytical tool, since it is one of the few techniques that allows a direct identification and quantification of all species present in solution as a function of time. However, the application of NMR spectroscopy for studying mixtures of molecules is often severally hampered by overlapping signals and difficulties in signal assignment. In a collaborative study with the group of Damien Jeannerat at the University of Geneva, we have shown that these problems can be completely solved by monitoring the dynamic system with a series of quick and highly resolved ^{1}H–^{13}C HSQC NMR experiments [48]. The experimental procedure relies on the reduction of the spectral width in the ^{13}C dimension of HSQC in order to enhance the resolution of signals *and* on a gradual increase in complexity of the system.

The dynamic network used to validate the use of ^{1}H–^{13}C HSQC NMR spectroscopy consisted of four hydrazones **11A**, **11B**, **11C**, and **11H**, which are in exchange due to the presence of the four hydrazides **A–C** and **D** (Figure 7.10a). The key trick in the experimental procedure is the sequential increase in complexity. So, the starting point was a 58 mM solution of hydrazone **11A** in CD$_3$OD, which has two signals in the "fingerprint" area (^{1}H: 7.8–9.2 ppm, ^{13}C: 142.6–147.4 ppm) corresponding to the *anti* and *syn* isomers with respect to the amide bond (**11A$_M$** and **11A$_m$**, Figure 7.10a). Addition of 1 equiv. of **A** and 2 equiv. of **B** results in the installment of an equilibrium between hydrazones **11A** and **11B**, evidenced by the appearance of two new signals in the "fingerprint" area (**11B$_M$** and **11B$_m$**, Figure 7.10b) and a concomitant decrease in intensity of signals **11A$_M$** and **11A$_m$**. Monitoring the signal intensities as a function of time yields the rate profile for equilibration (Section 7.1, Figure 7.10c), from which two second-order rate constants ($k_{11A \to 11B}$ and $k_{11B \to 11A}$) can be determined. After 4 hours, 2 equiv. of hydrazide **H** added resulting in the formation of hydrazone **11H** on the account of both **11A** and **11B**. After an additional 4 hours, 2 equiv. of hydrazide **C** added installs the complete network as given in Figure 7.10a. The sequential increase in complexity allows for an accurate determination of all rate constants by fitting each section separately, fixing the rate constants obtained from the previous sections. In this way 12 rate constants were obtained with a maximum error of 9.4% for $k_{11C \to 11H}$ ((3.0 ± 0.3) × 10^{-4} M^{-1} s^{-1}). An additional advantage of measuring

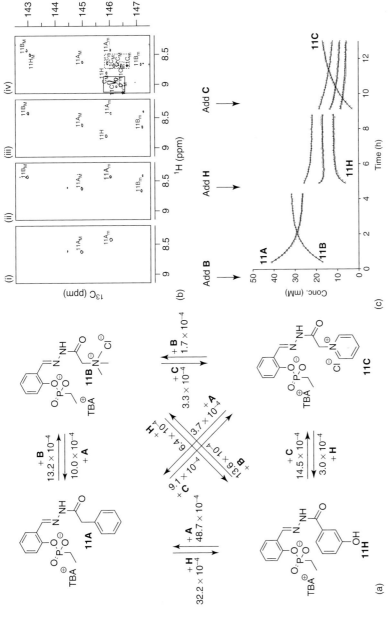

Figure 7.10 (a) Dynamic eight-component network containing four hydrazones **11A**, **11B**, **11C**, and **11H** and four hydrazides **A**, **B**, **C**, and **H** accounting for a total of 12 exchange reactions. Rate constants are calculated by fitting the kinetic profiles of Figure 7.10c. (b) $^{1}H-^{13}C$ HSQC NMR spectra (600 MHz, CD$_3$OD, 303 K) of the "fingerprint" area on different times ($t = 0$ (i), 4 (ii), 8 (iii), and 13 (iv) hours). The o- and p-protons of **11C** and **C** reside in the encircled areas in fingerprint area (iv). (c) Concentrations of hydrazones **11A** (□), **11B** (○), **11H** (△), and **11C** (×) as measured by $^{1}H-^{13}C$ HSQC NMR spectroscopy (600 MHz, CD$_3$OD, 303 K) throughout the mixing experiment. (c) At the beginning of each section, 2 equiv. of **B**, **H**, and **C** (with respect to the scaffold **11**) are added, respectively, and the evolution of the system is followed kinetically. Reproduced with permission from [48].

the kinetics of the system is the possibility to calculate the composition of the system at thermodynamic equilibrium via extrapolation of the kinetic profile. This significantly reduces the experimental time, especially in situations where the thermodynamic equilibrium is reached very slowly. In summary, this methodology allows the analysis of the complete kinetic and thermodynamic picture of an eight-component network involving 12 exchange reactions from a single mixing experiment.

7.6.2
UV/Vis Spectroscopy

The analysis of complex mixtures by NMR spectroscopy using the protocol described in the previous paragraph requires state-of-the-art NMR equipment (600 MHz with cryoprobe). For that reason we have very recently also developed an analytical protocol based on UV/Vis spectroscopy, which renders the parallel screening of hydrazides straightforward, fast, and cheap [49]. It should be remembered that the focal point of all our studies is the determination of the thermodynamic stability of a series of hydrazones with respect to a "neutral" reference hydrazone. The UV/Vis strategy is developed based on two general considerations. First, since all hydrazides are screened against the same reference hydrazide, it is more attractive to focus on the determination of the concentration of the reference rather than the target. Second, since no particular structural properties are requested for the reference, it can be chosen such that it has the ability to generate a unique signal. In fact, we decided to choose hydrazide **K** as a reference inspired by the observation of Herrmann, Lehn *et al.* [50] that reaction of **K** with *trans*-cinnamaldehyde **12** results in the formation of hydrazone **12K**, which has a highly characteristic UV/Vis absorption spectrum resulting from its conjugated push–pull system (Figure 7.11b). This allows for a quantification of the concentration of **12K** in the presence of other species, simply by measuring the absorbance at 348 nm.

A two-step protocol was used involving an initial competition between hydrazides **B** and **K** followed by a scavenging step in which all unreacted **K** is converted into hydrazone **12K** (Figure 7.11a). Mixing aldehyde **11** and hydrazides **B** and **K** in a 1 : 1 : 1 ratio (5 mM each) resulted in the quantitative formation of hydrazones **11B** and **11K**. As a result of the presence of an excess of hydrazides **B** and **K**, these hydrazones interconverted and reached a thermodynamic equilibrium in which **11B** and **11K** were present in a 68 : 32 ratio. Crucially, law of mass dictates that, at equilibrium, hydrazone **11B** has a concentration of 3.42 mM, which is equal to the "left-over" concentration of hydrazide **K**. Subsequently, the equilibrium was quenched by the addition of *trans*-cinnamaldehyde **12** (10 mM, 2 equiv.), which quantitatively converted hydrazides **B** and **K** into hydrazones **12B** and **12K**. In the final mixture, hydrazones **12K** and **11B** were present in an equimolar ratio as confirmed by ^{1}H NMR spectroscopy. Importantly, measurement of the absorbance of the mixture at 348 nm after a 100-fold dilution and comparison with a calibration curve for **12K** gave a concentration of 3.37 mM, which is in excellent agreement with

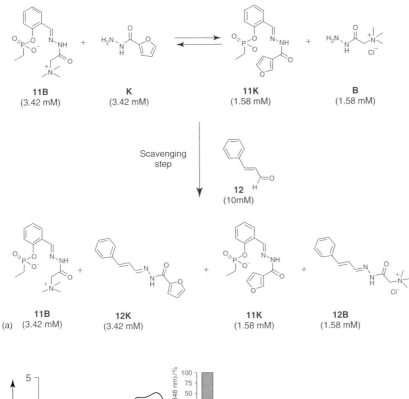

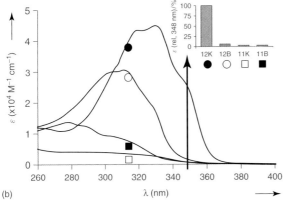

Figure 7.11 (a) Detection protocol for the analysis of the thermodynamic equilibrium between hydrazones **11B** and **11K** based on the quantitative conversion of **B** and **K** into **12B** and **12K**, respectively, using the scavenging agent **12**. The concentration of **12K** can be determined unequivocally from an absorbance measurement at 348 nm and allows determination of the concentration of all other species using mass balances. (b) UV/Vis absorption spectra for hydrazones **11B** (■), **11K** (□), **12A** (○), and **12K** (●) in a mixture of CD_3OD : DMSO-d_6 : CD_3CN (70 : 15 : 15) together with the relative molecular extinction coefficients at 348 nm (inset). Reproduced with permission from [49].

the concentration obtained for **12K** (and thus **11B**) by NMR analysis. In summary, the concentration of **11B** is determined in an indirect manner by measuring the concentration of the reporter molecule **12K**. The protocol has the advantage of being simple, fast (4 hours), and accurate, but the main advantage is the disconnection between the structure of the target and the output signal. In fact, reexamination of the thermodynamic stabilities of hydrazones **11A**, **11C**, **11G**, and **11I** returned the exact values as previously determined by ^{1}H NMR spectroscopy.

Additionally, we have shown that this approach can be used for the analysis of a complex chemical system. The possibility to extract the relative thermodynamic stabilities of hydrazones **11A**, **11B**, **11C**, **11G**, and **11I** from a mixture analysis is not only conceptually challenging, but also highly advantageous, since it eliminates the necessity of requiring building blocks in pure form. A 12-membered chemical system (six hydrazones and six hydrazides) composed of 15 different equilibrium reactions (Figure 7.12a) is installed upon adding a mixture of hydrazides **A**, **B**, **C**, **G**, **I**, and **K** to aldehyde **11**. Although complex, for screening purposes only the equilibria that involve reporter hydrazide **K** (which can be quantified using the scavenging reaction with **12**) are relevant. In fact, the system can be conceptually reduced to a simple thermodynamic equilibrium in which hydrazide **K** is in competition with all other hydrazides (**A**, **B**, **C**, **G**, and **I**) for hydrazone formation with aldehyde **11** (Figure 7.12b). Consequently, the amount of free hydrazide **K** in the system is directly related to the joined competitivity of the other hydrazides. Therefore, comparison of the amount of free **K** present in different hydrazide mixtures, selectively enriched in either one of the components **A**, **B**, **C**, **G**, and **I**, reveals the capacity of that given hydrazide to affect the overall "apparent" binding constant K_{system} and thus immediately reveals the relative thermodynamic stabilities of the corresponding hydrazones **11A**, **11B**, **11C**, **11G**, and **11**, respectively. This was illustrated by preparing five mixtures containing **11** and **K** in a constant 1 : 1 ratio (5 mM each), 0.2 equiv. of four of the five hydrazides **A**, **B**, **C**, **G**, and **I** and 0.6 equiv. of the fifth hydrazide (either **A**, **B**, **C**, **G**, or **I**). The mixtures were equilibrated overnight under the conditions described before, after which **12** (15 mM) was added. After 6 hours, the absorbance of each mixture was measured at 348 nm after dilution (Figure 7.12c, dark gray bars). A simple glance at the observed absorbance values is then sufficient to rank the hydrazones in terms of thermodynamic stability (**11I** > **11B** ∼ **11C** > **11A** ∼ **11G**). These experimental absorbance values were in excellent agreement with those calculated based on the individually obtained equilibrium constants (Figure 7.12c, light gray bars), which unambiguously validated the protocol for assay analysis.

7.7
Perspective

For several reasons dynamic covalent capture is a powerful tool to study molecular recognition. First, formation of the covalent bond makes that the noncovalent interactions occur intra- rather than intermolecularly, which significantly increases the

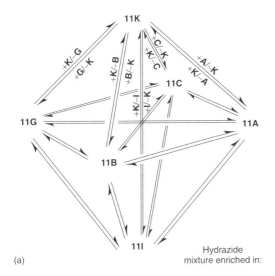

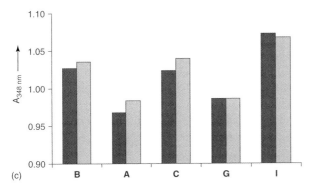

Figure 7.12 (a) Equilibria present in the system obtained from mixing **11** and hydrazides **A**, **B**, **C**, **G**, **I**, and **K**. For clarity reasons hydrazides are indicated only for the equilibria including reporter hydrazide **K**. (b) Interpretation of the chemical system as a thermodynamic equilibrium between hydrazone **11K** and hydrazones **1X** (X = A + B + C + G + I) that interconvert through hydrazide exchange. (c) Measured absorbance values (dark gray bars) at 348 nm of the five mixtures selectively enriched in either **A**, **B**, **C**, **G**, or **I** after scavenging with **12** and the calculated absorbance values (light gray bars) for the same mixtures based on the individually obtained equilibrium constants. Reproduced with permission from [49].

sensitivity. This allows the detection of weak interactions that would normally not have been observed, requiring very high concentrations. Second, the reversibility of the covalent bond ensures that the system is under thermodynamic control, which enables self-selection by the target. This is especially useful in case many compounds need to be tested for their potential to interact with a target. Examples in this chapter illustrate that dynamic covalent capture can indeed be successfully applied for detecting subtle molecular recognition events, such as the folding between peptide helices or transition state stabilization. Additionally, the reversible covalent bond flanking the recognition units can also serve to enforce the molecular complex. For example, this has led to the development of self-assembled systems in water, in which disulfide bonds give strength but in which the H bonds induce selectivity. Given the fact that self-selection plays a key role in dynamic covalent capture, it is the task of the chemist to identify which compounds are selected. This has led to the development of novel protocols that allow for a precise analysis of complex dynamic systems in a fast and straightforward manner using NMR, MS, or UV/Vis. Dynamic covalent capture is an emerging strategy. The fact that the concept is very general implies that it can be potentially applied for very different purposes. An important role of dynamic covalent capture is envisioned in areas where subtle noncovalent interactions are crucial, for example, in protein–protein interactions and in catalytic pathways.

Acknowledgments

The author is grateful to Prof. Paolo Scrimin for stimulating discussions and gratefully acknowledges the contributions of Giulio Gasparini, Giovanni Zaupa, Cristian Guarise, Marta Dal Molin, and Matteo Graziani. Financial support from the University of Padova (CPDA054893) and MIUR (PRIN2006) is acknowledged.

References

1. Prins, L.J. and Scrimin, P. (2009) *Angew. Chem. Int. Ed.*, **48**, 2288–2306.
2. Baker, B.R., Lee, W.W., Tong, E., and Ross, L.O. (1961) *J. Am. Chem. Soc.*, **83**, 3713–3714.
3. Evans, M.J. and Cravatt, B.F. (2006) *Chem. Rev.*, **106**, 3279–3301.
4. Phizicky, E.M. and Fields, S. (1995) *Microbiol. Rev.*, **59**, 94–123.
5. Li, X.Y. and Liu, D.R. (2004) *Angew. Chem. Int. Ed.*, **43**, 4848–4870.
6. Jackson, D.Y., King, D.S., Chmielewski, J., Singh, S., and Schultz, P.G. (1991) *J. Am. Chem. Soc.*, **113**, 9391–9392.
7. Rowan, S.J., Cantrill, S.J., Cousins, G.R.L., Sanders, J.K.M., and Stoddart, J.F. (2002) *Angew. Chem. Int. Ed.*, **41**, 898–952.
8. Corbett, P.T., Leclaire, J., Vial, L., West, K.R., Wietor, J.L., Sanders, J.K.M., and Otto, S. (2006) *Chem. Rev.*, **106**, 3652–3711.
9. Lehn, J.M. (1999) *Chem. Eur. J.*, **5**, 2455–2463.
10. Hoss, R. and Vogtle, F. (1994) *Angew. Chem. Int. Ed.*, **33**, 375–384.
11. Nowick, J.S., Chen, J.S., and Noronha, G. (1993) *J. Am. Chem. Soc.*, **115**, 7636–7644.
12. ten Cate, M.G.J., Crego-Calama, M., and Reinhoudt, D.N. (2004) *J. Am. Chem. Soc.*, **126**, 10840–10841.

13. Hirschberg, J., Brunsveld, L., Ramzi, A., Vekemans, J., Sijbesma, R.P., and Meijer, E.W. (2000) *Nature*, **407**, 167–170.
14. Brunsveld, L., Vekemans, J., Hirschberg, J., Sijbesma, R.P., and Meijer, E.W. (2002) *Proc. Natl. Acad. Sci. U.S.A.*, **99**, 4977–4982.
15. Li, M.F., Yamato, K., Ferguson, J.S., and Gong, B. (2006) *J. Am. Chem. Soc.*, **128**, 12628–12629.
16. Li, M.F., Yamato, K., Ferguson, J.S., Singarapu, K.K., Szyperski, T., and Gong, B. (2008) *J. Am. Chem. Soc.*, **130**, 491–500.
17. Wu, A.X. and Isaacs, L. (2003) *J. Am. Chem. Soc.*, **125**, 4831–4835.
18. Mukhopadhyay, P., Wu, A.X., and Isaacs, L. (2004) *J. Org. Chem.*, **69**, 6157–6164.
19. Dauer, W. and Przedborski, S. (2003) *Neuron*, **39**, 889–909.
20. Dobson, C.M. (2003) *Nature*, **426**, 884–890.
21. Chiti, F. and Dobson, C.M. (2006) *Annu. Rev. Biochem.*, **75**, 333–366.
22. Woll, M.G. and Gellman, S.H. (2004) *J. Am. Chem. Soc.*, **126**, 11172–11174.
23. Woll, M.G., Hadley, E.B., Mecozzi, S., and Gellman, S.H. (2006) *J. Am. Chem. Soc.*, **128**, 15932–15933.
24. Hadley, E.B. and Gellman, S.H. (2006) *J. Am. Chem. Soc.*, **128**, 16444–16445.
25. Saghatelian, A., Yokobayashi, Y., Soltani, K., and Ghadiri, M.R. (2001) *Nature*, **409**, 797–801.
26. Hadley, E.B., Witek, A.M., Freire, F., Peoples, A.J., and Gellman, S.H. (2007) *Angew. Chem. Int. Ed.*, **46**, 7056–7059.
27. Krishnan-Ghosh, Y. and Balasubramanian, S. (2003) *Angew. Chem. Int. Ed.*, **42**, 2171–2173.
28. Krishnan-Ghosh, Y., Whitney, A.M., and Balasubramanian, S. (2005) *Chem. Commun.*, 3068–3070.
29. Gennes, R.B. (1989) *Biomembranes: Molecular Structure and Function*, Springer-Verlag, New York.
30. Ono, A. and Freed, E.O. (2001) *Proc. Natl. Acad. Sci. U.S.A.*, **98**, 13925–13930.
31. Davidson, S.M.K. and Regen, S.L. (1997) *Chem. Rev.*, **97**, 1269–1279.
32. Regen, S.L. (2002) *Curr. Opin. Chem. Biol.*, **6**, 729–735.
33. Sugahara, M., Uragami, M., and Regen, S.L. (2002) *J. Am. Chem. Soc.*, **124**, 4253–4256.
34. Tokutake, N., Jing, B.W., Cao, H.H., and Regen, S.L. (2003) *J. Am. Chem. Soc.*, **125**, 15764–15766.
35. Erlanson, D.A., Braisted, A.C., Raphael, D.R., Randal, M., Stroud, R.M., Gordon, E.M., and Wells, J.A. (2000) *Proc. Natl. Acad. Sci. U.S.A.*, **97**, 9367–9372.
36. Wells, J.A., Erlanson, D., DeLano, W., Braisted, A., Raphael, D., Randal, M., Arkin, M., Raimundo, B., Oslob, J., Stroud, R., and Gordon, E. (2002) *FASEB J.*, **16**, A135–A135.
37. Erlanson, D.A., Wells, J.A., and Braisted, A.C. (2004) *Annu. Rev. Biophys. Biomol. Struct.*, **33**, 199–223.
38. Braisted, A.C., Oslob, J.D., Delano, W.L., Hyde, J., McDowell, R.S., Waal, N., Yu, C., Arkin, M.R., and Raimundo, B.C. (2003) *J. Am. Chem. Soc.*, **125**, 3714–3715.
39. Erlanson, D.A., Lam, J.W., Wiesmann, C., Luong, T.N., Simmons, R.L., DeLano, W.L., Choong, I.C., Burdett, M.T., Flanagan, W.M., Lee, D., Gordon, E.M., and O'Brien, T. (2003) *Nat. Biotechnol.*, **21**, 308–314.
40. Hajduk, P.J. and Greer, J. (2007) *Nat. Rev. Drug Discov.*, **6**, 211–219.
41. Obita, T., Muto, T., Endo, T., and Kohda, D. (2003) *J. Mol. Biol.*, **328**, 495–504.
42. Saitoh, T., Igura, M., Obita, T., Ose, T., Kojima, R., Maenaka, K., Endo, T., and Kohda, D. (2007) *EMBO J.*, **26**, 4777–4787.
43. Gasparini, G., Prins, L.J., and Scrimin, P. (2008) *Angew. Chem. Int. Ed.*, **47**, 2475–2479.
44. Gasparini, G., Martin, M., Prins, L.J., and Scrimin, P. (2007) *Chem. Commun.*, 1340–1342.
45. Godoy-Alcántar, C., Yatsimirsky, A.K., and Lehn, J.-M. (2005) *J. Phys. Org. Chem.*, **18**, 979–985.
46. Hochgürtel, M., Kroth, H., Piecha, D., Hofmann, M.W., Nicolau, C., Krause, S., Schaaf, O., Sonnenmoser, G., and Eliseev, A.V. (2002) *Proc. Natl. Acad. Sci. U.S.A.*, **99**, 3382–3387.

47. Zameo, S., Vauzeilles, B., and Beau, J.-M. (2005) *Angew. Chem. Int. Ed.*, **44**, 965–969.
48. Gasparini, G., Vitorge, B., Scrimin, P., Jeannerat, D., and Prins, L.J. (2008) *Chem. Commun.*, 3034–3036.
49. Gasparini, G., Bettin, F., Scrimin, P., and Prins, L.J. (2009) *Angew. Chem. Int. Ed.*, **48**, 4546–4550.
50. Levrand, B., Ruff, Y., Lehn, J.M., and Herrmann, A. (2006) *Chem. Commun.*, 2965–2967.

Part III
Chemical Reactions, Sustainable Processes, and Environment

8
Furfural and Furfural-Based Industrial Chemicals

Ana S. Dias, Sérgio Lima, Martyn Pillinger, and Anabela A. Valente

8.1
Carbohydrates for Life

In recent years we have been confronted with the reduction of fossil oil reserves, fluctuations of fossil fuel prices and the increase of CO_2 emissions, and the consequent problem of the greenhouse effect. These environmental, social, and economic problems have created the need for sustainable alternatives to fossil fuels and chemicals [1]. The use of plant biomass as starting material is one of the alternatives to decrease the dependency on fossil oil. The biomass can be transformed into energy, transportation fuels, various chemical compounds, and materials such as natural fibers by biochemical, chemical, physical, and thermal processes (Figure 8.1) [2–6]. However, when choosing the raw material, it is important to avoid the competition with food and feed applications and the consequent rise in prices.

Carbohydrates are among the most abundant organic compounds on earth and represent the major portion of the world's annually renewable biomass. Sources of carbohydrates include conventional forestry, by-products of wood processing (e.g., wood chips, pulp, and paper industrial residue), agricultural crops and surplus (e.g., corn stover, wheat, and rice straw), and plants (e.g., switchgrass) grown on degraded soils and algae. The bulk of the carbohydrate biomass comprises poly/oligosaccharides, such as hemicelluloses, cellulose, starch, inulin, and sucrose. In particular, lignocellulose plant matter is available in large quantities and is relatively cheap.

Cellulose and hemicellulose can be found in the cell wall of all plants cells. Cellulose is a linear polymer composed of β-D-glucopyranose (glucose) units forming microfibrils that give strength and resistance to the cell wall. The hemicellulose consists of a wide variety of polysaccharides (composed of pentoses, hexoses, hexuronic acids), which are interspersed with the microfibrils of cellulose, conferring consistency and flexibility to the structure of the cell wall [8].

The fermentation and the chemical conversion of carbohydrates into value-added compounds have received increasing interest in the last decade, and in a biorefinery different advantages may be taken from both processes [9–16]. Some of the most

Ideas in Chemistry and Molecular Sciences: Advances in Synthetic Chemistry. Edited by Bruno Pignataro
Copyright © 2010 WILEY-VCH Verlag GmbH & Co. KGaA, Weinheim
ISBN: 978-3-527-32539-9

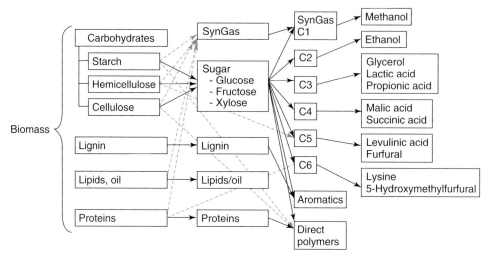

Figure 8.1 Bio-based products from the different biomass feedstocks. (Adapted from [7]).

important chemical transformations of carbohydrates are arguably the hydrolysis/dehydration of polysaccharides into the furan platform products, furfural and 5-hydroxymethylfurfural (HMF) [16, 17]. Furfural (Fur) has a wide industrial application profile and is considered as one of the top 30 building blocks that can be produced from biomass [7]. HMF is promising as a versatile, renewable furan chemical for the production of chemicals, polymers, and biofuels, similar to furfural [16, 18–20]. While Fur has been produced on an industrial scale for decades, the production of HMF has not reached industrial scale, to the best of our knowledge.

The hydrolysis/dehydration of polysaccharides into Fur and HMF may be promoted by Brönsted or Lewis acid catalysts. The industrial use of aqueous mineral acids as catalysts, such as sulfuric acid for furfural production, poses serious operational (corrosion), safety, and environmental problems (large amounts of toxic waste). Hence, it is desirable to replace conventional aqueous mineral acids by "green" nontoxic catalysts for converting sugars into Fur and HMF. The use of solid acids as catalysts may have several advantages over liquid acids, such as easier separation and reuse of the solid catalyst, longer catalyst lifetimes, toleration of a wide range of temperatures and pressures, and easier/safer catalyst handling, storage, and disposal. Several research groups have described approaches for converting hexoses (glucose and fructose) into HMF in the presence of solid acid catalysts with promising results [21–37]. Recently, to prevent the nonselective HMF decomposition, Avantium Technologies, a company based in Amsterdam, has reported a new approach to obtain an HMF derivative. In their work, the stable 5-(alkoxymethyl)furfural ether is formed from hexoses in the presence of an alcohol as solvent and an acid catalyst. When the alcohol is ethanol, the resulting 5-(ethoxymethyl)furfural) has an energy density of 31.3 MJ l^{-1}, which is as good as regular gasoline and diesel and significantly higher than ethanol. The encouraging results of the engine tests performed with a Citroen Berlingo has boosted interest in the development of furan products for application in transportation and aviation fuel/fuel additives and as bio-based polymers [16].

In this chapter, after an overview of the applications of Fur and the reaction mechanisms of dehydration/hydrolysis of polysaccharides into Fur, some of the most relevant results on the use of solid acid catalysts in the conversion of saccharides (in particular, xylose) into Fur are discussed.

8.2
Fur – Evolution over Nearly Two Centuries

Furfural was discovered in 1821 by Döbereiner, by the distillation of bran with dilute sulfuric acid [38, 39]. The resulting compound was first named furfurol (the name comes from the Latin word *furfur* that means bran cereal, while finishing *ol* means oil). Between 1835 and 1840, Emmet noted that the fur could be obtained from the majority of vegetable substances. The empirical formula of this product ($C_5H_4O_2$) was discovered by Stenhouse in 1840 and, in the year 1845, with the discovery of the aldehyde function in the molecule, it was named furfural (al for aldehyde). The fur molecule has an aldehyde group and a furan ring with aromatic character, and a characteristic smell of almonds. In the presence of oxygen, a colorless solution of Fur tends to become initially yellow, then brown, and finally black. This color is due to the formation of oligomers/polymers with conjugated double bonds formed by radical mechanisms and can be observed even at concentrations as low as 10^{-5} M [40].

The industrial production of fur was driven by the need of the United States of America to become self-sufficient during the First World War. Between 1914 and 1918, intensive exploration for converting agricultural wastes into industrially more valuable products was initiated. In 1921, the Quaker Oats company in Iowa initiated the production of Fur from oat hulls using "left over" reactors [40]. Over time, there was an increased industrial production of Fur and the discovery of new applications [41]. Currently, the annual world production of Fur is about 300 000 tons and, although there is industrial production in several countries, the main production units are located in China and in the Dominican Republic [42].

8.3
Applications of Furfural

The aldehyde group and furan ring furnish the Fur molecule with outstanding properties for use as a selective solvent [40, 41, 43]. Fur has the ability to form a conjugated double bond complex with molecules containing double bonds, and therefore is used industrially for the extraction of aromatics from lubricating oils and diesel fuels or unsaturated compounds from vegetable oils. Fur is used as a fungicide and nematocide in relatively low concentrations [40]. Additional advantages of Fur as an agrochemical are its low cost, safe and easy application, and its relatively low toxicity to humans. Despite the fact that Fur has an LD_{50} of 2330 mg kg^{-1} for dogs, man tolerates its presence in a wide variety of fruit juices, wine, coffee, and tea [40]. The highest concentrations of Fur are present in cocoa and coffee (55–255 ppm), in alcoholic beverages (1–33 ppm), and in brown bread (26 ppm) [44]. Most of the fur produced worldwide is converted through a hydrogenation process into furfuryl alcohol (FA), which is used for manufacturing polymers and plastics. Other furan compounds obtained from Fur include methylfuran and tetrahydrofuran. Fur and many of its derivatives can be used for the synthesis of new polymers based on the chemistry of the furan ring [41, 43, 45, 46].

8.4
Mechanistic Considerations on the Conversion of Pentosans into Furfural

Commercially, the pentosans (mainly xylan) present in the hemicellulose fraction of agricultural streams are hydrolyzed, using homogeneous acid catalysts in water, giving rise to pentose (xylose), which, by dehydration and cyclization reactions, leads to Fur with a theoretical mass yield of approximately 73% (Figure 8.2).

The hydrolysis of pentosans into pentoses in the presence of H_2SO_4 is faster than the dehydration of the pentose monomers into Fur [40, 41]. Hence, kinetic studies are generally focused on the rate limiting process, that is, the dehydration of pentoses. Xylose and arabinose are monomers found in pentosans, which can be converted into Fur, and some studies have shown that the dehydration of arabinose is slower than that of xylose [40, 47]. The concentration of xylose in the various raw materials is almost always much higher than that of arabinose. Considering these factors, it seems reasonable to investigate the kinetics of the dehydration process using xylose as substrate [21, 40, 42, 43, 45, 46, 48, 49]. In the dehydration and cyclization of xylose into fur, three molecules of water are released per molecule of fur produced. It is generally accepted that the xylose to Fur conversion involves a complex reaction mechanism consisting of a series of elementary steps. The two mechanisms presented have in common the fact that the furfural is formed from the xylopyranose ring and not from its open-chain aldehyde isomer (Figure 8.3 and 8.4). Considering the mechanism proposed by Zeitsch [40], the transformation of the pentose into Fur involves two eliminations in the positions 1,2 and one elimination in the position 1,4 (Figure 8.3). The

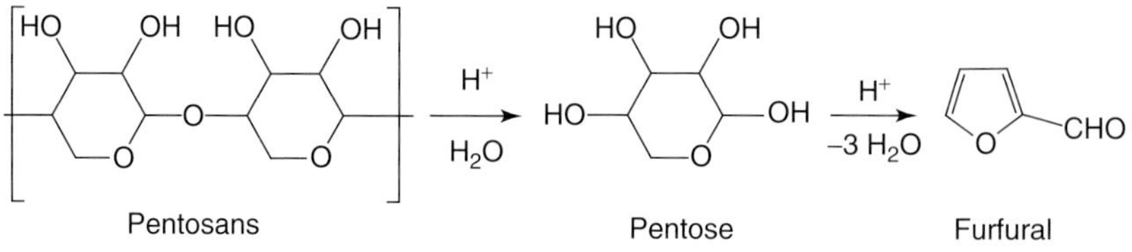

Figure 8.2 Net reaction of conversion of pentosans into furfural.

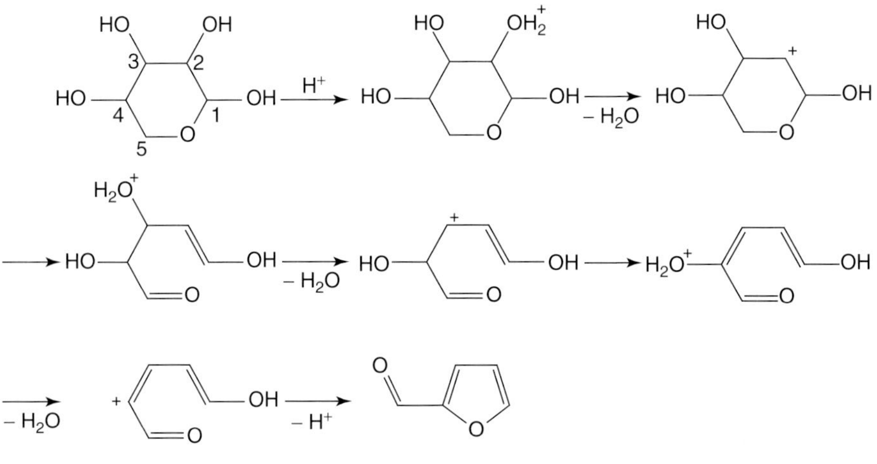

Figure 8.3 Mechanism of the dehydration of pentoses into furfural proposed by Zeitsch [40].

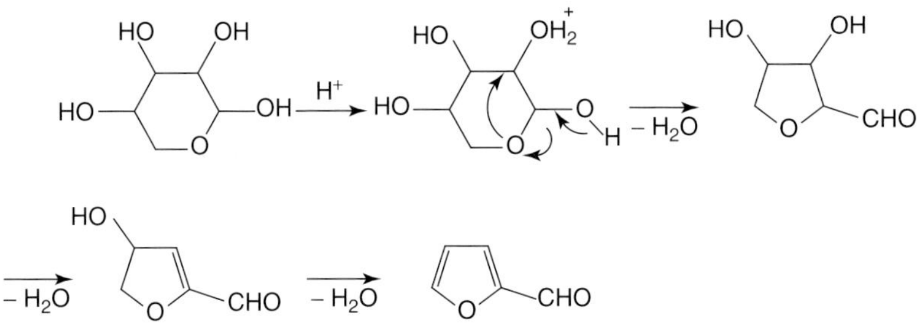

Figure 8.4 Reaction mechanism proposed by Antal et al. involving the protonation of the hydroxyl group in position C-2 [48].

1,2-eliminations imply the involvement of two neighboring carbon atoms and the formation of a double bond between them, while the 1,4-elimination involves two carbon atoms separated by two carbon atoms and the formation of the furan ring. Zeitsch summarizes the mechanism of conversion of the pentose into Fur in an acidic medium as a result of the transformation of hydroxyl groups of the pentose into H_2O^+ groups, leading to the liberation of water molecules with the formation of carbocations.

According to Antal et al. [48], there are two mechanistic alternatives to obtain Fur from D-xylose, depending on the hydroxyl group that is protonated first – the hydroxyl group at position 1 or 2 (Figure 8.4, only the mechanism resulting from the protonation of the hydroxyl group at the C2 position is shown). Both mechanisms involve the xylopyranose isomers, which lead to the formation of Fur by the loss of three molecules of water. A recent study of the xylose degradation using quantum mechanics modeling showed that the protonation of the hydroxyl group at position 2 is more favorable (requires less energy) than that of position 1 [50].

In acidic medium, the open-chain xylose undergoes isomerization into lyxose, which may be further dehydrated into Fur, albeit at a lower rate than that observed for the dehydration of xylose into Fur [48].

By-products formed in the xylose reaction may derive from the fragmentation of xylose, such as glyceraldehyde, glycolaldehyde, lactic acid, acetol [48]. On the other hand, as Fur is formed it can be transformed into higher molecular weight products by (i) condensation reactions between Fur and intermediates of conversion of xylose to furfural (and not directly with xylose) and (ii) Fur polymerization [40]. Aldol condensation between two molecules of Fur does not occur due to the absence of a carbon atom in H α position in relation to the carbonyl group [51]. The side reactions (i) and (ii) lead to oligomers and polymers and (i) is considered to be more relevant than (ii), although published characterization studies of the by-products formed are scarce [40]. The extent of these side reactions can be minimized by reducing the residence time of Fur in the reaction mixture and by increasing the reaction temperature [40, 49, 52]. If Fur is kept in the gas phase during the aqueous-phase reaction, it will not react with intermediates that are "nonvolatile". On the other hand, in a nonboiling system Fur yield increases with temperature possibly due to the entropy effect. The formation of by-products of high molecular weight results in a decrease of entropy and the change in Gibbs free energy (ΔG) becomes less negative: in the equation $\Delta G = \Delta H - T\Delta S$, the term $(-T\Delta S)$ becomes positive [52]. Increasing the temperature will eventually lead to $\Delta G \sim 0$, reached at the ceiling temperature (T_c). For $T > T_c$, fragmentation rather than the combination of molecules is favored. Another strategy for minimizing Fur losses is using a cosolvent immiscible with water to extract Fur from the aqueous phase (where the dehydration reaction of xylose takes place) as it is formed [21]. Extraction using supercritical CO_2 also enhances Fur yields [53–55].

The above mechanistic considerations for the homogeneous-phase conversion of xylose into Fur using H_2SO_4 as catalyst may also be considered for solid acid catalysts. Nevertheless, differences in product selectivity between homogeneous and heterogeneous catalytic processes are expected due to effects such as shape/size selectivity, competitive adsorption (related to hydrophilic/hydrophobic properties), and strength of the acid sites.

8.5
Production of Furfural

Industrially, Fur is directly produced from the lignocellulosic biomass in the presence of mineral acids, mainly sulfuric acid, under batch or continuous mode operation (Table 8.1). Attempts to improve Fur yields have been made by process innovation, although the use of mineral acids remains a drawback [40, 52, 56]. The cost and inefficiency of separating these homogeneous catalysts from the products makes their recovery impractical, resulting in large volumes of acid waste, which must be neutralized and disposed off. Other drawbacks include corrosion and safety problems. The production of Fur is therefore one of many industrial processes

Table 8.1 Industrial processes of furfural production.

Industrial process	Catalyst	Reaction type	Temperature (°C)
Quaker Oats	H_2SO_4	Batch	153
Chinese	H_2SO_4	Batch	160
Agrifurane	H_2SO_4	Batch	177–161
Quaker Oats	H_2SO_4	Continuous	184
Escher Wyss	H_2SO_4	Continuous	170
Rosenlew	Acids formed from the raw material	Continuous	180

where the replacement of the "toxic liquid" acid catalysts by alternative "green" catalysts is of high priority. Attempts have been made to develop heterogeneous catalytic processes for Fur production that offer environmental and economic benefits, but to the best of our knowledge none have been commercialized.

The acid properties of solid acids may be negatively affected by the presence of water in the reaction medium. Hence, one of the critical parameters in the choice of a stable, active heterogeneous catalyst is its tolerance toward water [57–64]. Several water-tolerant solid acids have been investigated in the conversion of saccharides into furan derivatives, including inorganic oxides and resins. Inorganic oxides have led to important improvements with respect to catalyst stability, recyclability, activity, and selectivity in comparison to conventional mineral acids and commercial acid ion exchange resins.

8.5.1
Crystalline Microporous Silicates

Conventional microporous zeolites, such as Faujasite HY and H-Modernite, seem quite promising, achieving selectivities to Fur in the range of 90–95% at xylose conversions between 30 and 40%, with water as solvent and in the presence of toluene as cosolvent, at 170 °C [21, 23]. However, xylose conversion has to be kept low in order to avoid significant drops in the Fur selectivity. Even then, the authors observed the formation of coke on the surface of the catalysts [21, 22, 46].

A novel microporous niobium silicate denoted as AM-11 was reported in 1998 and found to be a promising catalyst for gas-phase dehydration reactions, such as the conversion of *tert*-butanol to isobutene [65–67]. This solid contains octahedral niobium(V) and tetrahedral silicon, and the charge associated with framework niobium is balanced by Na^+ and NH_4^+ cations. Calcined AM-11 possesses a substantial amount of Brönsted and Lewis acidity [66]. Microporous AM-11 crystalline niobium silicates were studied as solid acid catalysts in the dehydration of xylose in water/toluene biphasic conditions (water and toluene (W/T)), at 140–180 °C. After 6 hours at 160 °C, xylose conversions of up to 90% and furfural yields of up to 50% were achieved, and the thermally regenerated catalysts could be reused without loss of activity or selectivity [68]. The calcined AM-11 (prepared in the NH_4^+ form) catalysts gave higher Fur yields at 6 hours (46% at 85% conversion) than the protonic form of commercial HY (Si/Al = 5; 39% yield at 94% conversion) and H-MOR (protonic form of zeolite mordenite, Si/Al = 6; 28% at 79% conversion), under identical reaction conditions. Zeolites and AM-11 materials are sufficiently stable to be used at elevated temperatures and to be regenerated (quite easily) by thermal treatments under air. This constitutes an important advantage in comparison to ion exchange resins as solid acid catalysts. The catalytic results obtained with the crystalline solid acids may be further optimized by, for example, using different solvent mixtures and compositions. The extensive studies carried out by the group of Dumesic and coworkers on the use of different solvent mixtures for the dehydration of sugars into HMF and Fur using mineral acids as catalysts at high temperatures give valuable insights on the solvent effects [18]. Preferably, the solvent(s) should have an excellent extracting capacity of the furan compound and should be used in minimal amounts, avoiding high dilution and long heating times. The reactor design is another important issue. For example, Moreau reported that a significant increase in HMF selectivity is obtained by simultaneous extraction of HMF with methyl isobutyl ketone (MIBK) circulating in a countercurrent manner in a continuous catalytic heterogeneous pulsed column reactor [26].

8.5.2
Functionalized Mesoporous Silicas

The application of mesoporous solid acids to convert sugars into furan derivatives may be advantageous in relation to microporous materials by avoiding diffusion limitations and fast catalyst deactivation. Micelle-templated mesoporous silicas are especially promising supports for liquid-phase acid catalysis because they have high specific surface area and pore volume, together with a regular pore structure and tuneable pore size, which enables rapid diffusion of reactants and products through the pores, thus minimizing consecutive reactions.

Heteropolyacids (HPAs) are promising candidates as green catalysts and are already used in several industrial processes, such as the hydration of olefins [57, 59, 69–72]. The advantages of HPAs in homogeneous liquid-phase catalysis are their low volatility, low corrosiveness, high flexibility, safety in handling, and generally high activity and selectivity compared to conventional mineral acids. Furthermore, side reactions such as sulfonation, chlorination, and nitration, which normally occur in the presence of mineral acids, are absent in the reactions catalyzed by HPAs. The Keggin-type HPAs are typically represented by the formula $H_{8-x}[XM_{12}O_{40}]$, where X is the heteroatom, x is its oxidation state and M is the addenda atom (Mo^{6+} or W^{6+}).

The HPAs $H_3PW_{12}O_{40}$ (PW), $H_4SiW_{12}O_{40}$ (SiW), $H_3PMo_{12}O_{40}$ (PMo), and $H_4SiMo_{12}O_{40}$ (SiMo) were investigated in the liquid-phase dehydration of D-xylose to Fur [73]. The catalytic results depend on the reaction temperature, type of solvent, and HPA composition. The most promising systems were the tungsten-containing HPAs used with either dimethyl sulfoxide (DMSO) or W/T as solvent: Fur yields

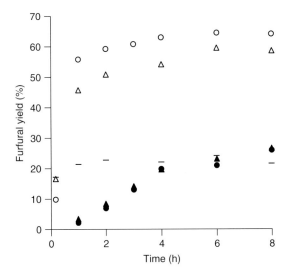

Figure 8.5 Dependence of furfural yield on reaction time using DMSO as the solvent and PW (○), SiW (Δ) or PMo (−) as the catalyst, or using W/T as the solvent and PW (●) or SiW (▲) as the catalyst, at 140 °C [73]. Copyright Elsevier (2005) with kind permission.

achieved within 8 hours at 140 °C were below 70% (Figure 8.5). The catalytic performance of the heteropolytungstate PW was on a par with that for sulfuric acid for the cyclodehydration of xylose into furfural, in homogeneous phase, using DMSO as solvent, at 140 °C. Kinetic studies showed that the initial reaction rate exhibits a first-order dependence on the initial concentration of xylose and a nonlinear dependence on the initial concentration of HPA.

Heterogenization of HPAs can facilitate product separation, catalyst recovery, and recycling [70, 71]. When supporting HPAs on ordered mesoporous silica, the immobilized species may interact more strongly with plain silica, due to the high dispersions achieved [74]. The most important and common HPAs for catalysis are the Keggin acids since they are the most stable and readily available ones. In particular, PW possesses the highest acid strength and thermal stability [71]. Complexation of PW with the hydroxyl groups of the hexagonally ordered mesoporous silica MCM-41 (Mobil Composition of Matter), for example, is thought to lead to $SiOH_2^+$ groups that can act as counterions for the polyanion [75–77].

Catalysts based on PW supported on silica have been used in the dehydration of xylose [78]. A series of composites comprising PW immobilized in micelle-templated silicas (e.g., MCM-41) with large unidimensional mesopores were prepared by either incipient wetness impregnation or immobilization in amino-functionalized silicas. These materials exhibited higher activity than the bulk HPA, and Fur yields after 4 hours were similar to those obtained with H_2SO_4 (58%), using DMSO as solvent, at 140 °C. Strong host–guest interactions and active site isolation for the materials with low HPA loadings (15 wt%)

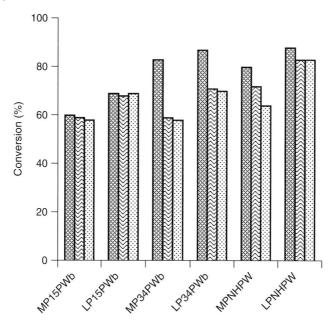

Figure 8.6 Xylose conversion after 4 hours reaction in the presence of the HPAs supported in medium pore (MP), large pore (LP), or in amino-fuctionalized silicas, in DMSO, at 140 °C: run 1 (diamonds), run 2 (waves), run 3 (dots) [78]. Copyright Elsevier (2006) with kind permission.

and the amino-functionalized supports appeared to benefit activity and stability with DMSO as solvent (Figure 8.6). High boiling DMSO requires difficult and energy-intensive isolation procedures to purify the target product and possible technical difficulties. Water is cleaner and cheaper, but xylose dehydration is sluggish with water as the solvent. As mentioned above, recent advances have shown that improved results are possible in biphasic systems consisting of W/T, allowing the in-situ extraction of Fur from the aqueous phase. However, for the supported heteropolyanion catalysts, a disadvantage is the significant leaching of the HPA from the support into the aqueous phase during reaction at high temperatures, which constitutes a major limitation for any potential industrial application. Similarly, the stability and reusability of the MCM-41-supported cesium salts of 12-tungstophosphoric acid ($Cs_xH_{3-x}PW_{12}O_{40}$) were better in DMSO than in W/T [79]. The latter materials showed no significant advantages over the corresponding PW-supported catalysts.

In order to increase the stability of the catalysts toward leaching, other materials were prepared taking into account the need for a covalent link between the active acid site and the support. Sulfonic acid–functionalized mesoporous silicas are active and selective catalysts for a number of reactions (Figure 8.7). The active sulfonic group is obtained postsynthetically by sulfonation reactions, or by the oxidation of thiol-functionalized silicas previously synthesized

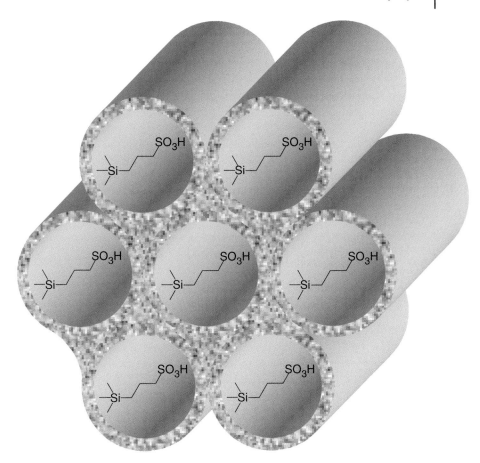

Figure 8.7 Representation of a sulfonic-functionalized mesoporous silica (MCM-41).

by a one-step sol gel or postmodification grafting route. The immobilization of (3-mercaptopropyl)trimethoxysilane (MPTS) in toluene onto MCM-41 with controlled water content resulted in a "coated" material (MCM-41-SHc) with a monolayer of MPTS moieties (7.1 wt% S), and a less covered "silylated" material (MCM-41-SHs) was obtained in dry conditions (4.5 wt% S). The oxidation of the mercaptopropyl groups by hydrogen peroxide in a water–methanol solution was complete, but resulted in a reduction of the sulfur contents and in the formation of disulfide and partially oxidized disulfide species. The performance of these materials is summarized in Table 8.2 [80]. For the synthesized materials, the highest yield was obtained with MCM-41-SO$_3$Hc as a catalyst (70% yield after 24 hours, at 140 °C, DMSO or W/T), which is higher than that attainable with commercial zeolites. The high furfural selectivity observed at high conversions may be explained by the presence of large unidimensional mesopores in the MCM-41 materials that promote the selective dehydration of xylose into furfural by allowing fast diffusion of furfural out of the catalyst as soon as it is formed. This diffusion

Table 8.2 Catalytic performance of sulfonic acid–functionalized materials in the dehydration of D-xylose[a] [80]. Copyright Elsevier (2005) with kind Permission.

Catalyst	S_{BET} ($m^2\ g^{-1}$)	V_p ($cm^3\ g^{-1}$)	H^+ (mequiv. g^{-1})[b]	TOF^c (mmol $g_{cat}^{-1}\ h^{-1}$)	Conversion[d] (%)	Selectivity[e] (%)
None	–	–	–	–	34/84	2/27
MCM-41	833	0.59	–	0.8	30/86	4/52
MCM-41-SO₃Hs	493	0.28	0.4	2.0 (5)	81/90	49/77
MCM-41-SO₃Hc	438	0.24	0.7	2.1 (3)	84/91	65/82
Hybrid-SO₃H	278	0.13	0.1	1.4 (14)	57/88	11/61
Amberlyst-15	–	–	4.6	2.2 (0.5)	87/90	68/70

[a] Reaction conditions: 1 ml DMSO, 30 mg xylose, 20 mg catalyst, 140 °C.
[b] Measured by titrating the solid with NaOH.
[c] Turnover frequency (TOF) calculated after 4 hours. In brackets the TOF values are expressed as millimole·per milliequivalent of H^+ per hour.
[d] Conversion after 4/24 hours.
[e] Selectivity to furfural after 4/24 hours.

effect avoids the extensive consecutive degradation reactions of furfural. Even so, these materials deactivate with long residence times, which is accompanied by the appearance of a brownish color during the reaction (coke formation). In fact, the progressive catalyst deactivation between recycling runs might be connected with the inefficient removal of the adsorbed by-products, which load the catalyst surface and lead to the decrease of the active sites that are active for the xylose dehydration by passivation effects. For better catalyst regeneration, these by-products must be removed with a thermal treatment (350 °C is needed for the coke removal), but the sulfonic acid groups are stable only up to 250 °C. This constitutes a major drawback for practical application. One approach that allows the thermal regeneration of the catalysts is to prepare materials without an organic component.

Early studies showed that Fur yields of about 60% at about 90% conversion could be achieved using conventional sulfated zirconia (SZ) or titania as solid acids and supercritical carbon dioxide as an extracting solvent, at 180 °C and 200 atm [53]. These results could be further improved by dispersing (per)sulfated zirconia ((P)SZ) on a mesoporous support with high specific surface area. Conventional SZ has a specific surface area usually in the range 80–100 $m^2\ g^{-1}$ and a lack of ordered mesoporosity and textural homogeneity, making it suitable for traditional vapor-phase reactions involving small molecules, but less amenable to liquid-phase reactions. On the other hand, the use of supercritical CO_2 as extracting solvent is greener than the use of organic solvents, but it may have economical drawbacks associated with process requirements and energy consumption.

Conventional (per)sulfated bulk zirconia, mesoporous SZ, and (P)SZ supported on an ordered mesoporous silica, MCM-41, with or without aluminum incorporation, were examined as acid catalysts for the dehydrocyclization of xylose into Fur in W/T, at 160 °C [81]. Fur yields of up to 50% could be achieved at >90% conversion with the mesostructured bulk and silica-supported zirconia catalysts, which was better than that achievable with H_2SO_4 (using approximately the same equivalent amount of sulfur). While these materials were stable toward zirconium leaching, loss of sulfur was observed in recycling runs, but in some cases no decrease in catalytic activity was observed. Of all the investigated materials, MCM-41-supported SZ with Al seemed to be the most attractive catalyst for aqueous-phase conversion of xylose, since it was the most stable to sulfur leaching and exhibited increasing activity and no significant loss of selectivity to Fur in three runs.

Niobium-containing materials such as hydrated niobium oxide, a water-tolerant solid acid catalyst, exhibit unique activity, selectivity, and stability for many different catalytic reactions [82, 83]. Ordered mesoporous MCM-41-type niobium silicates prepared by the doping of niobium in the micelle-templated silica, with Si/Nb molar ratios of either 25 or 50 (in the H^+ form), were found to be active catalysts for xylose dehydration in W/T and gave Fur yields consistently in the range of 34–39% (after 6 hours reaction at 160 °C) [68]. The niobium-containing mesoporous MCM-41-type catalysts exhibited higher activities than crystalline AM-11 materials (for the first catalytic runs), but were less selective to Fur at conversions above 80% [68]. Partial loss of activity in recycling runs and leaching of Nb from MCM-41 occurred during the reaction. Aluminum-containing MCM-41 catalysts gave similar results to those obtained with AM-11 at 160 °C in W/T at 6 hours reaction, and could be reused several times without loss of catalytic activity and selectivity and no metal leaching was detected [84] (Figure 8.8).

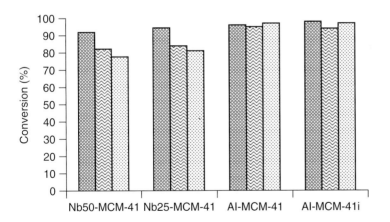

Figure 8.8 Xylose conversion after 6 hours reaction in the presence of the Nb- or Al-containing MCM-41 catalysts, in water/toluene at 160 °C: run 1 (diamonds), run 2 (waves), run 3 (dots). The i in Al-MCM-41i indicates postsynthesis impregnation of aluminum on MCM-41 (and not by hydrothermal conditions) [68, 84].

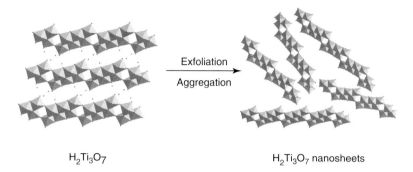

Figure 8.9 Representative scheme of the formation of the H$_2$Ti$_3$O$_7$ nanosheets after exfoliation and aggregation of the layered materials.

8.5.3
Transition Metal Oxide Nanosheets

Crystalline layered metal oxide cation exchangers, such as titanates, niobates, and titanoniobates, are potentially strong solid acids when in the H$^+$ form. However, the high charge density of the anionic sheets in these materials hinders the access of bulky substrate molecules to the acid sites. This problem has recently been addressed by exfoliating the layered metal oxides to give aggregates of nanosheets, where the two-dimensional sheet structure remains (Figure 8.9). The composites have much higher specific surface areas than the acid-exchanged layered precursors and function as strong solid acid catalysts, rivaling or even beating niobic acid that is a rare water-tolerant solid acid. It was found that they can be more active and somewhat more selective catalysts for the conversion of xylose into Fur than the microporous AM-11 crystalline niobium silicates, which in turn yielded more Fur than zeolites such as HY (the protonic form of Y-zeolite, with Si/Al = 5) and mordenite (Si/Al = 6), under similar reaction conditions (at 160 °C, in W/T) [85]. After 4 hours reaction, Fur yields of up to 55% were achieved. Furthermore, no metal leaching occurred and Fur yields remained practically the same in recycling runs (Figure 8.10).

8.6
Conclusion and Future Perspectives

Attempts have been made to convert (poly)saccharides into basic, versatile furan compounds, Fur and HMF, using heterogeneous catalytic routes and the published results obtained at the lab-scale seem quite promising and encouraging. The use of porous solid acids as catalysts instead of mineral acids for this reaction system may have several advantages, such as an easier separation of the catalyst from the products (e.g., by simple filtration), convenient regeneration (e.g., after thermal removal of coke), and the possibility to reuse the catalyst consecutively (avoiding treatments of effluent streams). Comparable or higher yields of the

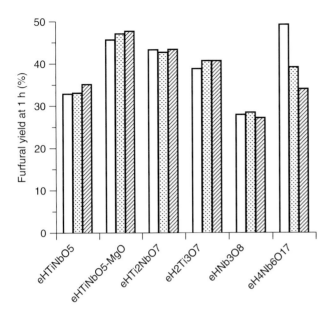

Figure 8.10 Furfural yield obtained in recycling runs after 1 hour reaction (run 1 – white bar, run 2 – dots, run 3 – hashed) over the exfoliated–aggregated nanosheet solid acid catalysts, at 160 °C [85]. Copyright Elsevier (2006) with kind permission.

target product may also be reachable. The results may be further optimized by fine-tuning the catalyst properties, such as acid–base (poor selectivity has been correlated to strong Brönsted acidity, and enhanced Lewis acidity seems favorable) and hydrophobic/hydrophilic properties (e.g., via dealumination, functionalization using organosilanes), and pore size distribution (e.g., by appropriate choice of template). Together with the adjustment of the reaction conditions, such as the composition of sugar, catalyst and solvent mixtures (types of solvents), temperature and residence times, and reactor design, these developments could open up valuable perspectives in the application of solid acid catalysts to the conversion of saccharides into basic furan derivatives. The use of water to dissolve saccharides instead of organic solvents, such as DMSO and dimethylformamide, is a more convenient, greener, and cheaper approach. These issues lead to stricter requirements in terms of catalytic properties: (i) the solid acid catalysts must be sufficiently tolerant toward water and impurities present in the raw materials, (hydro)thermally stable, and preferably readily prepared; (ii) high selectivity at high conversion for high substrate concentration is important for enhanced productivity. The use of extracting organic solvents to increase selectivity poses environmental concerns and future work in this direction must be considered carefully.

The use of heterogeneous catalysts for the production of Fur and HMF has not yet been implemented industrially, to the best of our knowledge. One critical factor is the choice of the biomass raw materials and its management/processing inputs in order to obtain liquid feed streams rich in saccharides for the heterogeneous catalytic hydrolysis/dehydration processes: the transformation of solid biomass using a solid catalyst would be subject to severe mass transfer limitations. Using di/oligo/polysaccharides as starting materials instead of the monosaccharides themselves to produce HMF and Fur in a one-pot process, thereby eliminating the separate hydrolysis step before the dehydration reaction, seems quite attractive. The design of versatile catalysts for mixed feed (fractions of cellulose, hemicellulose, starch) processing may make the process more cost competitive. Another important aspect that must be considered is the transport of biomass into the industrial plant.

Collaborative efforts between academia and industry will be crucial in developing competitive technologies for the production of HMF and Fur using a suitable reaction medium with readily prepared (with optimized properties) and sufficiently robust solid acids.

Acknowledgments

This work was partly funded by the FCT, POCTI, and FEDER (project POCI/QUI/56112/2004). The authors wish to express their gratitude to other colleagues at CICECO and also the University of Salamanca (their names appear in the references of A.S. Dias *et al.*) for their valuable collaborations, to Prof. C.P. Neto (Department of Chemistry) for helpful discussions, and Dr. F. Domingues (Department of Chemistry) for access to HPLC equipment. S.L. and A.S.D. are grateful to the FCT for grants. A.S.D. extends thanks to Dr. E. de Jong from Avantium Technologies for helpful discussions.

This article was published in Journal of Catalysis, Vol 244, Dias, A.S., Lima, S., Carriazo, D., Rives, V., Pillinger, M., Valente, A.A., Exfoliated titanate, niobate and titanoniobate nanosheets as solid acids for the liquid-phase dehydration of D-xylose into furfural, 230–237, Copyright Elsevier (2006).

References

1. Brown, R.C. (2003) *Biorenewable Resources – Engineering New Products from Agriculture*, 1st edn, Blackwell Publishing.
2. Huber, G.W., Iborra, S., and Corma, A. (2006) Synthesis of Transportation Fuels from Biomass: Chemistry, Catalysts, and Engineering. *Chem. Rev.*, **106** (9), 4044–4098.
3. Huber, G.W. and Dumesic, J.A. (2006) An overview of aqueous-phase catalytic processes for production of hydrogen and alkanes in a biorefinery. *Catal. Today*, **111** (1-2), 119–132.
4. Lichtenthaler, F.W. and Peters, S. (2004) Carbohydrates as green raw materials for the chemical industry. *C. R. Chim.*, **7**, 65–90.
5. Kamm, B., Kamm, M., Gruber, P.R., and Kromus, S. (2006) Biorefinery systems–an overview, in *Biorefineries–Industrial Processes and Products*, 1st edn, vol. **1** (eds P.R. Gruber and M. Kamm), Wiley-VCH Verlag GmbH, Weinheim, pp. 3–40.
6. Fernando, S., Adhikari, S., Chandrapal, C., and Murali, N. (2006) Biorefineries: current status, challenges, and

future direction. *Energy Fuels*, **20** (4), 1727–1737.
7. Werpy, T. and Peterson, G. (2004) Top Added Chemicals from Biomass. Volume I: Results of Screening Potential Candidates from Sugars and Synthesis Gas. US Department of Energy, www.eere.energy.gov/biomass/pdfs/35523.pdf (accessed 30 March 2009).
8. Spiridon, I. and Popa, V.I. (2005) Hemicelluloses: structure and properties, in *Polysaccharides-Structural Diversity and Functional Versatility*, 2 edn (ed. S. Dumitriu), Marcel Dekker, New York, pp. 475.
9. Kamm, B., Gruber, P.R., and Kamm, M. (eds) (2206) *Biorefineries–Industrial Processes and Products, Status Quo and Future Directions*, Vol. 1, Wiley-VCH Verlag GmbH, Weinheim.
10. Lin, Y.-C. and Huber, G. (2009) The critical role of heterogeneous catalysis in lignocellulosic biomass conversion. *Energy Environ. Sci.*, **2**, 68–80.
11. Stöcker, M. (2008) Biofuels and biomass-to-liquid fuels in the biorefinery: catalytic conversion of lignocellulosic biomass using porous materials. *Angew. Chem. Int. Ed.*, **47** (48), 9200–9211.
12. Dhepe, P.L.P.L. and Fukuoka, A.A. (2008) Cellulose conversion under heterogeneous catalysis. *ChemSusChem.*, **1** (12), 969–975.
13. Kamm, B., Gruber, P.R., and Kamm, M. (2006) *Biorefineries–Industrial Processes and Products, Status Quo and Future Directions*, Vol. 2, Wiley-VCH Verlag GmbH, Weinheim.
14. Lange, J.-P. (2007) Lignocellulose conversion: an introduction to chemistry process and economics, in *Catalysis for Renewables From Feedstock to Energy Production* (eds G. Centi and R.A. van Saten), Wiley-VCH, Weinheim, pp. 21–51.
15. Gallezot, P. (2007) Process options for the catalytic conversion of renewables into bioproducts, in *Catalysis for Renewables From Feedstock to Energy Production* (eds G. Centi and R.A. van Saten), Wiley-VCH, Weinheim, pp. 53–73.
16. Gruter, G.J. and de Jong, E. (2009) Furanics: novel fuel options from carbohydrates. *Biofuels Technol.*, **1** 10–17.
17. Chheda, J.N., Huber, G.W., and Dumesic, J.A. (2007) Liquid-phase catalytic processing of biomass-derived oxygenated hydrocarbons to fuels and chemicals. *Angew. Chem. Int. Ed.*, **46**, 7164–7183.
18. Román-Leshkov, Y., Chheda, J.N., and Dumesic, J.A. (2006) Phase modifiers promote efficient production of hydroxymethylfurfural from fructose. *Science*, **312**, 1933–1937.
19. Lewkowski, J. (2001) Synthesis, chemistry and applications of 5-hydroxymethylfurfural and its derivatives. *Arkivoc*, **2** (6) 17–54.
20. Binder, J.B. and Raines, R.T. (2009) Simple chemical transformation of lignocellulosic biomass into furans for fuels and chemicals. *J. Am. Chem. Soc.*, **131** (5), 1979–1985.
21. Moreau, C., Durand, R., Peyron, D., Duhamet, J., and Rivalier, P. (1998) Selective preparation of furfural from xylose over microporous solid acid catalysts. *Ind. Crops Prod.*, **7** (2-3), 95–99.
22. Moreau, C. (2006) Micro- and mesoporous catalysts for the transformation of carbohydrates, in *Catalysis for Chemical Synthesis* (ed E. Derouane), John Wiley & Sons, Ltd.
23. Moreau, C. (2002) Zeolites and related materials for the food and non food transformation of carbohydrates. *Agro Food Ind. Hi Tech*, **13** (1) 17–26.
24. Lourvanij, K. and Rorrer, G.L. (1993) Reactions of aqueous glucose solutions over solid-acid Y-zeolite catalyst at 110-160. degree-C. *Ind. Eng. Chem. Res.*, **32** (1), 11–19.
25. Moreau, C., Durand, R., Pourcheron, C., and Razigade, S. (1994) Preparation of 5-hydroxymethylfurfural from fructose and precursors over H-form zeolites. *Ind. Crops Prod.*, **3** (1-2), 85–90.
26. Moreau, C., Durand, R., Razigade, S., Duhamet, J., Faugeras, P., Rivalier, P., Ros, P., and Avignon, G. (1996) Dehydration of fructose to 5-hydroxymethylfurfural over H-mordenites. *Appl. Catal. A Gen.*, **145** (1-2), 211–224.

27. Seri, K., Inoue, Y., and Ishida, H. (2000) Highly efficient catalytic activity of lanthanide(III) ions for conversion of saccharides to 5-hydroxymethyl-2-furfural in organic solvents. *Chem. Lett.*, **29** (1), 22–23.
28. Ishida, H. and Seri, K. (1996) Catalytic activity of lanthanoide (III) ions for dehydration of -glucose to 5-(hydroxymethyl) furfural. *J. Mol. Catal. A Chem.*, **112** (2), L163–L165.
29. Seri, K., Inoue, Y., and Ishida, H. (2001) Catalytic activity of lanthanide(iii) ions for the dehydration of hexose to 5-hydroxymethyl-2-furaldehyde in water. *Bull. Chem. Soc. Jpn.*, **74** (6), 1145–1150.
30. Carlini, C., Giuttari, M., Galleti, A.M.R., Sbrana, G., Armaroli, T., and Busca, G. (1999) Selective saccharides dehydration to 5-hydroxymethyl-2-furaldehyde by heterogeneous niobium catalysts. *Appl. Catal. A: Gen.*, **183** (2), 295–302.
31. Armaroli, T., Busca, G., Carlini, C., Giuttari, M., Galletti, A.M.R. and Sbrana, G.G. (2000) Acid sites characterization of niobium phosphate catalysts and their activity in fructose dehydration to 5-hydroxymethyl-2-furaldehyde. *J. Mol. Catal. A Chem.*, **151** (1-2), 233–243.
32. Bebvenuti, F., Carlini, C., Patrono, P., Galleti, A.M.R., Sbrana, G., Massucci, M.A., and Galli, P. (2000) Heterogeneous zirconium and titanium catalysts for the selective synthesis of 5-hydroxymethyl-2-furaldehyde from carbohydrates. *Appl. Catal. A: Gen.*, **193** (1-2), 147–153.
33. Lansalot-Matras, C. and Moreau, C. (2003) Dehydration of fructose into 5-hydroxymethylfurfural in the presence of ionic liquids. *Catal. Commun.*, **4** (10), 517–520.
34. Carlini, C., Patrono, P., Galletti, A.M.R., and Sbrana, G. (2004) Heterogeneous catalysts based on vanadyl phosphate for fructose dehydration to 5-hydroxymethyl-2-furaldehyde. *Appl. Catal. A: Gen.*, **275** (1-2), 111–118.
35. Moreau, C., Finiels, A., and Vanoye, L. (2006) Dehydration of fructose and sucrose into 5-hydroxymethylfurfural in the presence of 1-H-3-methyl imidazolium chloride acting both as solvent and catalyst. *J. Mol. Catal. A Chem.*, **253** (1-2), 165–169.
36. Watanabe, M., Aizawa, Y., Iida, T., Nishimura, R., and Inomata, H. (2005) Catalytic glucose and fructose conversions with TiO2 and ZrO2 in water at 473 K: relationship between reactivity and acid–base property determined by TPD measurement. *Appl. Catal. A: Gen.*, **295** (2), 150–156.
37. Watanabe, M.M., Aizawa, Y.Y., Iida, T.T., Aida, T.M.T.M., Levy, C.C., Sue, K.K., and Inomata, H.H. (2005) Glucose reactions with acid and base catalysts in hot compressed water at 473 K. *Carbohydr. Res.*, **340** (12), 1925–1930.
38. International Furan Chemicals B.V. (2006) ©, Historical Overview and Industrial Development, http://www.furan.com/furfural, (accessed 30 March 2009).
39. Kamm, B., Kamm, M., Schimdt, M., Hirth, T., and Schulze, M. (2006) Lignocellulose-based chemical products and product family trees, in *Biorefineries–Industrial Processes and Products*, 1st edn, Vol. 2 (eds B. Kamm, P.R. Gruber, and M. Kamm), Wiley-VCH Verlag GmbH, Weinheim, pp. 97–149.
40. Zeitsch, K.J. (2000) *The Chemistry and Technology of Furfural and its Many By-products*, Sugar Series, Vol. 13, Elsevier, The Netherlands.
41. Hoydonckx, H.E., van Rhijn, W.M., van Rhijn, de Vos, D.E., and Jacobs, P.A. (2007) Furfural and derivatives. *Ullmann's Encyclopedia Ind. Chem.*, 1–29.
42. Win, D.T. (2005) Furfural–gold from garbage. *AU J. Technol.*, **8** (4), 185–190.
43. Sain, B., Chaudhuri, A., Borgohain, J.N., Baruah, B.P., and Ghose, J.L. (1982) Furfural and furfural-based industrial chemicals. *J. Sci. Ind. Res.*, **41**, 431.
44. International Programme on Chemical Safety (1999) Safety Evaluation of Certain Food Additives Furfural., WHO Food additives series: 42, World Health Organization, Geneva, First draft prepared by R. Kroes, Research Institute of Toxicology, Utrecht University, Utrecht, Netherlands, http://www.inchem.org, (accessed 30 March 2009).

45. Gandini, A. and Belgacem, M.N. (1997) Furans in polymer chemistry. *Prog. Polym. Sci.*, **22** (6), 1203–1379.
46. Moreau, C., Belgacem, M.N., and Gandini, A. (2004) Recent catalytic advances in the chemistry of substituted furans from carbohydrates and in the ensuing polymers. *Top. Catal*, **27** (1-4), 11–30.
47. Kootstra, A.M.J., Mosier, N.S., Scott, E.L., Beeftink, H.H., and Sanders, J.P.M. (2009) Differential effects of mineral and organic acids on the kinetics of arabinose degradation under lignocellulose pretreatment conditions. *Biochem. Eng. J.*, **43** (1), 92–97.
48. Antal, M.J., Leesomboon, T.Jr, Mok, W.S., and Richards, G.N. (1991) Mechanism of formation of 2-furaldehyde from D–xylose. *Carbohydr. Res.*, **217**, 71–86.
49. Root, D.F., Saeman, J.F., Harris, J.F., and Neill, W.K. (1959) Chemical conversion of wood residues, part II: kinetics of the acid-catalyzed conversion of xylose to furfural. *Forest Prod. J.*, **9** 158–165.
50. Nimlos, M.R., Qian, X., Davis, M., Himmel, M.E., and Nimlos, D.K.J. (2006) Energetics of xylose decomposition as determined using quantum mechanics modeling. *J. Phys. Chem. A*, **110** (42), 11824–11838.
51. Chheda, J.N. and Dumesic, J.A. (2007) An overview of dehydration, aldol-condensation and hydrogenation processes for production of liquid alkanes from biomass-derived carbohydrates. *Catal. Today*, **123** (1-4), 59–70.
52. Zeitsch, K.J. (2000) Furfural production needs chemical innovation. *Chem. Innov.*, **30** (4), 29–32.
53. Kim, Y.-C. and Lee, H.S. (2001) Selective synthesis of furfural from xylose with supercritical carbon dioxide and solid acid catalyst. *J. Ind. Eng. Chem.*, **7** (6), 424–429.
54. Sako, T., Sugeta, T., Nakazawa, N., Okubo, T., Sato, M., Taguchi, T., and Hiaki, T. (1991) Phase equilibrium study of extraction and concentration of furfural produced in a reactor using supercritical carbon dioxide extraction. *J. Chem. Eng. Jpn.*, **24** (4), 499–455.
55. Sako, T., Sugeta, T., Nakazawa, N., Okubo, T., Sato, M., Taguchi, T., and Hiaki, T. (1992) Kinetic study of furfural formation accompanying supercritical carbon dioxide extraction. *J. Chem. Eng. Japan*, **25** (4), 372–377.
56. Zeitsch, K.J. (2001) Gaseous acid catalysis: an intriguing new process. *Chem. Innov.*, **31** (1), 41–44.
57. Okuhara, T. (2002) Water-tolerant solid acid catalysts. *Chem. Rev.*, **102**, 3641–3666.
58. Li, L., Yoshinaga, Y., and Okuhara, T. (1999) Water-tolerant catalysis by Mo– Zr mixed oxides calcined at high temperatures. *Phys. Chem. Chem. Phys.*, **1** (20), 4913–4918.
59. Izumi, Y. (1997) Hydration/hydrolysis by solid acids. *Catal. Today*, **33** (4), 371–411.
60. Namba, S.S., Hosonuma, N.N., and Yashima, T.T. (1981) Catalytic application of hydrophobic properties of high-silica zeolites : I. Hydrolysis of ethyl acetate in aqueous solution. *J. Catal.*, **72** (1), 16–20.
61. Csicsery, S.M. (1986) Catalysis by shape selective zeolites-science and technology. *Pure Appl. Chem.*, **58** (6), 841–856.
62. Ishida, H. (1997) Liquid-phase hydration process of cyclohexene with zeolites. *Catal. Surv. Jpn.*, **1** (2), 241–245.
63. Kobayashi, S. and Hachiya, I. (1994) Lanthanide triflates as water-tolerant lewis acids. Activation of commercial formaldehyde solution and use in the aldol reaction of silyl enol ethers with aldehydes in aqueous media. *J. Org. Chem.*, **59** (13), 3590–3596.
64. Kobayashi, S. (1998) New types of Lewis acids used in organic synthesis. *Pure Appl. Chem.*, **70** (5), 1019–1126.
65. Rocha, J., Brandão, P., Pedrosa de Jesus, J.D., Philippou, A., and Anderson, M. (1999) Synthesis and characterisation of microporous titanoniobosilicate ETNbS-10. *Chem. Commun.*, 471–472.
66. Philippou, A., Brandão, P., Ghanbari-Siahkali, A., Dwyer, J., Rocha, J., and Anderson, M.W. (2001) Catalytic studies of the novel microporous niobium silicate AM-11. *Appl. Catal. A Gen.*, **207** (1-2), 229–238.
67. Brandão, P., Philippou, A., Rocha, J., and Anderson, M.W. (2002) Dehydration of alcohols by microporous niobium

silicate AM-11. *Catal. Lett.*, **80** (3-4), 99–102.

68. Dias, A.S., Lima, S., Brandão, P., Pillinger, M., Rocha, J., and Valente, A.A. (2006) Liquid-phase dehydration of D-Xylose over microporous and mesoporous niobium silicates. *Catal. Lett.*, **108** (3-4), 179–186.

69. Misono, M. and Nojiri, N. (1990) Recent progress in catalytic technology in Japan. *Appl. Catal.*, **64**, 1–30.

70. Kozhevnikov, I.V. (1998) Catalysis by heteropoly acids and multicomponent polyoxometalates in liquid-phase reactions. *Chem. Rev.*, **98** (1), 171–198.

71. Mizuno, N. and Misono, M. (1998) Heterogeneous catalysis. *Chem. Rev.*, **98** (1), 199–217.

72. Misono, M., Ono, I., Koyano, G., and Aoshima, A. (2000) Heteropolyacids. Versatile green catalysts usable in a variety of reaction media. *Pure Appl. Chem.*, **72** (7), 1305–1311.

73. Dias, A.S., Pillinger, M., and Valente, A.A. (2005) Liquid phase dehydration of D-xylose in the presence of Keggin-type heteropolyacids. *Appl. Catal. A: Gen.*, **285**, 126–131.

74. Molnár, Á., Beregszászi, T., Fudala, Á., Lentz, P., Nagy, J.B., Kónya, Z., and Kiricsi, I. (2001) The acidity and catalytic activity of supported acidic cesium dodecatungstophosphates studied by MAS NMR, FTIR, and catalytic test reactions. *J. Catal.*, **202** (2), 379–386.

75. Lefebvre, F. (1992) 31P MAS NMR study of H3PW12O40 supported on silica: formation of (SiOH2+)(H2PW12O40−). *J. Chem. Soc., Chem. Commun.*, **10**, 756.

76. Blasco, T., Corma, A., Martínez, A., and Martínez-Escolano, P. (19980) Supported heteropolyacid (HPW) catalysts for the continuous alkylation of isobutane with 2-butene: the benefit of using MCM-41 with larger pore diameters. *J. Catal.*, **177** (2), 306–313.

77. Ghanbari-Siahkali, A., Philippou, A., Dwyer, J., and Anderson, M.W. (2000) The acidity and catalytic activity of heteropoly acid on MCM-41 investigated by MAS NMR, FTIR and catalytic tests. *Appl. Catal. A: Gen.*, **192** (1), 57–69.

78. Dias, A.S., Pillinger, M., and Valente, A.A. (2006) Mesoporous silica supported 12-tungstophosphoric acid catalysts for the liquid phase dehydration of D-xylose. *Microporous Mesoporous Mater.*, **94**, 214–225.

79. Dias, A.S., Lima, S., Pillinger, M., and Valente, A.A. (2006) Acidic cesium salts of 12-tungstophosphoric acid as catalysts for the dehydration of xylose into furfural. *Carbohydr. Res.*, **34**, 2946–2953.

80. Dias, A.S., Pillinger, M., and Valente, A.A. (2005) Dehydration of xylose into furfural over micro-mesoporous sulfonic acid catalysts. *J. Catal.*, **229**, 414–423.

81. Dias, A.S., Lima, S., Pillinger, M., and Valente, A.A. (2007) Modified versions of sulfated zirconia as catalysts for the conversion of xylose to furfural. *Catal. Lett.*, **114** (3-4), 151–160.

82. Tanabe, K. and Okasaki, S. (1995) Various reactions catalyzed by niobium compounds and materials. *Appl. Catal. A: Gen.*, **133** (2), 191–218.

83. Ziolek, M. (2003) Niobium -containing catalysts – the state of the art. *Catal. Today*, **78** (1-4), 47–64.

84. Valente, A.A., Dias, A.S., Lima, S., Brandão, P., Pillinger, M., Plácido, H., and Rocha, J. (2006) Catalytic performance of microporous Nb and mesoporous Nb or Al silicates in the dehydration of D-xylose to furfural. IX Congreso Nacional de Materiales, Vigo, pp. 1203–1206.

85. Dias, A.S., Lima, S., Carriazo, D., Rives, V., Pillinger, M., and Valente, A.A. (2006) Exfoliated titanate, niobate and titanoniobate nanosheets as solid acid catalysts for the liquid-phase dehydration of D-xylose into furfural. *J. Catal.*, **244**, 230–237.

9
Multiple Bond-Forming Transformations: the Key Concept Toward Eco-Compatible Synthetic Organic Chemistry

Yoann Coquerel, Thomas Boddaert, Marc Presset, Damien Mailhol, and Jean Rodriguez

9.1
The Science of Synthesis

Chemistry is in part a descriptive science, which allows the discovery and the fundamental understanding of natural phenomena. But the most exciting part of chemical sciences is that it is certainly a creative experimental science. Indeed, chemical synthesis offers unique possibilities to create highly valuable substances. Synthetic chemists who have acquired a solid fundamental knowledge and developed practical skills place themselves at the frontier of the fascinating "molecular world." In this world, known molecules (i.e., naturally occurring molecules) can be man-made in a laboratory, which is already an outstanding performance. But of utmost importance, *de novo* molecules with unprecedented properties are also accessible by chemical synthesis. The only limitations to the exploration of the "molecular world" are the creativity of the chemist, and most importantly on a practical point of view, the current knowledge of the science of synthesis. In the current era, however, it is assumed that any three-dimensional molecular architecture, providing it is sufficiently stable, can be prepared by total synthesis (i.e., the laboratory construction of naturally occurring or designed molecules by chemical synthesis from simple starting materials) if the chemist has enough experience, knowledge, time, and money. This is a very comfortable idea. Although it might be true, the analysis of the total syntheses of some of the most highly complex and important molecules that are currently prepared, often with exceptional creativity and intellectual elegance, revealed that the desired target compound can actually be obtained at the expense of a huge number of elemental chemical transformations (sometimes over a hundred), which translates into prohibitive cost and time consumption, rendering the compound supply unpractical, and precluding its application at the benefit of mankind [1].

The discovery and control of fire, nearly 790 000 years ago, is certainly one of the most important revolution in the history of mankind [2]. Although it is not yet fully realized, the last 70 years have witnessed a "quiet revolution" in the

Ideas in Chemistry and Molecular Sciences: Advances in Synthetic Chemistry. Edited by Bruno Pignataro
Copyright © 2010 WILEY-VCH Verlag GmbH & Co. KGaA, Weinheim
ISBN: 978-3-527-32539-9

science of synthesis [3]. In 1828, Friedrich Wöhler reported the break-through synthesis of urea from silver isocyanate [4]. This was the very first synthesis of a molecule-of-life from inorganic materials, and this discovery marked both the death of the theory of vitalism and the birth of synthetic organic chemistry. Since this pioneering work, our knowledge of organic synthesis has grown considerably, with a dramatic acceleration since the early 1940s. This accumulated knowledge now translates into our capability to efficiently control the three-dimensional formation of virtually any single covalent chemical bond between, at least, the elements found in the restricted portion of the periodic table concerned with organic chemistry (C, H, O, N, P, S...) even on complicated substrates. We can highlight here that two recent Nobel Prizes in Chemistry were attributed to the development of methodologies in organic synthesis: the 2001 prize was jointly attributed to W. S. Knowles and R. Noyori, and K. B. Sharpless for their work on chirally catalyzed hydrogenation and oxidation reactions, respectively, and the 2005 prize to Y. Chauvin, R. H. Grubbs, and R. R. Schrock for their work on the catalytic olefin metathesis reaction.

The mastering of the chemical bond formation being established, we may ask ourselves what is the future of chemical synthesis? Predictions in sciences are of course uncertain, but some orientations, however, appear clearly. Certainly, the invention of new and improved methodologies to create a single covalent chemical bond will stay an important area of chemical research, as no one can reasonably think that the current tool-box of the synthetic chemist is full enough with the best and appropriate tools. But the next major evolution in synthetic organic chemistry might be the control of multiple bond-forming transformations (MBFTs). What better chemical transformation than the one which would allow the creation of multiple covalent bonds in the same operation while keeping the same level of control currently at hand in single bond-forming reactions? We should realize here that for billions of years, Nature's strategy for the biosynthesis of the complex molecules occurring in living organisms has been essentially based on such extremely efficient enzyme-mediated MBFTs. The most common examples of man-made controlled MBFTs are found in polymer synthesis, where *identical* covalent bonds are iteratively constructed between monomeric substrates to provide polymeric macromolecules. And there is no need here to convince the reader of the profound impact of polymer-based materials on our society [5]. The next evolution of chemical synthesis, ideally, would advance our knowledge to the control of chemical processes wherein a maximum of *different* covalent bonds are created in the same operation. In a previous excellent reflection on the topic, Paul A. Wender proposed to use the *ideal synthesis* as a calibration point for the evaluation of progresses in chemical synthesis, and gave the following definition: the *ideal synthesis* is "a synthesis in which the target molecule is made from readily available starting materials in one simple, safe, environmentally acceptable, and resource-effective operation that proceeds quickly and in quantitative yield" [6]. In recent years, and thanks to the rapid development of MBFTs, an increasing number of examples approaching the ideal synthesis have been reported, turning the concept close to a reality.

Before going any further, we should comment here on another aspect of contemporaneous synthetic chemistry. The synthetic chemical community and pharmaceutical companies have currently to respond to a double challenge: to continue to provide society with more and more complex molecules (essentially drugs) in an economically compatible manner, while taking into account the now required criterion of sustainable development. Sustainable synthetic chemistry is primarily concerned with the economy of raw material, energy, waste (often referred to as *atom-economy* [7]), time, and human resource in the development of a process. In the past two decades, catalysis has been proposed as a solution for sustainable chemistry, as it partially addresses some of the above issues. But all the above criteria can be easily respected if the overall number of synthetic transformations required to arrive at the target molecule can be significantly reduced. This latter concept, often referred to as *step-economy* [8], is the key to sustainable synthetic chemistry. This new area is the core of "green chemistry" [9], but might be better referred to as *eco-compatible chemistry* – we define eco-compatible as being both economically and ecologically compatible. *Eco-compatibility* can be measured by parameters such as selectivity and overall yield, of course, but mainly step count and the metric E-factor [10], as well as raw material, time, human resources, and energy requirements, and finally, the toxicity and hazard of the chemicals and the protocols involved. Step-economical synthetic approaches to well-defined complex molecular targets can be performed in only two ways: (i) a better strategic use of our current knowledge when planning the multiple step sequence of the synthesis, or (ii) the invention of new reactions, or better new MBFTs, that would simplify and shorten the synthetic sequence. If ingeniousness at the strategy level can indeed allow some shortening in a synthetic plan, we believe that the most important contributions to step-economical synthesis have resulted, and will do so in future, from advances in synthetic methodology. An eco-compatible synthetic plan should intend to maximize the use of MBFTs. Accordingly, when a complex molecule, or usually a family of complex molecules, is the target of a synthetic chemistry research program, a pleasing approach would start with the design of a daring and audacious synthetic strategy involving one or more unprecedented MBFTs, preferably under catalytic conditions, which after development should be validated through the total synthesis of the targeted molecule(s). This approach has proven highly rewarding in the past. More often, it occurs that a newly described MBFT opens-up new opportunities and triggers the elaboration of an eco-compatible synthetic plan, which is then turned to practice in target-oriented synthesis.

MBFTs can be divided into two classes: concerted, but not necessarily synchronized, reactions such as cycloadditions, and nonconcerted transformations such as *consecutive reactions*. In the following, we discuss the terminologies used to describe the four subclasses of nonconcerted MBFTs yet identified, and the differences between these conceptually distinct MBFTs are illustrated by a selection of examples chosen from our recent contributions.

9.2
Multiple Bond-Forming Transformations (MBFTs)

9.2.1
Consecutive Reactions

The descriptor *consecutive reactions* was introduced by Tietze in 1993 as one of the two subclasses of *sequential transformations* – herein denominated MBFTs [11]. It is regrettable that the terminology *consecutive reaction* was not adopted and used enough, often unwisely replaced by "one-pot," "tandem," or "cascade." In a consecutive reaction, the introduction of the reagent(s) and eventual additional solvent(s) and substrate(s) are performed in a stepwise manner in the same reaction mixture from which nothing is removed, resulting in a change of the reaction conditions during the overall transformation. In the cases where the sequence requires any kind of work-up of the intermediate product mixture, even one keeping the crude product in the same reactor (e.g., the removal of solvent and/or volatiles by simple evaporation), the reaction sequence can no longer be considered as a single transformation, but rather as a two-step sequence without purification of the intermediate in a *one-pot* operation.

9.2.2
Domino Reactions

Domino reactions have been well defined by Tietze as transformations that allow the formation of several covalent bonds in one process, the entire process being run without the addition of any further reagent (or solvent/catalyst) and without change in the reaction conditions, in which the latter transformation takes place at the functionalities obtained in the former bond-forming reaction [12]. The term *cascade reaction* is an acceptable synonym of "domino reaction," but the term *tandem reaction* often found in the literature to describe domino processes is not appropriate. Two things are important in the above definition: there should be no change in the reaction conditions, and the latter transformation must be a consequence of the former (i.e., the order of the different elemental transformations involved in a domino process is well defined). Strictly speaking, sequences involving even a limited and operationally simple change of the reaction conditions such as an elevation of temperature should not be denominated *domino reactions* but preferably *consecutive reactions*. A classification of domino reactions has been proposed by Tietze *et al.* based on the mechanism of the different bond-forming steps, with eight categories proposed for each individual step: cationic, anionic, radical, pericyclic, photochemical, transition metal–catalyzed, oxidative or reductive, and enzymatic. If the different steps of the domino process are of the same category, the reaction can be termed *homo domino* (e.g., a domino anionic/anionic reaction), while the terms *mixed domino* or *hetero domino* reaction would apply to a domino reaction involving mechanistically distinct elementary steps. The use of domino and consecutive reactions, embraced under the banner "cascade reactions," in the total synthesis of complex natural products has been highlighted recently [13].

9.2.3
Multicomponent Reactions

Multicomponent reactions (MCRs) can be defined as processes in which three or more reactants introduced simultaneously are combined through covalent bonds, regardless of their mechanistic nature, in a single chemical operation to produce products that incorporate substantial portions of all the reactants [14]. They are sometimes referred to as *multiple-component condensations* [15]. Most of the known MCRs are *domino reactions*, but in the cases where the protocol requires a change in the reaction conditions to promote the advanced elemental steps of the transformation, for example, an elevation of the temperature, or in the cases where the three or more components are assembled by elemental bond-forming operations taking place at independent functionalities, these MCRs are *consecutive reactions*. MCRs have been found most useful in the context of drug discovery for the preparation of large collections of molecules in a time- and cost-effective manner, and they are now key tools in industrial and academic research [16].

9.2.4
One-Pot Reactions

The term *one-pot* has been used with an always-increasing frequency since the mid-1970s, as an all-embracing term to describe several conceptually distinct eco-compatible transformations (Figure 9.1). If one-pot transformations are understood as MBFTs or sequences of transformations occurring in the same reaction vessel, the term *one-pot* encompasses under the same descriptor conceptually and practically very different transformations. Indeed, the three above-mentioned MBFTs are considered as single transformations, while a two-step sequence without purification of the intermediate in a *one-pot* operation is better regarded as two successive transformations performed in an eco-compatible manner. With this definition of one-pot transformations and according to the above, the commonly referred to and accepted definition of domino reactions [12], the frequently encountered association of the descriptors "one-pot" and "domino reaction" constitutes a pleonasm. In contemporaneous literature, the term *one-pot* is actually used more to underline the eco-compatibility of a protocol rather than to describe with a certain degree of precision the type of transformation involved. The overuse of "one-pot" as a descriptor complicates database searches and, more importantly, does not account for the relatively very different levels of efficiency and eco-compatibility that are reached, depending on the class of MBFTs involved. We recommend that the use of "one-pot" as a descriptor is restricted to MBFTs that require a work-up operation at one stage of the transformation – thus excluding domino and consecutive reactions as one-pot reactions – and to multistep sequences, without purification of the intermediates that are performed in the same reaction vessel.

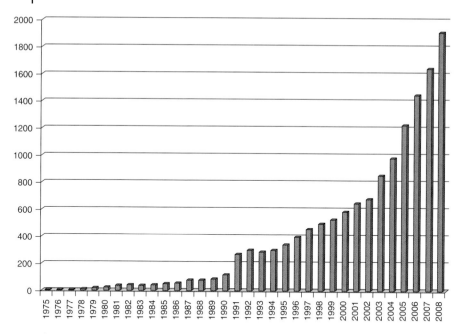

Figure 9.1 Occurrence of the descriptor "one-pot" in organic chemistry and multidisciplinary chemistry journals. The data were obtained from the ISI Web of Science (database = SCI EXPANDED) on 5 March 2009.

This more focused definition of one-pot reactions renders possible a comparison of the eco-compatibility of the different classes of nonconcerted MBFTs identified herein. Domino MCRs and consecutive MCRs are the most satisfying, provided the modifications of the reaction conditions in the latter are limited and operationally simple, for example, an elevation of the reaction temperature, while the eco-compatible benefit of one-pot reactions, though real, is reduced in comparison. A synoptic graphical representation of the definitions and terminologies used in this discussion can be found in Figure 9.2.

9.3
An Account of Our Recent Contributions

9.3.1
MBFTs Involving a Wolff Rearrangement

The Wolff rearrangement allows the synthesis of ketenes from α-diazo ketones by 1,2-sigmatropic transposition from the carbonyl group to the α-carbon atom [17]. In the case of 1,3-diketones, the rearrangement produces a reactive α-oxo ketene that can be engaged in a number of subsequent transformations, generally following a domino process [18]. The migratory aptitudes in the Wolff rearrangement are

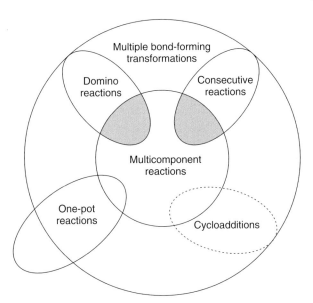

Figure 9.2 Synoptic overview of the various classes of nonconcerted MBFTs. Surfaces of the different domains and cross-domains have been chosen arbitrarily and are not representative of the relative importance of the corresponding reaction or any other criterion. The most eco-compatible MBFTs are highlighted in gray. A domain representing cycloadditions that are not part of a domino, consecutive, or one-pot reaction is delimited by a dashed-line for comparison.

influenced by the nature of the migrating group, the conformation of the reactant, and the dichotomy of concerted and nonconcerted reaction paths. In connection with our program on MBFTs involving 1,3-dicarbonyl compounds, we recently became interested in the Wolff rearrangement for the preparation of a variety of α-carbonylated cycloalkanones of type **2**, with modulated acidity of the activated α-proton and a functionalized Nu group (Scheme 9.1). Although some transition metal–catalyzed Wolff rearrangements have been described [19], the reaction usually occurs at high temperatures or requires some specific reactors for photoactivation (ultraviolet), precluding the presence of sensitive or reactive functional groups.

Microwave-assisted organic chemistry (MAOS) is no longer a laboratory curiosity and will soon be regarded as a common activation mode of chemical transformations. The synthetic community has started to realize the true benefits of microwave activation, and, as a testimony, a number of monographs and review articles have been made available in the past few years, on both the theory and the development of the technique [20]. A microwave effect can be observed in many reactions. Its origin can be from a purely thermal/kinetic effect according to the Arrhenius equation, as a consequence of the extremely rapid and internally localized heating of the reaction mixture, to a putative, highly controversial, microwave-specific nonthermal effect [21]. It is however universally recognized that polar substrates and solvents are microwave active, and owing to their high dipole moments, α-diazo-ketones fall into this category.

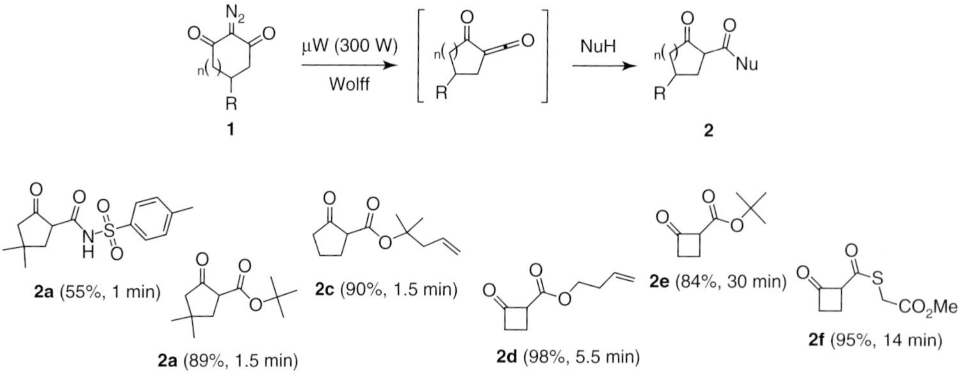

Scheme 9.1 Microwave-assisted Wolff rearrangement.

We have recently disclosed that a broad variety of cyclic 1,3-dicarbonyl compounds of type **2** can be prepared with high efficiency through a *mixed domino* microwave-assisted Wolff rearrangement of cyclic 2-diazo-1,3-diketones **1**, followed by the trapping of the resulting α-oxo ketene with a stoichiometric amount of alcohol, amine, or thiol (Scheme 9.1) [22]. The method has been found to be particularly of value for the preparation of otherwise difficultly available compounds. For example, the *p*-toluenesulfonyl β-ketoamide **2a**, which cannot be obtained by transamidation due to the poor nucleophilicity of *p*-toluenesulfonamide, was prepared in 55% yield. As illustrated by the efficient preparation of **2b** and **2c**, the method is also nicely suitable for the preparation of tertiary β-oxo esters, with regards to the few methods available for the synthesis of these sterically hindered esters [23]. Finally, the method also allows a straightforward access to strained functionalized cyclobutanones such as compounds **2d–f**. We may highlight here, that the eco-compatibility and the "ideal" character of this microwave-assisted domino Wolff rearrangement/ketene trapping are excellent, because no excess of substrate and no additives are used; only nitrogen gas is evolved, and no purification is required in most cases, the reaction time is very short, and the sequence requires a minimum of energy. Forthcoming developments of domino reactions involving a microwave-assisted Wolff rearrangement should include, in particular, the use of more functionalized nucleophiles allowing subsequent reactions in a prolonged domino process, and functionalized unsaturated substrates capable of cycloadditions.

9.3.2
MBFTs Involving Metals

The use of metal carbonyl complexes of unsaturated organic molecules is a well-established strategy for the functionalization of the organic moiety of the complex. Among these, the Pauson–Khand [2 + 2 + 1] cycloaddition of di(cobalt)hexa(carbonyl) alkyne complexes is a famous MBFT [24]. However, in the current era where the eco-compatibility of chemical processes is becoming an increasingly important criterion, the stoichiometric use of metals has to be avoided. Therefore, the Pauson–Khand reaction can now be performed efficiently with catalytic amounts of metallic species [25]. However, on a research laboratory scale, some of these stoichiometric organometallic complexes are sometimes used to prepare molecules that are otherwise difficult to access. That was the case for one of us (Y.C.) during his early Ph.D. thesis work when he got interested in the regio- and stereoselective synthesis of cyclohepta-1,3-dienes [26]. We required at the time, an efficient and scalable synthesis of the tricarbonyl(η^4-cycloheptadiene)iron complex (4), and two approaches were described: either the complexation of cycloheptadiene with $Fe(CO)_5$ [27], or the selective catalytic hydrogenation of the corresponding cycloheptatriene complex 3 [28]. They were both impractical, the first because of the prohibitive price of cycloheptadiene (about €100 g^{-1}), and the second due to high-pressure conditions required for the hydrogenation. We found that the expected cycloheptadiene complex 4 could be prepared directly in quantitative yield from cycloheptatriene (about €100 l^{-1}) and 3 equiv. of $Fe(CO)_5$ in the presence of a catalytic amount of sodium borohydride and iso-propanol by a *mixed domino reaction* (Scheme 9.2) [29]. The reaction most probably proceeds via initial complexation of the cycloheptatriene with excess $Fe(CO)_5$, and subsequent reduction of the resulting metal-free activated double-bond by hydrogen transfer from iso-propanol catalyzed by the trinuclear cluster hydride $NaHFe_3(CO)_{11}$, prepared *in situ* from $Fe(CO)_5$ and $NaBH_4$ (by an other domino process!). The interesting concept in this reaction is that the same reagent, namely $Fe(CO)_5$, is involved in both elemental steps of the domino reaction with a very different role: it is the source of the tricarbonyl(iron) moiety in the final complex and also the source of the organometallic catalyst of the reduction following its reaction with $NaBH_4$. It is worth noting now that even though the two newly incorporated hydrogen atoms in the final complex come from the iso-propanol solvent, there is not a substantial enough incorporation in the final product, and it cannot be considered as a third component of the reaction.

More recently, we got interested in the catalytic properties of the Pd/C–triethylamine system, and particularly, in the Pd/C-catalyzed conjugate reduction

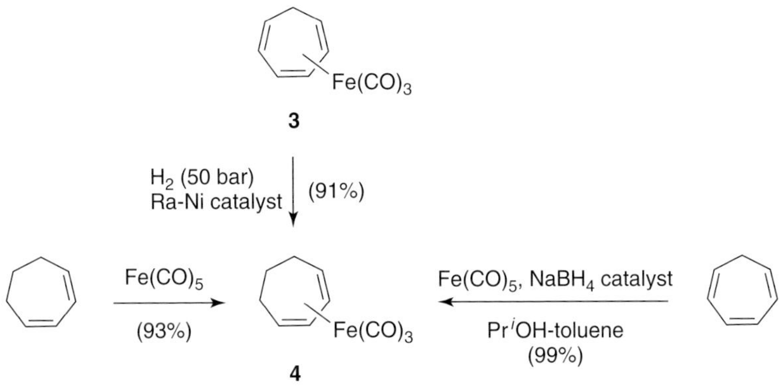

Scheme 9.2 Reductive complexation of cycloheptatriene.

of activated double bonds by hydrogen transfer from triethylamine [30]. In the course of our study with conjugated methyl esters, we observed and then optimized the formation of reduced ethyl amide by-products. The reaction of *trans*-methyl cinnamate (**5**) with 10 mol% Pd/C in the presence of 4 equiv. of triethylamine in toluene essentially gave the expected reduced ester **6** when the reaction was conducted at 140 °C for 16 hours, while the same mixture at 160 °C for four to five days afforded the corresponding reduced ethyl amide **7**, quantitatively [31]. From these results, we surmised that some ethylamine was generated *in situ* during or after the conjugate reduction reaction, and that the ethyl amide product would result from the efficient transamidation of the saturated methyl ester with ethylamine. It was rationalized that under the reaction conditions, triethylamine can undergo a Pd/C-catalyzed self-alkylation (or oligomerization) to produce ethylamine and longer-chain alkyl amines with the general formula $N[(C_2H_4)_xH][(C_2H_4)_yH][(C_2H_4)_zH]$. Thus, the saturated ethyl amide **7** is the product of a *domino* reduction/transamidation. Interestingly, triethylamine plays a dual role in this reaction: it is first the source of the two newly incorporated hydrogen atoms in the reduction step, and then the precursor of ethylamine used for transamidation. In the context of atom-economy, this reaction is a nice example of a particularly attractive concept, where under identical heterogeneous catalytic conditions, the same "bifunctional" reactant is used in two chemically distinct steps of a *hetero domino* process (Scheme 9.3).

9.3.3
MBFTs Involving Anions

In the context of a study on the chirality transfer from sulfoximines in MBFTs reactions, we became interested in the preparation of 2-sulfonymidoylmethyl-enetetrahydrofuranes of types **9** and **10**. Inspired by earlier work from our group and others, we have prepared a number of these chiral 2-sulfonymidoylmethyl-enetetrahydrofuranes by *consecutive reactions* involving the dianion of the corresponding S-methyl-sulfoximine **8** and α,ω-halogenoester bis-electrophiles [32]. The sequence is initiated by the chemoselective acylation of the sulfoximine dianion [33] to produce a stabilized β-oxo sulfoximine anion, which upon elevation of the reaction temperature, in turn, undergoes a regioselective O-cyclization to produce the target compounds **9** and **10**. As noted earlier, because the two elemental steps of this sequence are not conducted at the same temperature, the overall transformation is not a domino reaction but

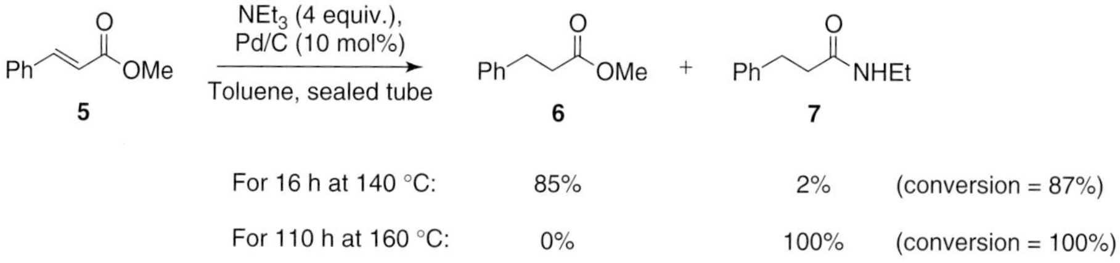

Ph—CH=CH—C(O)OMe **5**	NEt$_3$ (4 equiv.), Pd/C (10 mol%) → Toluene, sealed tube	Ph—CH$_2$CH$_2$—C(O)OMe **6** + Ph—CH$_2$CH$_2$—C(O)NHEt **7**	
	For 16 h at 140 °C:	85% 2%	(conversion = 87%)
	For 110 h at 160 °C:	0% 100%	(conversion = 100%)

Scheme 9.3 Unexpected catalytic properties of the Pd/C–NEt$_3$ system.

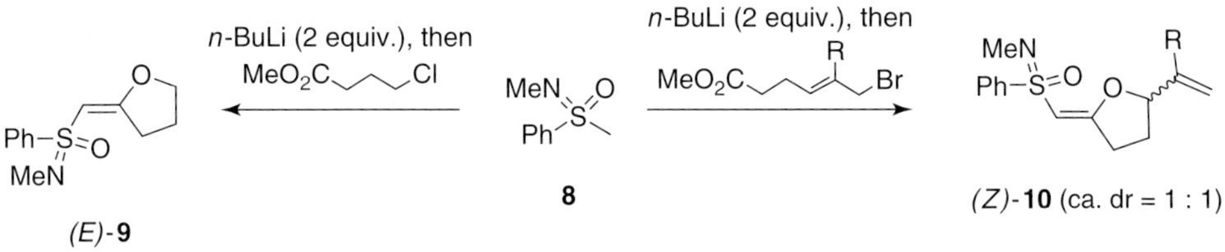

Scheme 9.4 Syntheses of 2-sulfonymidoylmethylenetetrahydrofuranes.

a consecutive reaction. Our first attempts of chirality transfer to prepare diastereoselectively 5-vinyl-2-sulfonymidoylmethylenetetrahydrofuranes using α,ω-haloesters bearing an allylic halide as the second electrophilic position revealed virtually no chiral induction of the sulfonimidoyl group, *(E)*-**10** (minor) and *(Z)*-**10** (major), both being isolated as near 1 : 1 mixtures of diastereomers. Our work in this series now concentrates on the exploitation of some of the 2-sulfonymidoylmethylenetetrahydrofuranes prepared very recently with this method as starting material in subsequent MBFTs. Overall, these studies are expected to result soon, in a general approach to cyclic polyethers featuring a two-step only iterative MBFTs strategy (Scheme 9.4).

In the course of a study aiming at the preparation of bicyclo[4.2.1]nonanes, we required an efficient synthetic access to the *cis*-α,γ-disubstituted β-oxo ester **13** from the α-monosubstituted β-oxo ester **11** [34]. We rapidly realized that the γ-alkylation of **11** through the reaction of its lithium enolate or enamine with allyl bromide was problematic, resulting in poor yields of the α,γ-bisallylated product **13**. In order to achieve better activation of the γ position for the second alkylation to proceed efficiently, we turned our attention to a 1,3-shift of the ester group. In early attempts, the treatment of **11** with potassium hydride in the presence of 5 equiv. of 18-crown-6 ether afforded the rearranged β-oxo ester **12** in moderate yield [35], which could then be allylated in a separate step (Scheme 9.5). Although this methodology allowed us to prepare the desired substrate, it was not efficient enough and too expensive on a preparative scale due to the use of a large excess of crown ether. We then looked for an alternative protocol for the 1,3-ester shift in **11** involving a domino retro-Dieckmann fragmentation–Dieckmann cyclization, as described

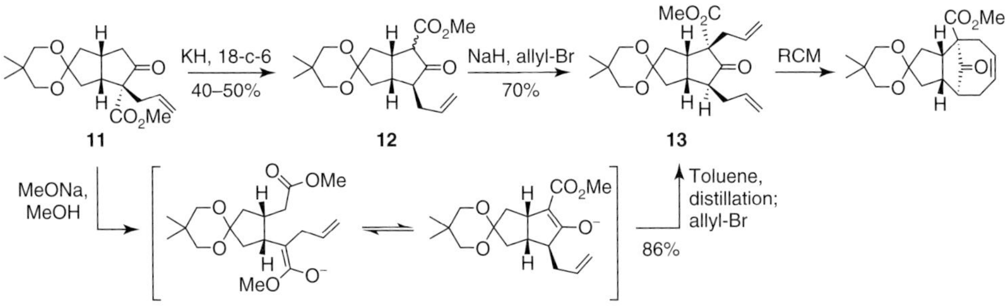

Scheme 9.5 Synthesis of bicyclo[4.2.1]nonanes.

by Isida and coworkers [36]. Of importance is the observation that the coupling of this 1,3-shift of the ester group with the direct alkylation of the resulting enolate was previously shown possible on two related examples from the Dauben and Paquette groups in a one-pot operation [37]. After some optimization studies, we have found that **13** could indeed be prepared very efficiently from **11** as follows: the treatment of **11** for 1 hour with sodium methoxide in refluxing methanol provided the transient shifted ester enolate, which, after removal of methanol by azeotropic distillation with toluene (Dean–Stark apparatus filled with 3 Å molecular sieves), was allowed to react with allyl bromide to furnish stereoselectively the desired 1,3-*cis* diallylated product **13** as the only detectable isomer. The method has proven very general and was applied successfully to the preparation of a number of analogs of **13**, which were converted to bicyclo[4.2.1]nonane compounds following a ring-closing olefin metathesis reaction. In this reaction, no intermediate crude product is isolated and thus, it is an MBFT, but unfortunately, a solvent switch from methanol to toluene has to be effected at one stage, and consequently the reaction should be best described as a *one-pot reaction* according to the definition proposed herein. The transformation **11** → **13** is, however, better performed under these one-pot conditions than the consensual two-step 1,3-shift/alkylation sequence, because of higher eco-compatibility.

In the continuation of our group's interest in the syntheses and applications of bicyclo[3.2.1]octane frameworks [38], we have recently developed a *homo domino MCR* for the stereoselective synthesis of a variety of seven-membered rings from simple substrates. The reaction is initiated by a base-promoted domino Michael-aldol sequence in methanol between an α-activated cyclopentanone **14** and an α, β-unsaturated aldehyde **15** to give the intermediate bicyclo[3.2.1]octanone **16**. Under the reaction conditions, the intermediate bicyclic product **16** can undergo an irreversible retro-Dieckmann fragmentation of the one-carbon bridge with methanol as a third component to give the corresponding seven-membered ring. The overall domino three-component reaction was named the MARDi cascade (an acronym for the Michael/aldol/retro-Dieckmann sequence) [39]. The excellent diastereoselectivity observed for compounds of type **17** was attributed to the totally reversible formation of the intermediate bicyclo[3.2.1]octane and a highly selective final fragmentation (three elemental steps in the domino process). The equally excellent diastereoselectivity obtained with products of type **18** is believed to originate from very different reasons: after the three-step domino sequence of the MARDi cascade, the intermediate seven-membered ring product undergoes an additional intramolecular lactonization and a final elimination step, thus affording the products of type **18** following a five-step domino process. In the sulfonyl series, the MARDi cascade is less diastereoselective, but of considerable importance; the relative configurations of the asymmetric carbon atoms on the northern part of the products is reversed (compare **19a** and **17**) [40]. This has been rationalized by the superior electron-withdrawing character and sterically demanding environment in phenysulfones than in primary esters, which induce a faster kinetic of the retro-Dieckmann fragmentation and a thermodynamically controlled final protonation. It should be noted that the reaction has also been applied successfully to the preparation of a number of heterocycles [41]. Overall, the MARDi cascade (up to five steps in the domino process) allows the regio- and stereocontrolled access to a variety of functionalized and substituted seven-membered rings. This MCR is a condensation and thus, no by-product except water, is formed when dehydration is observed. The substitution array can be diastereoselectively modulated by appropriate choice of the reaction partners, and the reaction allows the control of up to five newly created stereocenters and a complete chiral induction in the case of an optically active ketone precursor in a single operation. The eco-compatibility of the domino MCR is excellent, and only an enantioselective catalytic version remains desirable (Scheme 9.6).

In an early tentative to apply the MARDi cascade to the total synthesis of some natural products, we have prepared the tricyclic compound **22** exhibiting the ring system of the bioactive guaianolide sesquiterpenes (Scheme 9.7) [42]. A *domino multicomponent* MARDi cascade between methyl 2-oxo-cyclopentane carboxylate and cyclopentenecarboxaldehyde afforded the bicyclic products **20** in 49% yield as a

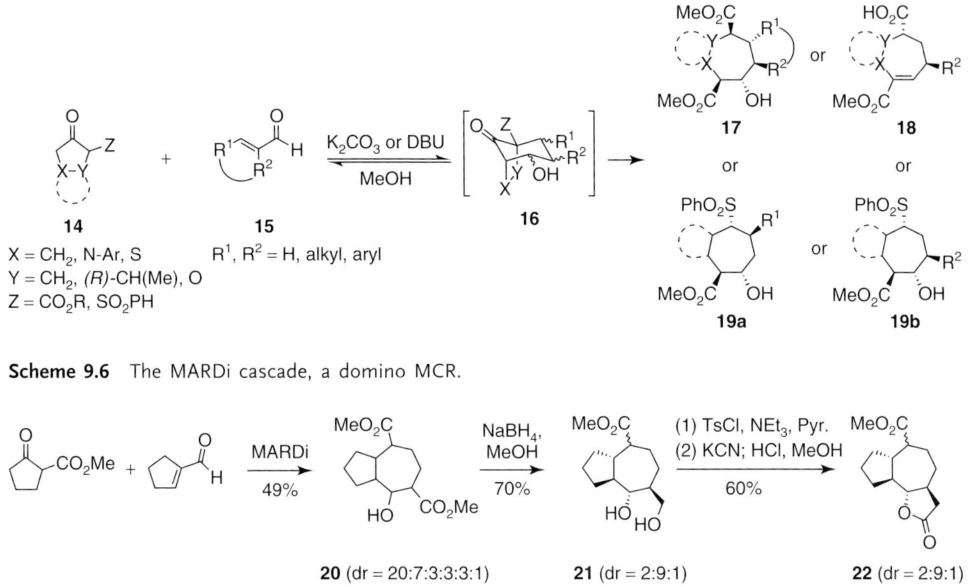

Scheme 9.6 The MARDi cascade, a domino MCR.

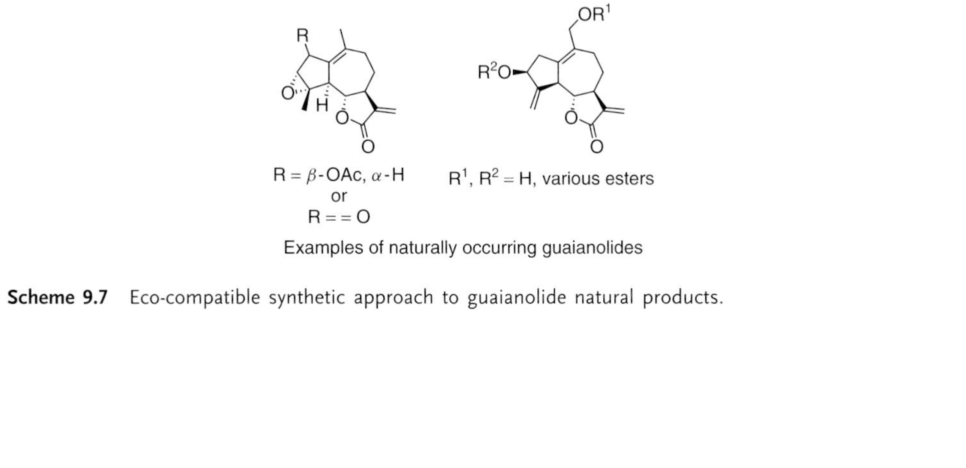

Scheme 9.7 Eco-compatible synthetic approach to guaianolide natural products.

mixture of six diastereomers (dr = 20 : 7 : 3 : 3 : 3 : 1). Because of the existence of an intramolecular activating hydrogen bond, only occurring in the two most abundant isomers of **20**, they could be selectively reduced with sodium borohydride to afford the diol **21** as only two diastereomers (dr = 2.9 : 1). The installation of the γ-lactone from the diol **21** by homologation at the primary alcohol position was realized in a straightforward manner by replacement of the corresponding tosylate with a cyano group, followed by hydrolysis and spontaneous cyclization in a *consecutive reaction*. Overall, compound **22**, which is closely related to naturally occurring guaianolides, has been obtained under eco-compatible conditions in four chemical operations from widely available substrates, involving a *domino MCR* and a *consecutive reaction*.

9.4
Conclusion

Synthetic chemistry offers unique opportunities for the exploration of the "molecular world," and this science has witnessed an explosive development over the past century, sometimes referred to as a *quiet revolution*. However, the next major evolution of chemical synthesis, which would nicely combine with the now required criteria of *eco-compatibility*, might be the control of MBFTs. On the basis of the previously introduced terminologies "*consecutive reaction*," "*domino reaction*," and "*MCR*," and a proposed restricted definition of "*one-pot reaction*," we have established a categorization of the nonconcerted MBFTs yet identified. Among these, the most desirable are enantioselective catalytic *domino MCRs* and *consecutive MCRs*; and, providing modifications of the reaction conditions in the latter are limited and operationally simple. Although the progresses in the development of MBFTs have been considerable in the past two decades, we are still very far from the efficiency of Nature. As illustrated above, through a brief account of our contributions, new MBFTs can be conceived and then put in practice only when a perfect control of the selectivity and a precise understanding of the mechanism and kinetics of each individual step is secured. The establishment and acceptance by the scientific community of adequate terminologies to describe emerging disciplines are essential for their development. The clarification of the terminologies discussed herein is expected to alter the perception of MBFTs by providing to this area of research a strong identity, and to accelerate its growth by facilitating analysis and categorization.

References

1. (a) Nicolaou, K.C. and Snyder, S.A. (2003) *Classics in Total Synthesis II*, Wiley-VCH Verlag GmbH, Weinheim; (b) Hudlicky, T. and Reed, J.W. (2007) *The Way of Synthesis*, Wiley-VCH Verlag GmbH, Weinheim.

2. Goren-Inbar, N., Alperson, N., Kislev, M.E., Simchoni, O., Melamed, Y., Ben-Nun, A., and Werker, E. (2004) *Science*, **304**, 725–727.

3. (a) Rocke, A.J. (1993) *The Quiet Revolution: Hermann Kolbe and the Science*

of *Organic Chemistry*, University of California Press, Berkeley; (b) Wilson, R.M. and Danishefsky, S.J. (2006) *J. Org. Chem.*, **71**, 8329–8351.
4. Wöhler, F. (1828) *Poggendorf's Ann.*, **12**, 253–256.
5. Hans-Georg, E. (2003) *An Introduction to Plastics*, Wiley-VCH Verlag GmbH, Weinheim.
6. (a) Wender, P.A. and Miller, B.L. (1993) in *Organic Synthesis: Theory and Applications*, vol. **2** (ed. T. Hudlicky), JAI Press, Greenwich, pp. 27–66; (b) Wender, P.A., Handy, S.T., and Wright, D.L. (1997) *Chem. Ind.*, 765–769.
7. Trost, B.M. (1991) *Science*, **254**, 1471–1477.
8. Wender, P.A., Baryza, J.L., Brenner, S.E., Clarke, M.O., Gamber, G.G., Horan, J.C., Jessop, T.C., Kan, C., Pattabiraman, K., and Williams, T.J. (2003) *Pure Appl. Chem.*, **75**, 143–155.
9. Sheldon, R.A., Arends, I., and Hanefeld, U. (2007) *Green Chemistry and Catalysis*, Wiley-VCH Verlag GmbH, Weinheim.
10. Sheldon, R.A. (1994) *Chemtech*, **24**, 38–47.
11. Tietze, L.F. and Beifuss, U. (1993) *Angew. Chem. Int. Ed.*, **105**, 137–170.
12. (a) Tietze, L.F., Brasche, G., and Gericke, K.M. (2006) *Domino Reactions in Organic Synthesis*, Wiley-VCH Verlag GmbH, Weinheim; (b) Tietze, L.F. (1996) *Chem. Rev.*, **96**, 115–136; See also: (c) Denmark, S.E. and Thorarensen, A. (1996) *Chem. Rev.*, **96**, 137–165.
13. Nicolaou, K.C., Edmonds, D.J., and Bulger, P.G. (2006) *Angew. Chem. Int. Ed.*, **45**, 7134–7186.
14. Zhu, J. and Bienaymé, H. (eds) (2005) *Multicomponent Reactions*, Wiley-VCH Verlag GmbH, Weinheim.
15. Armstrong, R.W., Combs, A.P., Tempest, P.A., Brown, S.D., and Keating, T.A. (1996) *Acc. Chem. Res.*, **29**, 123–131.
16. (a) Bienaymé, H., Hulme, C., Oddon, G., and Schmitt, P. (2000) *Chem. Eur. J.*, **6**, 3321–3329; (b) Isambert, N. and Lavilla, R. (2008) *Chem. Eur. J.*, **14**, 8444–8454; (c) Dömling, A. (2006) *Chem. Rev.*, **106**, 17–89; (d) Ramon, D.J. and Yus, M. (2005) *Angew. Chem. Int. Ed.*, **44**, 1602–1634; (e) Liéby-Müller, F., Simon, C., Constantieux, T., and Rodriguez, J. (2006) *QSAR Comb. Sci.*, **5-6**, 432–438.
17. Kirmse, W. (2002) *Eur. J. Org. Chem.*, 2193–2256.
18. Wentrup, C., Heilmayer, W., and Kollenz, G. (1994) *Synthesis*, 1219–1248.
19. For silver-catalyzed rearrangements, see: (a) Sudrik, S., Sharma, J., Chavan, V.B., Chaki, N.K., Sonawane, H.R., and Vijayamohanan, K.P. (2006) *Org. Lett.*, **8**, 1089–1092, and references therein; For rhodium-catalyzed rearrangements, see: (b) Lee, Y.R., Suk, J.Y., and Kim, B.S. (1999) *Tetrahedron Lett.*, **40**, 8219–8221.
20. For edited volumes on MAOS, see: (a) Loupy, A. (ed.) (2006) *Microwave in Organic Synthesis*, 2nd edn, Wiley-VCH Verlag GmbH, Weinheim; (b) Larhed, M. and Olafsson, K. (eds) (2006) *Topics in Current Chemistry*, Microwave Methods in Organic Synthesis, Vol. **266**, Springer, Berlin/Heidelberg; For representative reviews on MAOS and its applications, see: (c) Kappe, C.O. (2008) *Chem. Soc. Rev.*, **37**, 1127–1139; (d) Coquerel, Y. and Rodriguez, J. (2008) *Eur. J. Org. Chem.*, 1125–1132; (e) Razzaq, T. and Kappe, C.O. (2008) *ChemSusChem*, **1**, 123–132; (f) de la Hoz, A., Díaz-Ortiz, Á., and Moreno, A. (2005) *Chem. Soc. Rev.*, **34**, 164–178; (g) Kappe, C.O. (2004) *Angew. Chem. Int. Ed.*, **43**, 6250–6284; (h) Hayes, B.L. (2004) *Aldrichim. Acta*, **37** (2), 66–77; (i) Nüchter, M., Ondruschka, B., Bonrath, W., and Gum, A. (2004) *Green Chem.*, **6**, 128–141.
21. Herrero, M.A., Kremsner, J.M., and Kappe, C.O. (2008) *J. Org. Chem.*, **73**, 36–47.
22. (a) Presset, M., Coquerel, Y., and Rodriguez, J. (2009) *J. Org. Chem.*, **74**, 415–418; For pioneering studies on the microwave-assisted Wolff rearrangement, see: (b) Sudrik, S.G., Chavan, S.P., Chandrakumar, K.R.S., Pal, S., Date, S.K., Chavan, S.P., and Sonawane, H.R. (2002) *J. Org. Chem.*, **67**, 1574–1579; (c) O'Sullivan, O.C.M., Collins, S.G., and Maguire, A.R. (2008) *Synlett*, 659–662.

23. (a) Otera, J., Yano, T., Kawabata, A., and Nozaki, H. (1986) *Tetrahedron Lett.*, **27**, 2383–2386; See also: (b) Benetti, S., Romagnoli, R., De Risi, C., Spalluto, G., and Zanirato, V. (1995) *Chem. Rev.*, **95**, 1065–1114; (c) Otera, J. (1993) *Chem. Rev.*, **93**, 1449–1470.
24. Khand, I.U., Knox, G.R., Pauson, P.L., Watts, W.E., and Foreman, M.I. (1973) *J. Chem. Soc., Perkin Trans. 1*, 977–981.
25. Shibata, T. (2006) *Adv. Synth. Catal.*, **348**, 2328–2336.
26. Coquerel, Y. and Déprés, J.-P. (2006) Chemistry of seven-membered ring ligands in transition metal carbonyl complexes, in *Frontiers in Organometallic Chemistry*, Chapter 6 (ed. M.A.Cato), Nova Science Publishers, Hauppauge, pp. 123–154.
27. Pearson, A.J., Kole, S.L., and Ray, T. (1984) *J. Am. Chem. Soc.*, **106**, 6060–6074.
28. Coquerel, Y., Déprés, J.-P., Greene, A.E., Cividino, P., and Court, J. (2001) *Synth. Commun.*, **31**, 1291–1300.
29. Coquerel, Y. and Déprés, J.-P. (2002) *Chem. Commun.*, 658–659.
30. Coquerel, Y., Brémond, P., and Rodriguez, J. (2007) *J. Organomet. Chem.*, **692**, 4805–4808.
31. Coquerel, Y. and Rodriguez, J. (2008) *Arkivoc*, (xi), 227–237.
32. For preliminary results, see: Coquerel, Y. and Rodriguez, J. (2006) *Tetrahedron Lett.*, **47**, 8503–8506.
33. Müller, J.F.K. (2000) *Eur. J. Inorg. Chem.*, 789–799.
34. Michaut, A., Miranda García, S., Menéndez, J.C., Coquerel, Y., and Rodriguez, J. (2008) *Eur. J. Org. Chem.*, 4988–4998.
35. Habi, A. and Gravel, D. (1994) *Tetrahedron Lett.*, **35**, 4315–4318.
36. Sisido, K., Utimoto, K., and Isida, T. (1964) *J. Org. Chem.*, **29**, 2781–2782.
37. (a) Dauben, W.G. and Walker, D.M. (1982) *Tetrahedron Lett.*, **23**, 711–714; (b) Wright, J., Drtina, G.J., Roberts, R.A., and Paquette, L.A. (1988) *J. Am. Chem. Soc.*, **110**, 5806–5817.
38. (a) Filippini, M.-H., Faure, R., and Rodriguez, J. (1995) *J. Org. Chem.*, **60**, 6872–6882; (b) Filippini, M.-H. and Rodriguez, J. (1999) *Chem. Rev.*, **99**, 27–76.
39. Coquerel, Y., Filippini, M.-H., Bensa, D., and Rodriguez, J. (2008) *Chem. Eur. J.*, **14**, 3078–3092.
40. Coquerel, Y., Bensa, D., Moret, V., and Rodriguez, J. (2006) *Synlett*, 2751–2754.
41. Coquerel, Y., Bensa, D., Doutheau, A., and Rodriguez, J. (2006) *Org. Lett.*, **8**, 4819–4822.
42. Reboul, I., Boddaert, T., Coquerel, Y., and Rodriguez, J. (2008) *Eur. J. Org. Chem.*, 5379–5382.

10
Modeling of Indirect Phototransformation Reactions in Surface Waters

Davide Vione, Radharani Das, Francesca Rubertelli, Valter Maurino, and Claudio Minero

10.1
Introduction

The persistence of dissolved organic compounds in surface water bodies, including both natural organic molecules and man-made xenobiotics and pollutants, is influenced by their transformation kinetics due to abiotic and biotic [1].

Many organic pollutants such as polycyclic aromatic hydrocarbons, pesticides, pharmaceuticals, and their transformation intermediates are refractory to biological degradation. In such cases the abiotic transformation processes can represent major removal pathways from surface waters. Within the abiotic transformation reactions of xenobiotics, those induced by sunlight are receiving increasing attention nowadays because of their potential importance in the removal of the parent molecules and of the possible production of harmful secondary pollutants [2, 3]. Moreover, when considering the organic compounds of natural origin, biodegradation would often lead to biorefractory intermediates that undergo further biological processes with difficulty. Degradation by aquatic microorganisms would, for instance, yield biorefractory polysaccharides and peptides, the persistence of which in surface waters would be controlled by abiotic transformation processes (see below) [4]. The loading of dissolved organic matter (DOM) in surface waters would also include the contribution of organic compounds produced in soil [5], which reach the water bodies by leaching and surface runoff or through the underlying aquifers. This so-called pedogenic or allochthonous fraction of DOM is rich in humic and fulvic acids, which are able to absorb sunlight and take part in the abiotic transformation processes. The light-absorbing fraction of DOM is known as *colored dissolved organic matter* (CDOM) [6]. The CDOM can be transformed considerably upon absorption of sunlight, which decreases its absorbance. The resulting photobleaching is more marked for the allochthonous CDOM freshly released into the water body than for autochthonous (aquagenic) or aged allochthonous CDOM. Probably the aged allochthonous CDOM contains more photostable compounds as the photolabile ones have already undergone degradation [7]. The biorefractory organic compounds can

Ideas in Chemistry and Molecular Sciences: Advances in Synthetic Chemistry. Edited by Bruno Pignataro
Copyright © 2010 WILEY-VCH Verlag GmbH & Co. KGaA, Weinheim
ISBN: 978-3-527-32539-9

become bioavailable after some degree of abiotic processing, with the consequence that the combination of abiotic and biotic degradation can lead to the complete mineralization of organic matter [8].

In some cases, the removal of a compound from surface waters does not involve chemical transformation, but rather the transfer from the dissolved phase into suspended solids or sediment, or the volatilization from the surface water layer [9]. In this way, the molecule is simply shifted to a different phase, where it can be involved in transformation processes that can be very different from those in water [10]. Sometimes, the original molecule or its transformation intermediates can return back to the water body, as a consequence of dissolution from sediment or of wet and dry deposition from the atmosphere [11].

The abiotic transformation processes in surface waters include hydrolysis, oxidation mediated by dissolved species or by metal oxides such as Fe(III) and Mn(III,IV) (hydro)oxides, and light-induced reactions [12]. Hydrolysis will often produce bond cleavage, which in many cases results in the loss of a lateral functional chain. Hydrolytic reactions are usually acid- or base catalyzed, but at the pH values around neutrality that is typical of surface waters the effects of catalysis may be limited [13]. Among the oxidizing species dissolved in surface waters there are a number of reactive transients produced upon sunlight irradiation of photoactive compounds [14]. Under irradiation or even in the dark, Fe and Mn (hydr)oxides can be involved in charge-transfer processes with compounds that are able to form surface complexes. Salicylic and oxalic acid, as well as similar compounds and their derivatives, are examples of such molecules. Charge transfer would often result in the oxidation of organic molecules and the reduction of oxides, with formation of water-soluble Fe^{2+} and Mn^{2+}. Oxide dissolution is usually observed as a consequence of these processes [15, 16].

The photochemical processes that involve dissolved compounds in surface waters include the direct photolysis upon absorption of sunlight, the transformation sensitized by the triplet states of photoexcited CDOM (^{3}CDOM*, the precursors of which are often the humic and fulvic components of CDOM), and the reaction with photogenerated reactive transients [17]. The latter can be involved in the oxidation of organic substrates: it is the case of $^\bullet OH$, $CO_3^{-\bullet}$, and $^1O_2$. Other radical transients such as $^\bullet NO_2$, $Cl_2^{-\bullet}$, and $Br_2^{-\bullet}$ could be involved in the generation of secondary pollutants because they are nitrating and halogenating species [18].

It is noteworthy that the direct photolysis alone, although very important in defining the lifetime of many photolabile compounds in surface waters, will seldom lead to their complete removal. The process, which is sometimes very efficient, would rather yield a number of transformation intermediates. In some cases, they are even more harmful than the parent molecule [19, 20], and the complete detoxification of the xenobiotic species, for example, by mineralization, requires the reaction with oxidizing transients such as $^\bullet OH$, and/or microbial processing. In many cases, the mineralization is much slower than the primary step of phototransformation of the parent molecule [21].

An important issue concerning the photochemical reactions is the penetration of sunlight inside the water body. Absorption by CDOM plays a major role in decreasing the intensity of solar radiation in the water column; such an effect reduces the significance of the photochemical reactions when the water column depth increases, but would also protect the aquatic life from exposure to harmful UV radiation [22].

This chapter presents an overview of the main indirect photolysis processes in surface waters, and describes a modeling approach for the quantitative assessment of the weight of the transformation reactions induced by $^\bullet OH$, $CO_3^{-\bullet}$, and $^3CDOM^*$.

10.2
Indirect Photolysis Processes in Surface Waters

Indirect photochemical reactions involve dissolved molecules (substrates) that do not necessarily need to absorb sunlight on their own. These processes start as a result of the absorption of radiation by photoactive compounds, usually named *photosensitizers*, and imply the reaction of the substrates with the transient species that are produced by the light-excited sensitizers [23]. Important sensitizers in surface waters are CDOM, nitrate, and nitrite [24, 25]. We have recently shown that inorganic colloids such as Fe(III) oxides can play an important role under definite circumstances, for example, upon interaction with inorganic anions such as chloride [26], nitrite [27, 28], and carbonate [29]. Dissolved Fe(III) is also important because it can form complexes with many organic ligands, and the irradiation of the complexes Fe^{3+}-L can result in a ligand-to-metal electron transfer. The process generates Fe^{2+} and $L^{+\bullet}$, which is the first step toward the transformation of the ligand [30, 31]. In such cases, the photochemical reaction that initiates the process is a direct photolysis of the complex Fe^{3+}-L.

Generally speaking, the main reactive transients that are photochemically generated in surface waters are the radicals $^\bullet OH$ (hydroxyl) and $CO_3^{-\bullet}$ (carbonate), and the excited triplet states of DOM ($^3CDOM^*$) [32, 33]. Their main sources are CDOM, nitrate, and nitrite under irradiation; the formation of $CO_3^{-\bullet}$ also requires the oxidation of carbonate or bicarbonate [34, 35]. We have also shown that under definite circumstances the transients $^\bullet NO_2$, $Cl_2^{-\bullet}$, and $Br_2^{-\bullet}$ can play a significant role in the transformation reactions [36–38]. These species can be formed upon oxidation of nitrite, chloride, and bromide [39–41], and the formation of secondary pollutants such as aromatic nitro-, chloro-, and bromoderivatives is an important consequence of the cited processes [42, 43].

In some cases, knowledge concerning the generation and the reactivity of the transient species is just qualitative or semiquantitative or it is demonstrated in laboratory systems and the transfer of the results to the real environment is difficult. The latter problem is particularly true of the reactions induced by Fe(III) oxide colloids due to the insufficient knowledge about the speciation of Fe in the environmental waters [44]. However, in the case of the reactions induced by $^\bullet OH$, $CO_3^{-\bullet}$, and (to a lesser extent) by $^3CDOM^*$, there is sufficient data in the literature to attempt the modeling of the relevant reactions in surface waters. The formation and

10.2.1
Reactions Induced by •OH

The hydroxyl radical is generated in surface waters upon irradiation of CDOM, nitrate, and nitrite. The following light-induced reactions are involved (ISC = intersystem crossing) [32, 33, 45]:

$$CDOM + h\nu \longrightarrow {}^1CDOM^* - (ISC) \longrightarrow {}^3CDOM^* \qquad (10.1)$$

$$^3CDOM^* + H_2O \longrightarrow CDOM - H^\bullet + {}^\bullet OH \qquad (10.2)$$

$$NO_3^- + h\nu + H^+ \longrightarrow {}^\bullet NO_2 + {}^\bullet OH \qquad (10.3)$$

$$NO_2^- + h\nu + H^+ \longrightarrow {}^\bullet NO + {}^\bullet OH \qquad (10.4)$$

Figure 10.1 shows the absorption spectra of nitrate and nitrite, which have environmentally significant maxima in the UVB (305 nm) and UVA (360 nm), respectively. In contrast, the absorption of radiation by CDOM shows an exponential decay from the UV through the visible, but significant absorption is still operational above 400 and till 500 nm [46]. The spectra of nitrate and nitrite are compared in Figure 10.1 with the sunlight spectrum that reaches the ground, which is referred to midday on 15 June at 50°N latitude [47].

An important issue is that the penetration depth of sunlight in the column of surface waters shows a considerable wavelength trend. Generally speaking, the absorption of radiation is higher for lower wavelengths, and the visible radiation

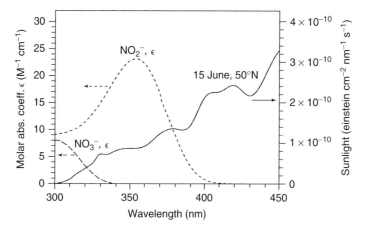

Figure 10.1 Molar absorption coefficients ε of nitrate and nitrite. Spectral photon flux of sunlight (15 June, 50°N, midday [47]).

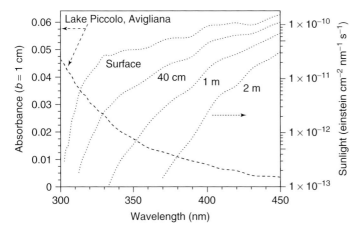

Figure 10.2 Absorption spectrum $A_1(\lambda)$ of the surface lake water from Lake Avigliana Piccolo (optical path length $b = 1$ cm, thus $A_1(\lambda)$ has units expressed per centimeter). Spectral photon flux of sunlight (15 June, 50°N, midday) as a function of the water column depth [50].

is therefore able to penetrate more deeply than the UV. In the UV range, it is the UVA that penetrates more deeply in the water column [22, 48, 49]. Figure 10.2 shows the absorption spectrum $A_1(\lambda)$ (optical path length $b = 1$ cm) of a lake water sample (Lake Avigliana Piccolo, NW Italy), together with the intensity of sunlight reaching different depths in the water column of the lake. It is clear from the figure that the absorption of radiation is higher at shorter wavelengths, and that the long-wavelength radiation penetrates more deeply into the water body [50].

The wavelength trend of sunlight absorption has considerable effects on the photochemical activity of the different sources of •OH radicals. We have shown that in the surface water layer, sufficiently thin to allow the absorption processes to be neglected, the average relative role of CDOM, nitrate, and nitrite as photochemical sources of •OH would be in the order CDOM > NO_2^- > NO_3^- [51–53]. Note that this is the average behavior of an elevated number of samples: the contribution of the different sensitizers can be very different in particular cases [54]. The attenuation of sunlight in the water column is expected to have a different impact on the photochemical •OH sources: nitrate has an absorption maximum in the UVB and its photochemistry would quickly decrease with depth. The photochemistry of nitrite that absorbs mostly in the UVA would undergo a somewhat lesser inhibition [52, 53], and CDOM that also absorbs in the visible is expected to be photochemically active in a larger part of the water column. Accordingly the reactivity order found in the water surface layer (CDOM > NO_2^- > NO_3^-) would be retained, with more marked relative differences, when the whole water column is considered.

Note that the Fe species might also give a contribution to the generation rate of •OH. The Fenton and photo-Fenton reactions [55] can be important in the photochemical redox cycling of Fe [56], and would also yield •OH with a higher efficiency under acidic conditions [57]. They can be important in surface water systems that are particularly rich in Fe [58]:

$$Fe^{3+} - L + h\nu \longrightarrow Fe^{2+} + L^{+\bullet} \qquad (10.5)$$

$$Fe^{2+} + H_2O_2 \longrightarrow Fe^3 + OH^- + {}^{\bullet}OH \qquad (10.6)$$

The direct photolysis of Fe(III) hydroxocomplexes to yield $^{\bullet}$OH is very efficient in acidic solution (Reaction 10.7), but the rate of $^{\bullet}$OH generation quickly decreases with increasing pH [59, 60]. Under neutral conditions, the yield of $^{\bullet}$OH upon irradiation of the Fe(III) (hydr)oxide colloids is very low [26, 27]:

$$Fe^{3+} - OH + h\nu \longrightarrow Fe^{2+} + {}^{\bullet}OH \qquad (10.7)$$

Overall, the role of the Fe species in the generation of $^{\bullet}$OH through the direct photolysis or the Fenton processes is likely to decrease with increasing pH, and could be limited in neutral to basic waters where the concentration of Fe is also low. A major limit to a proper understanding of the significance of the reported reactions is the insufficient knowledge about the speciation of Fe in surface waters [44]. However, Fe photochemistry could account for the elevated [$^{\bullet}$OH] in waters contaminated by acidic mine drainage [61]. Indeed, such waters are at the same time Fe-rich and acidic.

After photochemical formation, the hydroxyl radicals undergo very fast reaction: a steady-state condition for [$^{\bullet}$OH] is therefore quickly reached. The main $^{\bullet}$OH scavengers in freshwater are DOM, carbonate and bicarbonate, and nitrite to a lesser extent [24, 25, 34, 35, 62]. By studying lake water from Piedmont (NW Italy) we have found that, on average, DOM is able to scavenge over 90% of the photogenerated hydroxyl radicals [51, 53]. Estimates for the reaction rate constant between $^{\bullet}$OH and DOM vary between 2×10^4 and 5×10^4 l (mg C)$^{-1}$ s^{-1} [24, 51], and the fast scavenging often limits the steady-state [$^{\bullet}$OH] at levels around 10^{-16} M in the surface water layer under sunlight [24, 51, 63, 64].

The second-order rate constants between $^{\bullet}$OH and the dissolved organic compounds are often in the 10^9–10^{10} M^{-1} s^{-1} range [62]. The hydroxyl radical is therefore a very reactive but also little concentrated transient in surface waters. The combination of the two issues suggests that $^{\bullet}$OH can play a significant role in the phototransformation processes; in particular, in the water bodies where [$^{\bullet}$OH] is higher than average. It is required a high formation rate of $^{\bullet}$OH, or a low concentration of the scavengers [24, 51, 54]. Note that [$^{\bullet}$OH] is also dependent on the intensity of sunlight, and is therefore very variable in the diurnal as well as the seasonal cycle. Indeed, [$^{\bullet}$OH] is negligible during the night [65]. The diurnal and seasonal variability of the sunlight intensity has to be considered if one wants to carry out a proper modeling of the photochemistry of surface waters (see below).

Among the reactions that scavenge $^{\bullet}$OH, those involving DOM can give a contribution to the generation of organic peroxyl radicals (ROO$^{\bullet}$). These species take part somehow in the light-induced transformation processes in surface waters, but many important details of the relevant reactions are presently unknown [32, 66]. The reaction between $^{\bullet}$OH and carbonate or bicarbonate yields $CO_3^{-\bullet}$, which can further react with dissolved molecules and induce additional degradation processes (see the next section). Finally, the reaction between $^\bullet$OH and nitrite yields the nitrating agent $^\bullet NO_2$ [67, 68]. Although the fraction of $^\bullet$OH that reacts with nitrite is very small (around 1% of the total, [39, 52]), we have shown that the $^\bullet NO_2$ that is produced can be involved to a significant extent in the formation of aromatic nitroderivatives [39, 69–71]. Nitroaromatic compounds are often environmentally more persistent than the parent molecules [28, 72, 73]. They can be less toxic than the starting compounds, but may be characterized by a significant genotoxicity and/or mutagenicity [74–76].

10.2.2
Reactions Induced by $CO_3^{-\bullet}$

The main process that yields the carbonate radical in surface waters is the oxidation of the carbonate and bicarbonate anions by $^\bullet$OH (Reactions 10.8 and 10.9):

$$CO_3^{2-} + {}^\bullet OH \longrightarrow CO_3^{-\bullet} + OH^- \quad [k_8 = 3.9 \times 10^8 \text{ M}^{-1} \text{ s}^{-1}] \quad (10.8)$$

$$HCO_3^- + {}^\bullet OH \longrightarrow CO_3^{-\bullet} + H_2O \quad [k_9 = 3.9 \times 10^6 \text{ M}^{-1} \text{ s}^{-1}] \quad (10.9)$$

From Reactions 10.8 and 10.9, one gets that carbonate is more reactive toward $^\bullet$OH [62], but bicarbonate usually has a higher concentration in surface waters. The relative importance of carbonate versus bicarbonate as sinks of $^\bullet$OH is expected to increase with increasing pH. Results of comparison of different surface water samples (lake water from Piedmont, NW Italy, [51]) are shown in Figure 10.3. From the figure it is clear that carbonate is becoming a more important source of $CO_3^{-\bullet}$ above pH 8.5.

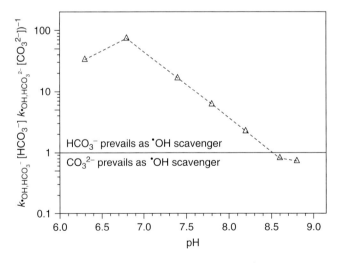

Figure 10.3 Relative contributions of $HCO_3^{-\bullet}$ and CO_3^{2-} to the scavenging of $^\bullet$OH in lake water, as a function of pH [51].

Another possible source of $CO_3^{-\bullet}$ is the oxidation of CO_3^{2-} by the photochemically excited triplet states of CDOM ($^3CDOM^*$):

$$CO_3^{2-} + {}^3CDOM^* \longrightarrow CO_3^{-\bullet} + CDOM^{-\bullet} \tag{10.10}$$

Reaction 10.10 is likely to be less important than Reactions 10.8 and 10.9 to produce the carbonate radical, and its relative weight should be around 10% or lower [77]. Considering that Reactions 10.8 and 10.9 prevail over Reaction 10.10 as sources of $CO_3^{-\bullet}$, and that the reaction between $^\bullet OH$ and carbonate or bicarbonate accounts for 10% or less of the total scavenging of $^\bullet OH$ (90% or more of $^\bullet OH$ is scavenged by DOM; [51]), it can be concluded that the generation rate of $CO_3^{-\bullet}$ would be significantly lower than that of $^\bullet OH$ in surface waters.

A further issue concerning $CO_3^{-\bullet}$ is that it is less reactive than $^\bullet OH$ toward most organic compounds [62, 77–79]. However, the combination of lower formation rate and lower reactivity does not necessarily imply that the role of $CO_3^{-\bullet}$ in surface water photochemistry is unimportant. Indeed, DOM is the main natural scavenger of both $^\bullet OH$ and $CO_3^{-\bullet}$ [24, 77]. The rate constant between $^\bullet OH$ and DOM is in the 10^4 l (mg C)$^{-1}$ s^{-1} range [24, 51], and estimates for the corresponding rate constant between $CO_3^{-\bullet}$ and DOM vary between 40 and 280 ± 90 l (mg C)$^{-1}$ s^{-1} [77, 80]. The difference of two to three orders of magnitude in the rate constants of $^\bullet OH$ and $CO_3^{-\bullet}$ with DOM can more than compensate for the lower generation rate of $CO_3^{-\bullet}$ compared to $^\bullet OH$. Moreover, unlike $^\bullet OH$, $CO_3^{-\bullet}$ is not scavenged by additional water components such as inorganic carbon [81].

A major consequence of the cited processes is that $CO_3^{-\bullet}$ reaches steady-state concentration values around 10^{-14} M in surface waters under summertime irradiation conditions [77], compared with 10^{-16}–10^{-15} M for [$^\bullet OH$]. The higher steady-state concentration of $CO_3^{-\bullet}$ could or could not compensate for its lower reactivity compared to $^\bullet OH$, depending on the substrate under consideration. The carbonate radical is unlikely to contribute much to the transformation of hard-to-oxidize molecules. In contrast, it could be involved to a very significant extent in the degradation of electron-rich phenols, aromatic amines, and sulfur-containing compounds [77, 81–83].

The importance of the carbonate radical in the transformation reactions would also depend on the ecosystem variables. The formation of $CO_3^{-\bullet}$ requires the reaction between $^\bullet OH$ and carbonate or bicarbonate, while DOM scavenges both $^\bullet OH$ (therefore inhibiting the formation of $CO_3^{-\bullet}$) and $CO_3^{-\bullet}$ after it is formed (inhibiting the reactions between $CO_3^{-\bullet}$ and other dissolved molecules). Accordingly, the carbonate radical is expected to play a more important role in the surface waters with a high ratio of inorganic versus organic carbon. Under the latter circumstances, a higher fraction of $^\bullet OH$ would react with carbonate and bicarbonate to produce $CO_3^{-\bullet}$, and there would be comparatively less DOM to scavenge $CO_3^{-\bullet}$ after its formation [84].

10.2.3
Reactions Induced by $^3CDOM^*$

The irradiation of the photoactive components of CDOM causes electron transitions from the fundamental state to the first excited singlet state ($^1CDOM^*$). $^1CDOM^*$ can decay back to the fundamental state by thermal deactivation or fluorescence emission, undergo some transformation reactions, or an ISC to the first excited triplet state, $^3CDOM^*$ (Reaction 10.1 and Figure 10.4) [32].

The excited triplet states of molecules with photosensitizing properties, such as quinones and aromatic carbonyls, are able to react with dissolved substrates via electron or hydrogen abstraction, or simply by energy transfer [85, 86]. All these reactions can induce the degradation of the substrates. The relevant processes that can involve $^3CDOM^*$ and an organic substrate HS are depicted below [87]:

$$^3CDOM^* + HS \longrightarrow CDOM-H^{\bullet} + S^{\bullet} \qquad (10.11)$$

$$^3CDOM^* + HS \longrightarrow CDOM^{-\bullet} + S^{+\bullet} \qquad (10.12)$$

$$^3CDOM^* + HS \longrightarrow CDOM + HS^* \qquad (10.13)$$

The reactions between $^3CDOM^*$ and the dissolved substrates are in competition with the thermal and radiative deactivation of $^3CDOM^*$ to the fundamental state, and with the reaction between $^3CDOM^*$ and O_2 to yield singlet oxygen, $^1O_2$ [88]. In oxygenated surface waters, $^3CDOM^*$ has an average lifetime of 2 μs or lower, which corresponds to an average deactivation rate constant $k^* = 5 \times 10^5$ s^{-1} or higher [89].

The importance of the processes involving $^3CDOM^*$ is enhanced by the considerable absorption of radiation by CDOM. Indeed, CDOM accounts almost completely for the absorption of sunlight UV by surface waters [6]. In contrast, a limiting factor to the reactivity of $^3CDOM^*$ is its deactivation to the fundamental state, with rate constant k^*. The deactivation is likely to be considerably faster than the

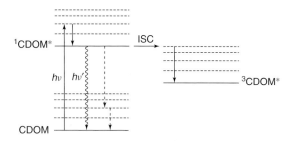

Figure 10.4 Schematic of the processes that follow the absorption of radiation by CDOM. Radiation absorption is labeled as $h\nu$. The continuous lines represent the pathways that lead to $^3CDOM^*$. The dashed lines represent nonradiative deactivation processes. $h\nu\prime$ is referred to fluorescence emission. ISC = intersystem crossing.

reaction between ^{3}CDOM* and the water-dissolved compounds [66, 90], and would therefore be the main sink for ^{3}CDOM* in the vast majority of cases.

The reactions induced by ^{3}CDOM* have been shown to play a substantial role in the photodegradation of phenylurea pesticides in Lake Greifensee (Switzerland, [91]), and are probably the main transformation pathway for electron-rich phenols in surface waters [66].

10.2.4
Other Reactions

CDOM can induce phototransformation processes directly, via the reactions of its excited triplet states (^{3}CDOM*) with dissolved compounds, or indirectly as a consequence of the generation of singlet oxygen (^{1}O$_2$, Reaction 10.14, [88]). In Reaction 10.16, HS represents a dissolved molecule. Additionally the light-induced transformation reactions of CDOM can yield peroxy radicals ROO$^\bullet$, which can show a certain reactivity [32, 66]:

$$^3\text{CDOM}^* + \text{O}_2 \longrightarrow \text{CDOM} + {}^1\text{O}_2 \tag{10.14}$$

$$^1\text{O}_2 \longrightarrow \text{O}_2 \tag{10.15}$$

$$\text{HS} + {}^1\text{O}_2 \longrightarrow \text{Products} \tag{10.16}$$

Singlet oxygen can lose the surplus energy because of collisions with the solvent molecules [23]. Such a deactivation process (Reaction 10.15) has rate constant $k_{15} = 2.5 \times 10^5$ s^{-1} [92], and is in competition with Reaction 10.16 that involves dissolved compounds. There is evidence that Reaction 10.15 prevails to a large extent over Reaction 10.16 as a sink of ^{1}O$_2$ [23]. As a consequence ^{1}O$_2$ cannot be accumulated in solution, and its ability to induce the degradation of organic substrates depends on their reactivity toward ^{1}O$_2$. It is likely that ^{1}O$_2$ plays an important role in inducing the degradation of photolabile amino acids [93], but its importance in the transformation of dissolved pollutants is still highly controversial [91, 94]. Very little is also known regarding the importance of ROO$^\bullet$ in surface water photochemistry. The relevant reactions could simulate those of ^{3}CDOM* [32], but it is generally accepted that ^{3}CDOM* is more important than ROO$^\bullet$ in the phototransformation processes that take place in surface waters [90].

An interesting class of photochemical reactions is initiated by Fe(III) (hydr)oxide colloids. They are semiconductor oxides, and the absorption of visible radiation promotes electrons to the conduction band of the semiconductor, leaving electron vacancies (holes) in the valence band [95]. Conduction band electrons are reducing species and can, for instance, interact with oxygen. In contrast, valence band holes are oxidizing species and can extract electrons from a number of compounds adsorbed onto the semiconductor surface [96, 97]. The general processes that take place on the semiconductor oxides under irradiation are presented in Figure 10.5.

Among the species that can be oxidized at the surface of Fe(III) (hydr)oxides there are organic ligands that can form surface complexes [15] and, as shown by us,

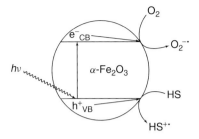

Figure 10.5 Schematic of the processes that follow the absorption of radiation by hematite (α-Fe_2O_3). HS represent a dissolved molecule, $e_{CB}^{-\bullet}$ is an electron of the conduction band, and h_{VB}^{+} is a hole of the valence band.

dissolved anions such as carbonate [29], nitrite [27, 36], chloride [26], and bromide [98]. Oxalate has been shown to undergo thermal and photoinduced oxidation on the surface of hematite, with parallel dissolution of the oxide via the formation of Fe(II) [15]. We have shown that the oxidation of CO_3^{2-} by irradiated hematite yields the oxidizing agent $CO_3^{-\bullet}$ [29], and the oxidation of nitrite yields $^{\bullet}NO_2$ that is a nitrating agent in the aqueous solution [27, 28, 36, 69, 99]. The irradiation of Fe(III) (hydr)oxides in the presence of chloride and bromide results in the generation of the radicals $Cl^{\bullet}$ and $Br^{\bullet}$ [26, 98], which are soon transformed into $Cl_2^{-\bullet}$ and $Br_2^{-\bullet}$ [100, 101] (Reactions 10.17 and 10.18):

$$Cl^{\bullet} + Cl^{-} \rightleftarrows Cl_2^{-\bullet} \tag{10.17}$$

$$Br^{\bullet} + Br^{-} \rightleftarrows Br_2^{-\bullet} \tag{10.18}$$

The radical $Cl_2^{-\bullet}$ is an oxidizing and chlorinating agent, and can, for instance, chlorinate aromatic hydrocarbons [26, 37]. The radical $Br_2^{-\bullet}$ is a less powerful oxidant [42, 78], but a very efficient brominating agent for aromatic compounds [38, 41].

Overall, a major obstacle to the modeling of the photochemical reactions involving Fe(III) (hydr)oxides is the complex and still insufficiently understood speciation of Fe in surface waters [44]. In contrast, there are sufficient data to undertake the modeling of the indirect photolysis reactions initiated by $^{\bullet}OH$, $CO_3^{-\bullet}$, and $^3CDOM^{*}$. A modeling approach that we have recently developed aims at the assessment of the reactivity of surface waters as a function of the chemical composition of water, the absorption spectrum of water, and the depth of the water column. Indeed, photochemical reactions are generally more important in shallow waters that can be thoroughly illuminated [102]. In contrast, the bottom part of deep water bodies receives practically no light, and no photoinduced reactivity is expected there. Our model of surface water photochemistry is described in Section 10.3.

10.3
Modeling the Photochemistry of Surface Waters

10.3.1
Modeling the Absorption Spectrum of Lake Water

A major issue in the context of surface water photochemistry is represented by the absorption spectrum of lake water, which influences the intensity of radiation absorption of the different photoactive species. Here, the water absorbance will be considered as a first approximation, neglecting the scattering processes for the sake of simplicity. Indeed, all the photosensitizers compete with one another for solar irradiance, with different results that depend on the features of the water body (e.g., chemical composition and average depth). The absorption spectrum can be expressed as the water absorbance over an optical path length of 1 cm (hereafter $A_1(\lambda)$, the units of which are expressed per centimeter). The most significant spectrum is that of water taken from the surface layer: it is the most illuminated compartment, and also the one where the highest absorbed photon flux and the highest rate of the photochemical reactions are observed [23].

The use of the spectrum of actual water sampled from the relevant water body is certainly recommended as input datum for the model, but in some cases it would be interesting to foresee the possible photochemical fate of a pollutant independent of the particular water body, with the purpose of assessing the general degree of photolability of the compound as a function of the ecosystem variables (e.g., chemical composition of water and water column depth). For this purpose one needs that the variables are not linked to a particular case, but only represent a plausible set of values for actual ecosystems. This a priori approach can work if it is possible to find a way to simulate the water spectrum as a function of the chemical composition, even with an unavoidable loss of accuracy.

Considering that most of the absorption of sunlight UV in surface waters is carried out by CDOM [8, 103], a possible parameter upon which the simulated spectrum could be based is the nonpurgeable organic carbon (NPOC) that includes both the colored and the noncolored fraction of the organic matter. The NPOC is the most suitable way to measure the amount of DOM in waters rich in carbonate and bicarbonate, which would interfere with the measurement of the dissolved organic carbon (DOC). In the case of NPOC the water sample is acidified, and the inorganic carbon is eliminated as gas-phase CO_2 by purging with a gentle flow of CO_2-free air. It follows the measurement of total dissolved carbon [104]. Some volatile organic compounds can be lost in this procedure: we have shown that on average DOC $\approx$ 1.3 NPOC in lake water [105].

The approach based on NPOC is justified by the fact that there is a good correlation between surface water absorbance in the UV (254 and 285 nm) and the values of the NPOC [4, 106]. Interestingly, in [106] we considered 26 lake water samples from NW Italy, and obtained $A_{254\,nm} = (1.40 \pm 0.15) \times 10^{-2}$ NPOC ($\mu \pm \sigma$, absorbance expressed per centimeter and NPOC in milligrams of carbon per liter) and $A_{285\,nm} = (8.57 \pm 1.09) \times 10^{-3}$ NPOC, the intercept being in both cases insignificantly different from 0. Additionally, the absorption spectra of CDOM (and of surface waters as a consequence) can be approximated reasonably well with an exponential-like trend with wavelength [107, 108]. These considerations prompt for the fitting of the experimental absorption spectra with a general empirical equation of the following form:

$$\frac{A_1(\lambda)}{NPOC} = B \cdot e^{-k \cdot \lambda} \quad (10.19)$$

Note that it has recently been possible to quantify the difference between the absorption spectrum of lake water and a true exponential function, and to relate the shift to the photodegradation processes undergone by CDOM [109]. These findings could open up the way for a future use of more precise functions than (Eq. 10.19) to describe the spectrum of lake water.

Figure 10.6 reports the absorption spectra ($A_1(\lambda)$ NPOC^{-1} vs λ) of various filtered lake water samples (n = 9, [51, 52, 105]). The use of filtered water is motivated by simplicity issues, and by the fact that relatively small differences were observed in the relevant wavelength interval between the filtered and the unfiltered samples. The fitting of each spectrum with Eq 10.19 yields different values of B and k. On average, one finds that B = 0.45 ± 0.041 (mg C)$^{-1}$ cm^{-1} and k = 0.015 ± 0.002 nm^{-1} ($\mu \pm \sigma$). Note that the units of $A_1(\lambda)$ are expressed per centimeter.

Figure 10.6 also reports Eq. (10.20) (obtained from Eq. 10.19 by substituting the fitting values of B and k) with its error bounds (bold curves) superposed to the experimental absorption spectra. Upon application of Eq. (10.20), the absorption spectrum $A_1(\lambda)$ can be obtained approximately from the NPOC value of the surface

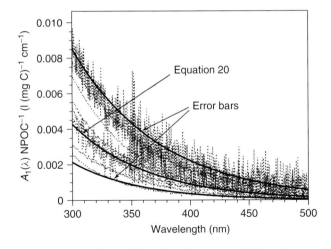

Figure 10.6 Simulation of the absorption spectra of filtered lake water with Eq. 10.20. The lake water spectra [51, 52, 105] are normalized for the NPOC. The three bold curves represent Eq. 10.20 and its error bounds ($\mu \pm \sigma$).

water sample.

$$\frac{A_1(\lambda)}{\text{NPOC}} = (0.45 \pm 0.04) \cdot e^{-(0.015 \pm 0.002)\cdot \lambda} \qquad (10.20)$$

10.3.2
Mass versus Concentration Approach in Photochemistry Models

It was previously shown that CDOM, nitrate, and nitrite are among the main compounds involved in the production of reactive transient species in surface waters. The assessment of their role in surface water photochemistry requires the calculation of the intensity of radiation absorption by each compound in the whole column of the aquatic systems. At the moment the modeling of the photochemical reactions cannot include the Fe species because insufficient data are available on their speciation and reactivity.

A possible approach could be to look for a correlation between the concentration of the photoactive species for which modeling is possible ($[NO_3^-]$, $[NO_2^-]$, NPOC for DOM) and the intensity of radiation absorption. The latter is directly linked to the photochemical activity [53]. However, a major obstacle to this methodology is that the intensity of radiation absorption varies with the depth of the water column, and integration on the whole column is therefore required. Another problem with this approach is that one needs to know the whole column trend of the concentration values of the photoactive compounds, while most of the photochemical reactivity is usually concentrated in a mixed surface layer of constant chemical composition [28, 50].

To overcome the problems mentioned above, we have recently devised an alternative, approximated but more workable approach. The idea is to calculate the intensity of radiation absorption and the rate of production of the reactive species in the whole volume $V = Sd$. Here, d is the depth of a mixed water column having approximately constant chemical composition (average depth for thoroughly mixed water bodies, and mixing layer depth for the stratified lakes). S is a standard surface area (Figure 10.7). We have assumed $S = 12.6 \, \text{cm}^2$ to allow for a direct comparison with the results of irradiation experiments of surface water samples [51]. Both the intensity of radiation absorption by the compound i, P_a^i, and the rate of production of the transient species j, R_j, should therefore be expressed within the volume V. This means that P_a^i will be in units of einstein per second (1 einstein = 1 mole of photons) and R_j in moles per second, instead of the more common units of einstein per liter per second and moles per liter per second. The actual value of P_a^i will depend on the concentration of i, and in this case it will be used the value of $[i]$ found in the surface water layer. The reason behind this choice is that the majority of the photochemical reactivity is concentrated in the surface layer where most of the absorption of radiation takes place, and that is usually thoroughly mixed (i.e., has constant chemical composition) both in rivers and in lakes [110]. The modeling results will be best applicable to the mixing layer in the case of deep, stratified lakes, and to the whole water column for shallow, thoroughly mixed water bodies.

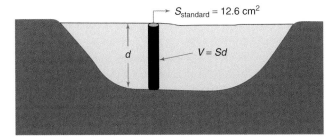

Figure 10.7 Schematic of the approach adopted for the modeling. One is interested to compute what happens inside the volume $V = Sd$, where $S = 12.6$ cm² is a standard surface area. Concerning d, it is the average depth for thoroughly mixed lakes or the mixing layer depth for stratified lakes. Note that the absorbed photon flux will be expressed in units of einstein per second instead of einstein per liter per second, and the reaction rates in moles per second instead of molarity per seconds.

To achieve the goal of relating all the important quantities to the volume $V = Sd$, consider that the sunlight radiation density reaching the ground ($q°(\lambda)$) is usually expressed in units of einstein per square centimeter per second per nanometer [47]. For our purposes it will be sufficient to multiply such a value for the standard surface $S = 12.6$ cm². The integration over wavelength of $p°(\lambda) = Sq°(\lambda)$ will give units of einstein per second as required.

10.3.3
Radiation Absorption by Photoactive Water Components

Here, only absorption is considered in a simplified approach that neglects the scattering of radiation. A major issue in the calculation of the intensity of radiation absorption by a molecule Q in a mixture is that the absorbance A_Q is the same (at equal concentration of Q) in the mixture or when Q is alone in the solution. In contrast, the absorbed photon flux density p_a^Q and the related fraction of radiation absorption (f_Q) are lower in the mixture, because of competition for absorption between Q and the other species. Moreover, for two species Q and R at wavelength λ, the ratio of the absorbance values is equal to the ratio of the absorbed photon flux densities: $A_Q(\lambda) A_R(\lambda)^{-1} = p_a^Q(\lambda) \left[p_a^R(\lambda)^{-1} \right]$ [111]. For a water column depth d (expressed in centimeters) at wavelength λ, the absorbance of nitrate, nitrite, CDOM, and the total absorbance of the water column (A_{tot}) can be expressed as follows:

$$A_{tot}(\lambda) = A_1(\lambda) d \qquad (10.21)$$

$$A_{NO_3^-}(\lambda) = \varepsilon_{NO_3^-}(\lambda) d \left[NO_3^- \right] \qquad (10.22)$$

$$A_{NO_2^-}(\lambda) = \varepsilon_{NO_2^-}(\lambda) d \left[NO_2^- \right] \qquad (10.23)$$

$$A_{CDOM}(\lambda) = A_{tot}(\lambda) - A_{NO_3^-}(\lambda) - A_{NO_2^-}(\lambda) \qquad (10.24)$$

$A_1(\lambda)$ is the absorbance of water over an optical path length of 1 cm (units expressed per centimeter) and ε represents a molar absorption coefficient (units are expressed per molarity per centimeter). Note that $A_1(\lambda)$ could be the actually measured absorption spectrum of water, or could be obtained by modeling based on the value of NPOC as shown in Section 3.1 (Eq. 10.20). Let $p°(\lambda)$ be the incident photon flux density of sunlight (in einstein per second per nanometer over the surface $S = 12.6 \text{ cm}^2$). In the Lambert–Beer approximation, the total photon flux density absorbed by water is

$$p_a^{tot}(\lambda) = p°(\lambda)\left(1 - 10^{-A_{tot}(\lambda)}\right) \tag{10.25}$$

For the photon flux density absorbed by CDOM, nitrate, and nitrite at wavelength λ, one has to consider that the photon flux densities are proportional to the values of the absorbance:

$$p_a^{CDOM}(\lambda) = p_a^{tot}(\lambda) A_{CDOM}(\lambda) [A_{tot}(\lambda)]^{-1} \tag{10.26}$$

$$p_a^{NO_3^-}(\lambda) = p_a^{tot}(\lambda) A_{NO_3^-}(\lambda) [A_{tot}(\lambda)]^{-1} \tag{10.27}$$

$$p_a^{NO_2^-}(\lambda) = p_a^{tot}(\lambda) A_{NO_2^-}(\lambda) [A_{tot}(\lambda)]^{-1} \tag{10.28}$$

Finally, the total photon flux absorbed by the species i (P_a^i, with i = CDOM, NO_3^-, NO_2^-), expressed in einstein per second, is the integral over wavelength of $p_a^i(\lambda)$

$$P_a^i = \int_\lambda p_a^i(\lambda) d\lambda \tag{10.29}$$

Table 10.1 reports the values of P_a^{CDOM}, P_a^{NO3-}, and P_a^{NO2-} for several surface waters, based on the chemical composition of the surface layer [51] and on the water column depth d. The absorption spectra of the filtered water samples ($A_1(\lambda)$ [51, 52]) and the sunlight spectrum adopted for the calculations ($p°(\lambda)$, with 22 W m^{-2} irradiance in the UV, [47]) are shown in Figure 10.8. The molar absorption coefficients of nitrate and nitrite ($\varepsilon(\lambda)$) are reported in Figure 10.1. Note that a sunlight UV irradiance of 22 W m^{-2} could be found at mid latitude (45°N) in a summer sunny day (SSD) such as 15 July, at 10 a.m. or 15 p.m. solar time [51].

10.3.4
Generation and Reactivity of •OH upon Irradiation of CDOM, Nitrate, and Nitrite

The generation of •OH by the relevant photosensitizers in surface waters is initiated by the absorption of radiation. It is therefore reasonable that the generation rate of •OH by the compound i, $R_{•OH}^i$, is proportional to P_a^i.

The previous section showed how to calculate the absorbed photon fluxes of CDOM, nitrate, and nitrite from $A_1(\lambda)$, [NO_3^-], [NO_2^-], and d. It is then necessary to derive the relationships between $R_{•OH}^{CDOM}$, $R_{•OH}^{NO_3^-}$, $R_{•OH}^{NO_2^-}$, and the corresponding absorbed photon fluxes. Figure 10.9a reports the trend of $R_{•OH}$ versus P_a for nitrate and nitrite under simulated sunlight, based on the previous

Table 10.1 Parameters of photochemical significance in the lake water samples under consideration [51, 52].

	Av. Piccolo	Candia	Av. Grande	Rouen
NPOC (mg C l^{-1})	5.1	5.4	5.0	0.63
NO$_3^-$ (M)	1.9×10^{-5}	1.6×10^{-6}	9.6×10^{-6}	1.9×10^{-5}
NO$_2^-$ (M)	1.2×10^{-6}	1.5×10^{-7}	1.4×10^{-6}	3.7×10^{-7}
HCO$_3^-$ (M)	4.0×10^{-4}	1.1×10^{-3}	3.6×10^{-3}	2.4×10^{-5}
CO$_3^{2-}$ (M)	1.1×10^{-6}	6.1×10^{-6}	4.8×10^{-5}	2.4×10^{-9}
d (m)	7.7	5.9	19.5	2.0
V (l)	9.7	7.4	24.6	2.5
P_a^{tot} (einstein s^{-1})	3.2×10^{-7}	2.5×10^{-7}	3.2×10^{-7}	1.9×10^{-7}
P_a^{CDOM} (einstein s^{-1})	3.2×10^{-7}	2.5×10^{-7}	3.2×10^{-7}	1.9×10^{-7}
P_a^{NO3-} (einstein s^{-1})	1.1×10^{-11}	1.9×10^{-12}	1.3×10^{-11}	9.1×10^{-11}
P_a^{NO2-} (einstein s^{-1})	6.3×10^{-11}	2.0×10^{-11}	2.3×10^{-10}	1.1×10^{-10}
$R_{\bullet OH}^{CDOM}$ (mol s^{-1})	$(9.6 \pm 1.3) \times 10^{-12}$	$(7.5 \pm 1.0) \times 10^{-12}$	$(9.6 \pm 1.3) \times 10^{-12}$	$(5.7 \pm 0.8) \times 10^{-12}$
$R_{\bullet OH}^{NO3-}$ (mol s^{-1})	$(4.8 \pm 0.2) \times 10^{-13}$	$(8.2 \pm 0.3) \times 10^{-14}$	$(5.6 \pm 0.2) \times 10^{-13}$	$(3.9 \pm 0.2) \times 10^{-12}$
$R_{\bullet OH}^{NO2-}$ (mol s^{-1})	$(7.3 \pm 0.2) \times 10^{-12}$	$(2.3 \pm 0.1) \times 10^{-12}$	$(2.7 \pm 0.1) \times 10^{-11}$	$(1.3 \pm 0.1) \times 10^{-11}$
$R_{\bullet OH}^{tot}$ (mol s^{-1})	$(1.7 \pm 0.2) \times 10^{-11}$	$(9.9 \pm 1.1) \times 10^{-12}$	$(3.7 \pm 0.2) \times 10^{-11}$	$(2.2 \pm 0.2) \times 10^{-11}$
$\Sigma_i \; k_{Si} [S_i]$ (s^{-1})	2.7×10^5	2.8×10^5	3.2×10^5	3.5×10^4
($\tau_P^{\bullet OH}$)$_{SSD}$ Diuron	600 ± 70	810 ± 100	830 ± 100	15 ± 2
Fenuron	430 ± 50	580 ± 70	590 ± 70	11 ± 1
Atrazine	1000 ± 100	1400 ± 200	1400 ± 200	26 ± 3
Aniline	210 ± 30	290 ± 30	290 ± 30	5.5 ± 0.6
Phenolate	310 ± 40	420 ± 50	430 ± 50	8.0 ± 0.9
4-HOBz	350 ± 40	480 ± 60	490 ± 60	9.1 ± 1.1

Note that d is the average depth and $V = 1000 \; Sd$ is the volume of the water column with surface area $S = 1.26 \times 10^{-3}$ m^2. The data in the upper section of the table, which are referred to the water surface layer, are the basis to derive those of the lower section. The four lakes are all located in NW Italy, within 45°01'–45°20' N and 7°10'–7°55' E.

4-HOBz, 4-hydroxybenzoate; SSD, summer sunny days equivalent to 15 July at 45°N latitude.

data from our group [52]. The formation rate of •OH has been measured with the reaction benzene $+$•OH $\to$ phenol (95% yield, [112]). Irradiation took place under a solar simulator (22 W m^{-2} UV irradiance) inside cylindrical Pyrex glass cells with surface area $S = 12.6$ cm^2. This explains the adoption of that value of S and of 22 W m^{-2} UV irradiance as standard conditions in our model. Note that in Figure 10.9a P_a is expressed in einstein per second, and $R_{\bullet OH}$ in moles per second. The calculation of P_a for nitrate and nitrite was based on the already reported Eqs. 10.25–10.28, adopting as $p°(\lambda)$ the emission spectrum of the lamp used (which simulates summertime sunlight) instead of that of sunlight itself.

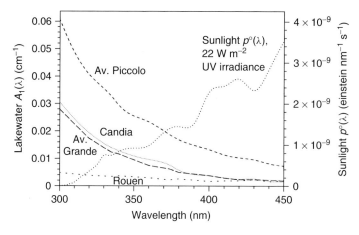

Figure 10.8 Absorption spectra $A_1(\lambda)$ of four lake water samples (surface layer) [51, 52, 105]. Spectral photon flux of sunlight, corresponding to 22 W m^{-2} UV irradiance.

The derivation of $R_{\bullet OH}^{CDOM}$ versus P_a^{CDOM} is more complicated because CDOM is not a species of definite chemical composition. Nevertheless, for various lake water samples it has been possible to find a statistically significant correlation between the value of NPOC (that measures DOM) and the formation rate of $^{\bullet}$OH unaccounted for by nitrate and nitrite ($R_{\bullet OH}^{tot} - R_{\bullet OH}^{NO_3^-} - R_{\bullet OH}^{NO_2^-}$) [51]. Such a quantity would measure the generation rate of $^{\bullet}$OH by CDOM, $R_{\bullet OH}^{CDOM}$.

Figure 10.9b reports $R_{\bullet OH}^{CDOM} = R_{\bullet OH}^{tot} - R_{\bullet OH}^{NO_3^-} - R_{\bullet OH}^{NO_2^-}$ as a function of P_a^{CDOM}, calculated as for the previous section (Eq. 10.26) but adopting the lamp emission spectrum as $p°(\lambda)$. From linear least-square fitting of the data reported in Figure 10.9a and b, one gets the following expressions for the generation rate of $^{\bullet}$OH from CDOM, nitrate, and nitrite:

$$R_{\bullet OH}^{CDOM} = (3.0 \pm 0.4) \times 10^{-5} \, P_a^{CDOM} \tag{10.30}$$

$$R_{\bullet OH}^{NO_3^-} = (4.3 \pm 0.2) \times 10^{-2} \, P_a^{NO_3^-} \tag{10.31}$$

$$R_{\bullet OH}^{NO_2^-} = (1.2 \pm 0.1) \times 10^{-1} \, P_a^{NO_2^-} \tag{10.32}$$

$$R_{\bullet OH}^{tot} = R_{\bullet OH}^{CDOM} + R_{\bullet OH}^{NO_3^-} + R_{\bullet OH}^{NO_2^-} \tag{10.33}$$

The error bounds represent one standard deviation ($\pm\sigma$). Table 10.1 reports $R_{\bullet OH}$ for various surface waters, calculated on the basis of the corresponding values of P_a for CDOM, nitrate, and nitrite. Note that P_a was calculated assuming a sunlight spectrum $p°(\lambda)$ characterized by 22 W m^{-2} UV irradiance (Figure 10.8), thus also the values of $R_{\bullet OH}$ are referred to the same irradiance.

Once generated, the $^{\bullet}$OH radicals react quickly with many dissolved compounds. A steady state is almost suddenly reached, where the rate of consumption of $^{\bullet}$OH

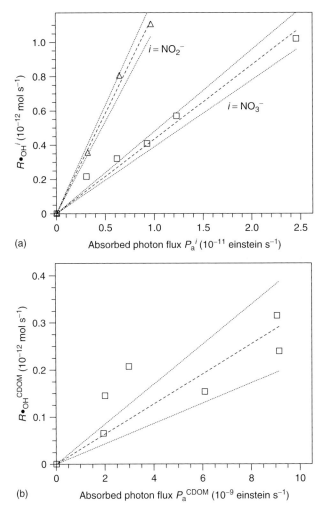

Figure 10.9 (a) Formation rate of •OH by nitrate and nitrite, as a function of the respective absorbed photon fluxes. (b) Formation rate of •OH attributable to CDOM, as a function of its absorbed photon flux. The figures report a recalculation of the experimental data of [51, 52]. The dashed lines represent the linear least-square fitting and the dotted lines are the 95% confidence bounds.

is equal to the rate of production. Let S_i be a generic scavenger molecule and k_{S_i} its second-order rate constant for the reaction with •OH. At the steady state the following relationship holds:

$$R^{tot}_{\bullet OH} = \sum_i k_{S_i} [^\bullet OH][S_i] = [^\bullet OH] \sum_i k_{S_i}[S_i] \tag{10.34}$$

The main scavengers of hydroxyl radicals in surface freshwater are DOM, HCO_3^-, CO_3^{2-}, and NO_2^- [23]. In contrast, in seawater the main role is played by bromide [81]. From the literature rate constants for reaction with •OH

[62], and empirically derived relationships in the case of DOM [51], it is possible to express the scavenging rate constant for •OH in freshwater as follows:

$$\sum_i k_{S_i}[S_i] = 5 \times 10^4 \text{NPOC} + 8.5 \times 10^6 [\text{HCO}_3^-] + 3.9 \times 10^8 [\text{CO}_3^{2-}]$$
$$+ 1.0 \times 10^{10} [\text{NO}_2^-] \qquad (10.35)$$

Here, NPOC is expressed in milligrams per carbon per liter, and the concentration values are in molarity. $\sum_i k_{S_i}[S_i]$ has units expressed per second. Consider a generic dissolved molecule P, with second-order reaction rate constant $k_{P,\bullet OH}$ with •OH. In the very vast majority of the environmental cases, it will be $k_{P,\bullet OH}[P] \ll \sum_i k_{S_i}[S_i]$. The rate of P degradation for reaction with •OH ($R_P^{\bullet OH}$ is given by the fraction of $R_{\bullet OH}^{tot}$ that is involved in the degradation of P, namely:

$$R_P^{\bullet OH} = R_{\bullet OH}^{tot} \cdot \frac{k_{P,\bullet OH}[P]}{\sum_i k_{S_i}[S_i]} \qquad (10.36)$$

Equation 10.36 describes a first-order decay kinetics, with rate constant $k_P = R_P^{\bullet OH} V^{-1} [P]^{-1} = R_{\bullet OH}^{tot} k_{P,\bullet OH} (V \sum_i k_{S_i}[S_i])^{-1}$. $V = Sd$ is the volume of solution contained in a cylinder of surface $S = 12.6 \text{ cm}^2$ and height d, which is the average depth for a thoroughly mixed water body and the mixing layer depth for a stratified lake (Figure 10.7). Note that if [P] is expressed in molarity, V in liters, and $R_P^{\bullet OH}$ in moles per second, the units of k_P will be expressed per second. The volume V has to be included in the expression of k_P to obtain compatibility between the mass approach (adopted to simplify the absorbed light calculations) and the kinetic treatment of the results.

For a first-order kinetics it is possible to calculate the half-life time of P, $(t_P^{\bullet OH})_{\frac{1}{2}} = \ln2(k_P)^{-1}$. If $R_{\bullet OH}^{tot}$ is expressed in moles per second and is referred to 22 W m^{-2} sunlight UV irradiance, if $\sum_i k_{S_i}[S_i]$ is in per second, and $k_{P,\bullet OH}$ in per molar per second, then the units of $(t_P^{\bullet OH})_{\frac{1}{2}}$ will be seconds of steady irradiation under 22 W m^{-2} sunlight UV. A major issue is that the outdoor sunlight intensity is not constant, and it has been shown that the energy reaching the ground in a sunny summer day (15 July, 45°N latitude) corresponds to 10 hour irradiation at 22 W m^{-2} UV irradiance [51]. It is therefore possible to convert $(t_P^{\bullet OH})_{\frac{1}{2}}$ in units of SSD (taking 15 July at 45°N latitude as reference), by dividing it for 10 hour = 3.6 × 10^4 seconds. The resulting half-life time for reaction with •OH, $(\tau_P^{\bullet OH})_{SSD}$, will therefore be expressed as follows:

$$\left(\tau_P^{\bullet OH}\right)_{SSD} = \frac{\ln2 V \sum_i k_{S_i}[S_i]}{3.6 \times 10^4 R_{\bullet OH}^{tot} k_{P,\bullet OH}} = 1.9 \times 10^{-5} \frac{V \sum_i k_{S_i}[S_i]}{R_{\bullet OH}^{tot} k_{P,\bullet OH}} \qquad (10.37)$$

Table 10.1 reports the values of $(\tau_P^{\bullet OH})_{SSD}$ for a number of substrates (the herbicides, diuron, fenuron, and atrazine, and miscellaneous organic compounds, aniline,

Table 10.2 Reaction rate constants of the compounds of interest with respect to the reactive species considered in the model ($^{\bullet}OH$, $CO_3^{-\bullet}$, $^3CDOM^*$) [62, 77, 78, 90].

	$k_{P,^{\bullet}OH}$ $(M^{-1}s^{-1})$	$k_{P,CO_3^{-\bullet}}$ $(M^{-1}s^{-1})$	$k_{P,BP*}$ $(M^{-1}s^{-1})$	$k_{P,MAP*}$ $(M^{-1}s^{-1})$	$k_{P,^{\bullet}OH}$ $(k_{P,CO_3^{-\bullet}})^{-1}$
Diuron	5×10^9	8×10^6	5.2×10^8	9×10^6	620
Fenuron	7×10^9	1×10^7	2.0×10^9	8.1×10^7	700
Atrazine	3×10^9	4×10^6	n/a	n/a	750
Aniline	1.4×10^{10}	5×10^8	n/a	n/a	28
Phenolate	9.6×10^9	2.5×10^8	n/a	n/a	38
4-HOBz	8.5×10^9	1×10^8	n/a	n/a	85

In the case of $^3CDOM^*$, the table reports the rate constants for the reaction with the excited triplet states of two different model molecules for CDOM, benzophenone ($k_{P,BP*}$) and 3'-methoxyacetophenone ($k_{P,MAP*}$). The ratio of the rate constants with $^{\bullet}OH$ and with $CO_3^{-\bullet}$, is also reported.
n/a = not available.

phenolate, and 4-hydroxybenzoate). The values of $k_{P,^{\bullet}OH}$ for each substrate are reported in Table 10.2.

In Table 10.1 it can be observed that there is both variability of $(\tau_p^{\bullet OH})_{SSD}$ for the same substrate in different water bodies and variability among different substrates in the same water body. The degradation kinetics for the reaction with $^{\bullet}OH$ can thus be expected to depend on the ecosystem variables, about as much as it depends on the intrinsic reactivity of the substrate with $^{\bullet}OH$. Note that in Eq. (10.37) the ecosystem-related quantities (ability to produce and consume $^{\bullet}OH$, depth of the water column) are expressed by the product of $(R_{\bullet OH}^{tot})^{-1} V \sum_i k_{Si}[S_i]$, while the reactivity of the substrate is expressed by $k_{P,^{\bullet}OH}$.

Finally, note that when $(\tau_p^{\bullet OH})_{SSD}$ is high, the reaction with $^{\bullet}OH$ is very likely not the main removal process for P in the relevant water body. Under such circumstances the actual degradation kinetics of P could be much faster than foreseen, because of the contribution of additional processes.

10.3.5
Formation and Reactivity of $CO_3^{-\bullet}$ in Surface Waters

As already mentioned in Section 10.2.2, the carbonate radical can be formed upon oxidation of carbonate and bicarbonate by $^{\bullet}OH$, and upon carbonate oxidation by $^3CDOM^*$. The main sink of $CO_3^{-\bullet}$ in surface freshwater is DOM, and the estimates for the second-order rate constant of the reaction between $CO_3^{-\bullet}$ and DOM vary between 40 and 280 $\pm$ 90 l (mg C)$^{-1}$ s^{-1} [77, 80]. Note that in these cases DOM is quantified as NPOC. For our purposes we will assume $k_{CO3-\bullet,DOM} = 10^2$ l (mg C)$^{-1}$ s^{-1} as a reasonable intermediate value of the literature data. The calculations of the generation rates of $CO_3^{-\bullet}$ by $^{\bullet}OH$ and by $^3CDOM^*$ will be carried out separately. Within our model the first process mainly depends on $R_{\bullet OH}^{tot}$, and the second process mainly on P_a^{CDOM}.

10.3.5.1 Oxidation of HCO_3^- and CO_3^{2-} by $^{\bullet}OH$

The generation rate of $CO_3^{-\bullet}$ from the reaction of carbonate and bicarbonate with $^{\bullet}OH$ is given by the generation rate of $^{\bullet}OH$ times the fraction of the hydroxyl radicals that react with HCO_3^- or CO_3^{2-}. From the literature rate constants of carbonate and bicarbonate with $^{\bullet}OH$ [62] one gets the following relationship:

$$R_{CO_3^{-\bullet}}^{\bullet OH} = R_{\bullet OH}^{tot} \frac{8.5 \times 10^6 [HCO_3^-] + 3.9 \times 10^8 [CO_3^{2-}]}{\sum_i k_{S_i}[S_i]} \quad (10.38)$$

Here, $\sum_i k_{S_i}[S_i]$ is expressed by Eq. (10.35). Note that the generation rate of $CO_3^{-\bullet}$ via this route is only a fraction (often less than 10%) of the generation rate of $^{\bullet}OH$.

10.3.5.2 Oxidation of CO_3^{2-} by $^3CDOM^*$

The following are the chain of reactions that lead to the production of $CO_3^{-\bullet}$ from $^3DOM^*$:

$$CDOM + h\nu \longrightarrow {}^1CDOM^* - (ISC) \longrightarrow {}^3CDOM^* \quad (10.39)$$

$$^3CDOM^* \longrightarrow CDOM + heat \quad (10.40)$$

$$^3CDOM^* + CO_3^{2-} \longrightarrow CDOM^{-\bullet} + CO_3^{-\bullet} \quad (10.41)$$

From Reaction 10.41 one gets that the initial formation rate of $CO_3^{-\bullet}$ is $R_{CO_3^{-\bullet}} = k_{41} [CO_3^{2-}] [^3CDOM^*]$. The initial formation rate of $^3CDOM^*$ is proportional to P_a^{CDOM} (Reaction 10.39). Considering that Reaction 10.40 is likely to be the main transformation pathway for $^3CDOM^*$ [66], [$^3CDOM^*$] would be proportional to $P_a^{CDOM} k_{40}^{-1}$, and therefore to P_a^{CDOM} because k_{40} is constant. In lake Greinfensee (Switzerland, 47°N), for a water column depth $d = 1$ m, 30 W m^{-2} sunlight UV irradiance, and $[CO_3^{2-}] = 1 \times 10^{-5}$ M, it has been evaluated that $[^3CDOM^*] = 8 \times 10^{-15}$ M and $R_{CO_3^{-\bullet}} = 1 \times 10^{-14}$ M s^{-1} [77]. In our model the rates should be expressed in units of moles per second; if $d = 1$ m and $S = 0.0013$ m^2, one gets $V = 1.3$ l and $R_{CO_3^{-\bullet}}^{CDOM} = VR_{CO_3^{-\bullet}} = 1.3 \times 10^{-14}$ mol s^{-1}. From NPOC = 3.5 mg C l^{-1} [77] and from Eqs. (10.20–10.26 and 10.29), under the reasonable hypothesis that $P_a^{tot} \approx P_a^{CDOM}$, one gets $P_a^{CDOM} = 2.0 \times 10^{-7}$ einstein s^{-1}. Considering that $3CDOM^*] \propto P_a^{CDOM}$, from the values of $R_{CO_3^{-\bullet}}^{CDOM}$ and $[CO_3^{2-}]$ one gets the following relationship:

$$R_{CO_3^{-\bullet}}^{CDOM} = 6.5 \times 10^{-3} [CO_3^{2-}] P_a^{CDOM} \quad (10.42)$$

P_a^{CDOM} is expressed in einstein per second, $[CO_3^{2-}]$ in molarity, and $R_{CO_3^{-\bullet}}^{CDOM}$ in moles per second. We will also consider that $R_{CO_3^{-\bullet}}^{tot} = R_{CO_3^{-\bullet}}^{\bullet OH} + R_{CO_3^{-\bullet}}^{CDOM}$. Other possible sources of $CO_3^{-\bullet}$ in surface waters exist [29], but at the moment insufficient data are available to allow their precise modeling.

10.3.5.3 Reactivity of $CO_3^{-\bullet}$ in Surface Waters

Assume $R_{CO_3^{-\bullet}}^{tot} = R_{CO_3^{-\bullet}}^{\bullet OH} + R_{CO_3^{-\bullet}}^{CDOM}$ as the total formation rate of $CO_3^{-\bullet}$ in surface waters. It can be calculated from $R_{\bullet OH}^{tot}$, P_a^{CDOM}, and the water chemical composition (Eqs. 10.38 and 10.42). The rates of formation and scavenging of $CO_3^{-\bullet}$ would be equal under the steady-state conditions that are typically reached in sunlit waters. Accordingly, it would be $R_{CO_3^{-\bullet}}^{tot} = k_{CO_3^{-\bullet},DOM}$ NPOC $= 10^2$ NPOC. Let P be a dissolved molecule, with a second-order rate constant $k_{P,CO_3^{-\bullet}}$ for the reaction with $CO_3^{-\bullet}$. In the very vast majority of the cases it will be $k_{P,CO_3^{-\bullet}} \cdot [P] \ll 10^2$ NPOC. In analogy to the case of $\bullet OH$, the degradation rate of P for reaction with $CO_3^{-\bullet}$, $R_P^{CO_3^{-\bullet}}$, is given by $R_{CO_3^{-\bullet}}^{tot}$ times the fraction of $CO_3^{-\bullet}$ that reacts with P:

$$R_P^{CO_3^{-\bullet}} = R_{CO_3^{-\bullet}}^{tot} \cdot \frac{k_{P,CO_3^{-\bullet}} \cdot [P]}{k_{CO_3^{-\bullet},DOM} \text{NPOC}} \quad (10.43)$$

Also in this case one gets a first-order kinetics for the degradation of P, with degradation rate constant $k_P = R_P^{CO_3^{-\bullet}} V^{-1} [P]^{-1} = R_{CO_3^{-\bullet}}^{tot} \cdot k_{P,CO_3^{-\bullet}} \cdot (Vk_{CO_3^{-\bullet},DOM} \text{NPOC})^{-1}$. The half-life time $(t_P^{CO_3^{-\bullet}})_{\frac{1}{2}} = \ln 2 (k_P)^{-1}$ would be referred to the same irradiation conditions as for $R_P^{CO_3^{-\bullet}}$. If the conditions are 22 W m^{-2} sunlight UV irradiance, the same conversion factor can be applied between steady irradiation and our standard SSD (15 July, 45°N latitude), which was already discussed for $\bullet OH$ in Section 10.3.4. Accordingly, the half-life time of P in SSD for reaction with $CO_3^{-\bullet}$ would be expressed as follows:

$$\left(\tau_P^{CO_3^{-\bullet}}\right)_{SSD} = \frac{\ln 2 V k_{CO_3^{-\bullet},DOM} \text{NPOC}}{3.6 \times 10^4 R_{CO_3^{-\bullet}}^{tot} \cdot k_{P,CO_3^{-\bullet}}}$$

$$= 1.9 \times 10^{-5} \frac{V k_{CO_3^{-\bullet},DOM} \text{NPOC}}{R_{CO_3^{-\bullet}}^{tot} \cdot k_{P,CO_3^{-\bullet}}} \quad (10.44)$$

Table 10.3 reports the values of P_a^{CDOM}, $R_{CO_3^{-\bullet}}^{tot}$, and $(\tau_P^{CO_3^{-\bullet}})_{SSD}$ for various substrates in different water bodies. The table also reports $R_{CO_3^{-\bullet}}^{\bullet OH}$ and $R_{CO_3^{-\bullet}}^{CDOM}$, and it is $R_{CO_3^{-\bullet}}^{\bullet OH} \gg R_{CO_3^{-\bullet}}^{CDOM}$ in all the cases. P_a^{CDOM} and the formation rates of $CO_3^{-\bullet}$ are referred to 22 W m^{-2} sunlight UV irradiance. The values of $k_{P,CO_3^{-\bullet}}$ for the different substrates are reported in Table 10.2.

In Table 10.3, the half-life times for reaction with $CO_3^{-\bullet}$ are compared with those for reaction with $\bullet OH$. It can be seen that the reaction with $\bullet OH$ is usually more important, but $CO_3^{-\bullet}$ is able to outcompete $\bullet OH$ in a number of cases in which the transformation kinetics is fast (i.e., $(\tau_P^{CO_3^{-\bullet}})_{SSD}$ is low). In such cases photochemistry can be a very significant degradation process, and the reaction with $CO_3^{-\bullet}$ could therefore control the persistence of the relevant compound. However, for certain substrates the carbonate radical has no chances to induce degradation to a higher extent than $\bullet OH$. When considering the values of $k_{P,\bullet OH}$ and $k_{P,CO_3^{-\bullet}}$

Table 10.3 Half-life times for reaction with $CO_3^{-\bullet}$ in the lake water samples under consideration.

		Av. Piccolo	Candia	Av. Grande	Rouen
NPOC (mg C l^{-1})		5.1	5.4	5.0	0.63
NO_3^- (M)		1.9×10^{-5}	1.6×10^{-6}	9.6×10^{-6}	1.9×10^{-5}
NO_2^- (M)		1.2×10^{-6}	1.5×10^{-7}	1.4×10^{-6}	3.7×10^{-7}
HCO_3^- (M)		4.0×10^{-4}	1.1×10^{-3}	3.6×10^{-3}	2.4×10^{-5}
CO_3^{2-} (M)		1.1×10^{-6}	6.1×10^{-6}	4.8×10^{-5}	2.4×10^{-9}
d (m)		7.7	5.9	19.5	2.0
V (l)		9.7	7.4	24.6	2.5
p_a^{CDOM} (einstein s^{-1})		3.2×10^{-7}	2.5×10^{-7}	3.2×10^{-7}	1.9×10^{-7}
$R_{\bullet OH}^{tot}$ (mol s^{-1})		$(1.7 \pm 0.2) \times 10^{-11}$	$(9.9 \pm 1.1) \times 10^{-12}$	$(3.7 \pm 0.2) \times 10^{-11}$	$(2.2 \pm 0.2) \times 10^{-11}$
$\sum_i k_{Si} [S_i]$ (s^{-1})		2.7×10^5	2.8×10^5	3.2×10^5	3.5×10^4
$R_{CO_3^{-\bullet}}^{\bullet OH}$ (mol s^{-1})		$(2.4 \pm 0.3) \times 10^{-13}$	$(4.1 \pm 0.5) \times 10^{-13}$	$(5.7 \pm 0.3) \times 10^{-12}$	$(1.3 \pm 0.1) \times 10^{-13}$
$R_{CO_3^{-\bullet}}^{CDOM}$ (mol s^{-1})		2.3×10^{-15}	9.9×10^{-15}	1.0×10^{-13}	3.0×10^{-18}
$R_{CO_3^{-\bullet}}^{tot}$ (mol s^{-1})		$(2.4 \pm 0.3) \times 10^{-13}$	$(4.2 \pm 0.5) \times 10^{-13}$	$(5.8 \pm 0.3) \times 10^{-12}$	$(1.3 \pm 0.1) \times 10^{-13}$
Diuron	$(\tau_p^{CO_3^{-\bullet}})_{SSD}$	$(4.8 \pm 0.6) \times 10^4$	$(2.2 \pm 0.2) \times 10^4$	$(5.0 \pm 0.3) \times 10^3$	$(2.9 \pm 0.3) \times 10^3$
	$(\tau_p^{\bullet OH})_{SSD}$	600 ± 70	810 ± 100	830 ± 100	15 ± 2
Fenuron	$(\tau_p^{CO_3^{-\bullet}})_{SSD}$	$(3.9 \pm 0.5) \times 10^4$	$(1.8 \pm 0.2) \times 10^4$	$(4.0 \pm 0.2) \times 10^3$	$(2.3 \pm 0.2) \times 10^3$
	$(\tau_p^{\bullet OH})_{SSD}$	430 ± 50	580 ± 70	590 ± 70	11 ± 1
Atrazine	$(\tau_p^{CO_3^{-\bullet}})_{SSD}$	$(9.7 \pm 1.1) \times 10^4$	$(4.5 \pm 0.5) \times 10^4$	$(1.0 \pm 0.1) \times 10^4$	$(5.8 \pm 0.5) \times 10^3$
	$(\tau_p^{\bullet OH})_{SSD}$	1000 ± 100	1400 ± 200	1400 ± 200	26 ± 3
Aniline	$(\tau_p^{CO_3^{-\bullet}})_{SSD}$	770 ± 90	360 ± 40	81 ± 4	46 ± 4
	$(\tau_p^{\bullet OH})_{SSD}$	210 ± 30	290 ± 30	290 ± 30	5.5 ± 0.6
Phenolate	$(\tau_p^{CO_3^{-\bullet}})_{SSD}$	$(1.5 \pm 0.2) \times 10^3$	720 ± 80	160 ± 10	93 ± 8
	$(\tau_p^{\bullet OH})_{SSD}$	310 ± 40	420 ± 50	430 ± 50	8.0 ± 0.9
4-HOBz	$(\tau_p^{CO_3^{-\bullet}})_{SSD}$	$(3.9 \pm 0.5) \times 10^3$	1800 ± 200	400 ± 20	230 ± 20
	$(\tau_p^{\bullet OH})_{SSD}$	350 ± 40	480 ± 60	490 ± 60	9.1 ± 1.1

See also all the notes and observations reported in Table 10.1 caption. The half-life times with $CO_3^{-\bullet}$ are compared to those with $^\bullet OH$, reported from Table 10.1. SSD = summer sunny days equivalent to 15 July at 45°N latitude.

(Table 10.2), one gets that it should be $k_{P,\bullet OH}(k_{P,CO_3^{-\bullet}})^{-1} < 100$ for $CO_3^{-\bullet}$ to be more important than $^\bullet OH$ as a reactive species for the degradation of P in at least some water bodies. Favorable environmental conditions for the degradation processes induced by $CO_3^{-\bullet}$ are elevated values of carbonate and bicarbonate, and low NPOC.

10.3.6
Formation and Reactivity of $^3CDOM^*$ in Surface Waters

The absorption of sunlight by some photoactive components of CDOM leads through ISC to the formation of the corresponding excited triplet states, $^3CDOM^*$. These species can undergo thermal deactivation or reaction with dissolved O_2, with a global rate constant $k^* \approx 5 \times 10^5$ s^{-1} [66], or react with other dissolved molecules and induce their transformation. It is reasonable to correlate the generation rate of $^3CDOM^*$ with P_a^{CDOM}. Referring again to the surface water of Lake Greifensee ($d = 1$ m, $V = 1.3$ l, 30 W m^{-2} sunlight UV irradiance), one has $[^3CDOM^*] = 8 \times 10^{-15}$ M [77] and $P_a^{CDOM} = 2.0 \times 10^{-7}$ einstein s^{-1}. With $k* \approx 5 \times 10^5$ s^{-1}, one can infer that thermal deactivation or the reaction with O_2 is a more important quenching pathway for $^3CDOM^*$ than the reactions with other dissolved molecules [90]. Under steady-state conditions, one can assume that the production rate of $^3CDOM^*$ is equal to the rate of consumption, thus $R_{3CDOM*} = k*[^3CDOM^*] = 4 \times 10^{-9}$ Ms^{-1}. Considering that R_{3CDOM*} would be proportional to $P_a^{CDOM} = 2.0 \times 10^{-7}$ einstein s^{-1} one gets:

$$R_P^{3CDOM*} = 0.02 P_a^{CDOM} \tag{10.45}$$

Assume a molecule P, with reaction rate constant $k_{P,3CDOM*}$ with $^3CDOM^*$. In most cases it will be $k_{P,3CDOM*}[P] \ll k*$. The degradation rate of P for reaction with $^3CDOM^*$, R_P^{3CDOM*}, is given by R_{3CDOM*} times the fraction of $^3CDOM^*$ that reacts with P, which yields the following equation:

$$R_P^{3CDOM*} = R_{3CDOM*} \frac{k_{P,3CDOM*}[P]}{k*} \tag{10.46}$$

The first-order degradation rate constant of P is $k_P = R_P^{3CDOM*} V^{-1} [P]^{-1}$ (note that both R_P^{3CDOM*} and R_{3CDOM*} are expressed in moles per second, [P] in molarity, V in liters, and k_P in per second). From the first-order kinetics one gets the half-life time $(t_P^{3CDOM*})_{\frac{1}{2}} = \ln 2 (k_P)^{-1}$. It is referred to continuous sunlight irradiation with 22 W m^{-2} irradiance, like for R_{3CDOM*}. The usual conversion gives the lifetime for reaction with $^3CDOM^*$ in SSD equivalents (15 July, 45°N latitude) as follows:

$$\left(\tau_P^{3CDOM*}\right)_{SSD} = \frac{\ln 2 V k*}{3.6 \times 10^4 R_{3CDOM*} k_{P,3CDOM*}}$$

$$= 9.5 \times 10^{-4} \frac{V k*}{P_a^{CDOM} k_{P,3CDOM*}} \tag{10.47}$$

The main problem concerning the modeling of the reactivity of $^3CDOM^*$ in surface waters is the measurement of $k_{P,3CDOM*}$. Actually, CDOM is not a species of definite chemical composition, and a major problem arises to find a suitable model material as the basis for the determination of $k_{P,3CDOM*}$. Canonica et al. [90] have adopted benzophenone (BP) and 3'-methoxyacetophenone (MAP) as photosensitizers for this purpose, and have measured the values of $k_{P,3BP*}$ and of $k_{P,3MAP*}$ for a series of phenylurea herbicides. The choice of BP and MAP has been motivated by the photosensitizing properties of aromatic ketones [86], and by the presence of this class of compounds within CDOM. A problem with the use of BP and MAP as model molecules is that $k_{P,3BP*} > k_{P,3MAP*}$, with large differences for some substrates [90], which makes the determination of $k_{P,3CDOM*}$ not univocal.

Figure 10.10 reports examples of plots of $(\tau_P^{3CDOM*})_{SSD}$ versus $k_{P,3CDOM*}$ for phenuron and diuron in the water bodies that were considered in this study. The value of $k_{P,3CDOM*}$ was set to vary in the interval between the rate constant obtained with MAP and that obtained with BP ($k_{P,^3MAP*}$ and $k_{P,3BP*}$, see Table 10.2 and [90]). Note the large variability of $(\tau_P^{3CDOM*})_{SSD}$, depending on the chosen model photosensitizer. Whenever possible on the plots the value of $(\tau_P^{•OH})_{SSD}$, taken from Table 10.1, for each relevant water body is also shown. When $(\tau_P^{•OH})_{SSD}$ is not indicated, this means that it is higher than $(\tau_P^{3CDOM*})_{SSD}$ obtained with either BP or MAP. Interestingly, it was $(\tau_P^{CO_3^{-•}})_{SSD} \gg (\tau_P^{3CDOM*})_{SSD}$ in all the cases, suggesting that the reaction with $CO_3^{-•}$ would not be a significant sink for phenuron and diuron in the water bodies under consideration. Probably, $CO_3^{-•}$ would not be a significant sink for phenuron and diuron, in general, because of the very elevated values found for $(\tau_P^{CO_3^{-•}})_{SSD}$ compared to $(\tau_P^{•OH})_{SSD}$ or $(\tau_P^{3CDOM*})_{SSD}$ in all the cases under consideration. As far as $(\tau_P^{•OH})_{SSD}$ is considered, it is often included within the range of the likely values of $(\tau_P^{3CDOM*})_{SSD}$. Generally speaking, the following considerations can be made: (i) The relative role of $•OH$ compared to $^3CDOM*$ is likely to be higher for diuron than for fenuron. (ii) The relative role of $•OH$ would increase in shallow, clear water bodies with a limited DOM content, and

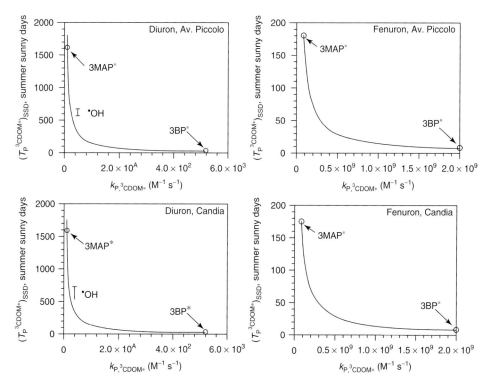

Figure 10.10 Assessment of the half-life time of the phenylurea herbicides diuron and fenuron in lake water. Because the rate constants with $^3CDOM*$ are not known, the estimate is reported based on the rate constants [86] of the two compounds with the excited triplet states of 3′-methoxyacetophenone ($^3MAP*$) and benzophenone ($^3BP*$).

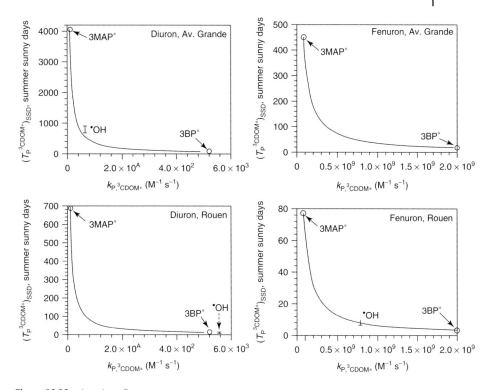

Figure 10.10 (continued)

decrease in the opposite case. Indeed, the role of $^{\bullet}OH$ is most significant for Lake Rouen. (iii) $^{3}CDOM^{*}$ would be more important than $^{\bullet}OH$ for the degradation of phenylurea herbicides if its reactivity resembles that of $^{3}BP^{*}$, and probably of comparable importance if the reactivity is similar to $^{3}MAP^{*}$.

The main rationale behind point (ii) is that the CDOM is a major photochemical source of $^{\bullet}OH$, but not the only one because nitrite and nitrate also contribute. In contrast, CDOM is the only source of $^{3}CDOM^{*}$. Additionally, DOM is the main sink for $^{\bullet}OH$ but presumably a minor sink for $^{3}CDOM^{*}$. Moreover, CDOM would absorb a higher fraction of radiation compared to nitrate and nitrite in deep water bodies. A further observation that can be made concerning point (iii) is that the reaction with $^{3}CDOM^{*}$ is already known to be the main removal pathway for electron-rich phenols in surface waters [66], and a major transformation route for phenylurea herbicides in Lake Greifensee, Switzerland [91].

Note that the present model cannot take into account the variability of DOM/CDOM in different environments. Indeed, Eq. (10.45) that correlates $R_{3CDOM^{*}}$ with P_{a}^{CDOM} is just an approximation, and the proportionality factor is very likely to be different for different water bodies. A similar variability would also affect $k_{P,3CDOM^{*}}$, which has been partially considered in Figure 10.10 by letting $k_{P,3CDOM^{*}}$ vary between $k_{P,3MAP^{*}}$ and $k_{P,3BP^{*}}$. Additional research is needed to assess the ecosystem-dependent variability of the photochemistry of CDOM.

10.4
Conclusions and Perspectives

A modeling approach was presented to describe the indirect photolysis processes that can take place in surface waters, with particular emphasis on the reactions that involve $^{\bullet}OH$, $CO_3^{-\bullet}$, and $^3CDOM^*$. The model allows the assessment of the lifetime of dissolved compounds in water bodies, including hazardous xenobiotics, as far as the indirect photochemical processes are concerned. The main applications are the assessment of the self-decontamination potential of the aquatic systems and the prediction of the importance of the photochemical transformation processes. Additionally, in the cases in which it is known that a certain photochemical pathway yields secondary pollutant(s), the probability of formation of these compounds can be assessed.

It is possible to calculate the rate of photon absorption by CDOM, nitrate, and nitrite in the whole column of water bodies under definite irradiation conditions (22 W m^{-2} sunlight UV irradiance), based on the chemical composition of water, its absorption spectrum, and the water column depth. Radiation absorption calculations are the basis for the assessment of the indirect photochemical reactivity induced by the transient species under consideration. A comparison between $^{\bullet}OH$ and $CO_3^{-\bullet}$ shows that the carbonate radical cannot be a significant sink of difficult-to-oxidize compounds under any circumstance. In contrast, $CO_3^{-\bullet}$ can play a significant role for sufficiently reactive compounds (with $k_{P,CO_3^{-\bullet}} \cdot (k_{P,^{\bullet}OH})^{-1} > 0.01$) if the environmental conditions are favorable. Note that with increasing reactivity of the substrate P, both $k_{P,CO_3^{-\bullet}}$ and $k_{P,^{\bullet}OH}$ increase and the ratio $k_{P,CO_3^{-\bullet}} \cdot (k_{P,^{\bullet}OH})^{-1}$ generally increases as well. Favorable conditions that are important for the reactions induced by $CO_3^{-\bullet}$ are an elevated ratio of inorganic to organic carbon. Indeed, $CO_3^{-\bullet}$ is generated upon oxidation of carbonate and bicarbonate by $^{\bullet}OH$ (and only to a lesser extent upon oxidation of carbonate by $^3CDOM^*$), and it is mainly scavenged by DOM. An elevated amount of DOM would both consume $CO_3^{-\bullet}$ and keep its formation rate low through the scavenging of $^{\bullet}OH$.

There are indications from both the literature [91] and this work (Figure 10.10) that $^3CDOM^*$ could be more important than $^{\bullet}OH$ toward the degradation of phenylurea herbicides, at least in water bodies that are at the same time relatively deep and quite rich in CDOM. Indeed, the formation rate of $^3CDOM^*$ would decrease more slowly with depth compared to $^{\bullet}OH$. Moreover, organic matter produces $^3CDOM^*$ upon irradiation but does not contribute to a significant extent to its scavenging, while $^{\bullet}OH$ is photoproduced also (but not exclusively) by organic material but it is mainly scavenged by DOM.

The model can be easily extended to the direct photolysis processes, which would require the absorption spectrum of the substrate P and its photolysis quantum yield as input data. Note that $^{\bullet}OH$ is produced by the direct photolysis of nitrate and nitrite, thus the direct photolysis rates of the two ions have to be implicitly calculated to derive $R_{^{\bullet}OH}^i$. Another interesting issue deals with the reactivity of singlet oxygen ($^1O_2$), which is photogenerated by CDOM (Reactions 10.1 and 10.14). We have recently found that $R_{1O2} = [(1.1 \pm 0.3) \times 10^{-2}]P_a^{DOM}$, and

$^1O_2$ could be more important than $^\bullet OH$ for the degradation of compounds with $k_{P,\bullet OH}(k_{P,1O2})^{-1} < 100$ [113].

The correlation between chemical composition, water column depth, and photochemical reactivity may allow the inference of the possible effects of climate change on the sunlight-driven reactions in surface waters, first of all linked to the modifications of the DOM levels because of global warming [114, 115]. Additionally, the evaporative concentration of water bodies that can take place during the hot season in areas subject to desertification [116, 117] can have considerable impact on photochemistry, because both the chemical composition of water and the depth of the water column would be affected.

Acknowledgment

Financial support by INCA Consortium, Università di Torino – Ricerca Locale and PNRA – Progetto Antartide is gratefully acknowledged.

References

1. Bucheli-Witschel, M. and Egli, T. (2001) *FEMS Microbiol. Rev.*, **25**, 69–106.
2. Grabner, G. and Richard, C. (2005) in *Environmental Photochemistry Part II (The Handbook of Environmental Chemistry)*, Vol. 2M (eds P. Boule, D. Bahnemann, and P.J.K. Robertson), Springer, Berlin, pp. 161–192.
3. Pagni, R.M. and Dabestani, R. (2005) in *Environmental Photochemistry Part II (The Handbook of Environmental Chemistry)*, Vol. 2M (eds P. Boule, D. Bahnemann, and P.J.K. Robertson), Springer, Berlin, pp. 193–219.
4. Oliveira, J.L., Boroski, M., Azevedo, J.C.R., and Nozaki, J. (2006) *Acta Hydrochim. Hydrobiol.*, **34**, 608–617.
5. Bracchini, L., Cozar, A., Dattilo, A.M., Loiselle, S.A., Tognazzi, A., Azza, N., and Rossi, C. (2006) *Chemosphere*, **63**, 1170–1178.
6. Zumstein, J. and Buffle, J. (1989) *Water Res.*, **23**, 229–239.
7. Loiselle, S.A., Bracchini, L., Cozar, A., Dattilo, A.M., Tognazzi, A., and Rossi, C. (2009) *J. Photochem. Photobiol. B Biol.*, **95**, 129–137.
8. Brinkmann, T., Hörsch, P., Sartorius, D., and Frimmel, F.H. (2003) *Environ. Sci. Technol.*, **37**, 3004–3010.
9. Acharid, A., Sadiki, M., Elmanfe, G., Derkaoui, N., Olier, R., and Privat, M. (2006) *Langmuir*, **22**, 8790–8799.
10. Muller, B. and Heal, M.R. (2001) *Chemosphere*, **45**, 309–314.
11. Bucheli, T., Müller, S., Heberle, S., and Schwarzenbach, R. (1998) *Environ. Sci. Technol.*, **32**, 3457–3464.
12. Kuivila, K.M. and Jennings, B.E. (2007) *Int. J. Environ. Anal. Chem.*, **87**, 897–911.
13. Comoretto, L., Arfib, B., and Chiron, S. (2007) *Sci. Total Environ.*, **380**, 124–132.
14. Fenner, K., Canonica, S., Escher, B.I., Gasser, L., Spycher, S., and Tuelp, H.C. (2006) *Chimia*, **60**, 683–690.
15. Siffert, C. and Sulzberger, B. (1991) *Langmuir*, **7**, 1627–1634.
16. Xyla, A.G., Sulzberger, B., Luther, G.W., Hering, J.G., Van Cappellen, P., and Stumm, W. (1992) *Langmuir*, **8**, 95–103.
17. Czaplicka, M. (2006) *J. Hazard. Mater.*, **134**, 45–59.
18. Vione, D., Maurino, V., Minero, C., Pelizzetti, E., Harrison, M.A.J., Olariu, R.I., and Arsene, C. (2006) *Chem. Soc. Rev.*, **35**, 441–453.
19. Lam, M.E. and Mabury, S.A. (2005) *Aquatic. Sci.*, **67**, 177–188.

20. Vogna, D., Marotta, R., Andreozzi, R., Napolitano, A., and D'Ischia, M. (2004) *Chemosphere*, **54**, 497–505.
21. Rafqah, S., Mailhot, G., and Sarakha, M. (2006) *Environ. Chem. Lett.*, **4**, 213–217.
22. Dattilo, A.M., Decembrini, F., Bracchini, L., Focardi, S., Mazzuoli, S., and Rossi, C. (2005) *Annal. Chim.*, **95**, 177–184.
23. Hoigné, J. (1990) in *Aquatic Chemical Kinetics* (ed W. Stumm), John Wiley & Sons, Inc., New York, pp. 43–70.
24. Brezonik, P.L. and Fulkerson-Brekken, J. (1998) *Environ. Sci. Technol.*, **32**, 3004–3010.
25. Vaughan, P.P. and Blough, N.V. (1998) *Environ. Sci. Technol.*, **32**, 2947–2953.
26. Chiron, S., Minero, C., and Vione, D. (2006) *Environ. Sci. Technol.*, **40**, 5977–5983.
27. Vione, D., Maurino, V., Minero, C., and Pelizzetti, E. (2002) *Environ. Sci. Technol.*, **36**, 669–676.
28. Chiron, S., Minero, C., and Vione, D. (2007) *Environ. Sci. Technol.*, **41**, 3127–3133.
29. Chiron, S., Barbati, S., Khanra, S., Dutta, B.K., Minella, M., Minero, C., Maurino, V., and Vione, D. (2009) *Photochem. Photobiol. Sci.*, **8**, 91–100.
30. Kari, F. and Giger, W. (1995) *Environ. Sci. Technol.*, **29**, 2814–2827.
31. Meunier, L., Laubscher, H., Hug, S.J., and Sulzberger, B. (2005) *Aquatic. Sci.*, **67**, 292–307.
32. Richard, C. and Canonica, S. (2005) in *Environmental Photochemistry Part II (The Handbook of Environmental Chemistry)*, vol. **2M** (eds P. Boule, D. Bahnemann, and P.J.K. Robertson), Springer, Berlin, pp. 299–323.
33. Vione, D., Maurino, V., Minero, C., and Pelizzetti, E. (2005) in *Environmental Photochemistry Part II (The Handbook of Environmental Chemistry)*, Vol. 2M (eds P. Boule, D. Bahnemann, and P.J.K. Robertson), Springer, Berlin, pp. 221–253.
34. Huang, J. and Mabury, S.A. (2000) *Chemosphere*, **41**, 1775–1782.
35. Lam, M.W., Young, C.J., and Mabury, S.A. (2005) *Environ. Sci. Technol.*, **39**, 513–522.
36. Vione, D., Maurino, V., Minero, C., Borghesi, D., Lucchiari, M., and Pelizzetti, E. (2003) *Environ. Sci. Technol.*, **37**, 4635–4641.
37. Vione, D., Maurino, V., Minero, C., Calza, P., and Pelizzetti, E. (2005) *Environ. Sci. Technol.*, **39**, 5066–5075.
38. Vione, D., Maurino, V., Cucu Man, S., Khanra, S., Arsene, C., Olariu, R.I., and Minero, C. (2008) *ChemSusChem*, **1**, 197–204.
39. Minero, C., Maurino, V., Pelizzetti, E., and Vione, D. (2007) *Environ. Sci. Pollut. Res. Int.*, **14**, 241–243.
40. Khanra, S., Minero, C., Maurino, V., Pelizzetti, E., Dutta, B.K., and Vione, D. (2008) *Environ. Chem. Lett.*, **6**, 29–34.
41. Das, R., Dutta, B.K., Maurino, V., Vione, D., and Minero, C. (2009) *Environ. Chem. Lett.*, **7**, 337–342 doi: 10.1007/s10311-008-0176-8.
42. Minero, C., Maurino, V., Pelizzetti, E., and Vione, D. (2006) *Environ. Sci. Pollut. Res. Int.*, **13**, 212–214.
43. Minero, C., Bono, F., Rubertelli, F., Pavino, D., Maurino, V., Pelizzetti, E., and Vione, D. (2007) *Chemosphere*, **66**, 650–656.
44. Cullen, J.T., Bergquist, B.A., and Moffett, J.W. (2006) *Mar. Chem.*, **98**, 295–303.
45. Wayne, R.P. (2005) in *Environmental Photochemistry Part II (The Handbook of Environmental Chemistry)*, vol. 2M (eds P. Boule, D. Bahnemann, and P.J.K. Robertson), Springer, Berlin, pp. 1–47.
46. Caplanne, S. and Laurion, I. (2008) *Aquatic. Sci.*, **70**, 123–133.
47. Frank, R. and Klöpffer, W. (1988) *Chemosphere*, **17**, 985–994.
48. Bracchini, L., Cozar, A., Dattilo, A.M., Falcucci, M., Gonzales, R., Loiselle, S., and Hull, V. (2004) *Chemosphere*, **57**, 1245–1255.
49. Bracchini, L., Loiselle, S., Dattilko, A.M., Mazzuoli, S., Cozar, A., and Rossi, C. (2004) *Photochem. Photobiol.*, **80**, 139–149.
50. Vione, D., Minero, C., Maurino, V., and Pelizzetti, E. (2007) *Annal. Chim.*, **97**, 699–711.

51. Vione, D., Falletti, G., Maurino, V., Minero, C., Pelizzetti, E., Malandrino, M., Ajassa, R., Olariu, R.I., and Arsene, C. (2006) *Environ. Sci. Technol.*, **40**, 3775–3781.
52. Minero, C., Chiron, S., Falletti, G., Maurino, V., Pelizzetti, E., Ajassa, R., Carlotti, M.E., and Vione, D. (2007) *Aquatic. Sci.*, **69**, 71–85.
53. Minero, C., Lauri, V., Maurino, V., Pelizzetti, E., and Vione, D. (2007) *Annal. Chim.*, **97**, 685–698.
54. Vione, D., Casanova, I., Minero, C., Duncianu, M., Olariu, R.I., and Arsene, C. (2008) *Rev. Chim.*, **60**, 123–126
55. Fisher, J.M., Reese, J.G., Pellechia, P.J., Moeller, P.L., and Ferry, J.L. (2006) *Environ. Sci. Technol.*, **40**, 2200–2205.
56. Southworth, B.A. and Voelker, B.M. (2003) *Environ. Sci. Technol.*, **37**, 1130–1136.
57. Chang, C.Y., Hsieh, Y.H., Cheng, K.Y., Hsieh, L.L., Cheng, T.C., and Yao, K.S. (2008) *Water Sci. Technol.*, **58**, 873–879.
58. White, E.M., Vaughan, P.P., and Zepp, R.G. (2003) *Aquatic. Sci.*, **65**, 402–414.
59. Mazellier, P., Mailhot, G., and Bolte, M. (1997) *New J. Chem.*, **21**, 389–397.
60. King, D.W., Aldrich, R.A., and Charneeki, S.E. (1993) *Mar. Chem.*, **44**, 105–120.
61. Allen, J.M., Lucas, S., and Allen, S.K. (1996) *Environ. Toxicol. Chem.*, **15**, 107–113.
62. Buxton, G.V., Greenstock, C.L., Helman, W.P., and Ross, A.B. (1988) *J. Phys. Chem. Ref. Data*, **17**, 513–886.
63. Zepp, R.G., Hoigné, J., and Bader, H. (1987) *Environ. Sci. Technol.*, **21**, 443–450.
64. Takeda, K., Takedoi, H., Yamaji, S., Ohta, K., and Sakugawa, H. (2004) *Anal. Sci.*, **20**, 153–158.
65. Arakaki, T., Fujimura, H., Hamdun, A.M., Okada, K., Kondo, H., Oomori, T., Tanahara, A., and Taira, H. (2005) *J. Oceanogr.*, **61**, 561–568.
66. Canonica, S. and Freiburghaus, M. (2001) *Environ. Sci. Technol.*, **35**, 690–695.
67. Fischer, M. and Warneck, P. (1996) *J. Phys. Chem.*, **100**, 18749–18756.
68. Mack, J. and Bolton, J.R. (1999) *J. Photochem. Photobiol. A Chem.*, **128**, 1–13.
69. Vione, D., Maurino, V., Minero, C., and Pelizzetti, E. (2005) *Environ. Sci. Technol.*, **39**, 7921–7931.
70. Vione, D., Maurino, V., Minero, C., and Pelizzetti, E. (2001) *Chemosphere*, **45**, 893–902.
71. Vione, D., Maurino, V., Minero, C., and Pelizzetti, E. (2001) *Chemosphere*, **45**, 903–910.
72. Vione, D., Minero, C., Housari, F., and Chiron, S. (2007) *Chemosphere*, **69**, 1548–1554.
73. Chiron, S., Comoretto, L., Rinaldi, E., Maurino, V., Minero, C., and Vione, D. (2009) *Chemosphere*, **74**, 599–604.
74. Papaefthimiou, C., Cabral, M.D., Micailidou, C., Viegas, C.A., Sa-Correia, I., and Theophilidis, G. (2004) *Environ. Toxicol. Chem.*, **23**, 1211–1218.
75. Heng, Z.C., Ong, T., and Nath, J. (1996) *Mutat. Res.*, **368**, 149–155.
76. Chiron, S., Barbati, S., De Méo, M., and Botta, A. (2007) *Environ. Toxicol.*, **22**, 222–227.
77. Canonica, S., Kohn, T., Mac, M., Real, F.J., Wirz, J., and Von Gunten, U. (2005) *Environ. Sci. Technol.*, **39**, 9182–9188.
78. Neta, P., Huie, R.E., and Ross, A.B. (1988) *J. Phys. Chem. Ref. Data*, **17**, 1027–1228.
79. Canonica, S. and Tratnyek, P.G. (2003) *Environ. Toxicol. Chem.*, **22**, 1743–1754.
80. Larson, R.A. and Zepp, R.G. (1988) *Environ. Toxicol. Chem.*, **7**, 265–274.
81. Bouillon, R.C. and Miller, W.L. (2005) *Environ. Sci. Technol.*, **39**, 9471–9477.
82. Huang, J.P. and Mabury, S.A. (2000) *Environ. Toxicol. Chem.*, **19**, 1501–1507.
83. Huang, J.P. and Mabury, S.A. (2000) *Environ. Toxicol. Chem.*, **19**, 2181–2188.
84. Vione, D., Maurino, V., Minero, C., Carlotti, M.E., Chiron, S., and Barbati, S. (2009) *C. R. Chimie*, **12**, 865–871 doi: 10.1016/j.crci.2008.09.024.
85. Alegría, A., Ferrer, A., Santiago, G., Sepúlveda, E., and Flores, W. (1999) *J. Photochem. Photobiol. A Chem.*, **127**, 57–65.

86. Anastasio, C., Faust, B.C., and Rao, C.J. (1997) *Environ. Sci. Technol.*, **31**, 218–232.
87. Patai, S. (1974) *The Chemistry of the Quinonoid Compounds*, John Wiley & Sons, Inc., New York.
88. Halladja, S., Ter Halle, A., Aguer, J.P., Boulkamh, A., and Richard, C. (2007) *Environ. Sci. Technol.*, **41**, 6066–6073.
89. Canonica, S., Jans, U., Stemmler, K., and Hoigné, J. (1995) *Environ. Sci. Technol.*, **29**, 1822–1831.
90. Canonica, S., Hellrung, B., Müller, P., and Wirz, J. (2006) *Environ. Sci. Technol.*, **40**, 6636–6641.
91. Gerecke, A.C., Canonica, S., Muller, S.R., Scharer, M., and Schwarzenbach, R.P. (2001) *Environ. Sci. Technol.*, **35**, 3915–3923.
92. Rodgers, M.A.J. and Snowden, P.T. (1982) *J. Am. Chem. Soc.*, **104**, 5541–5543.
93. Boreen, A.L., Edhlund, B.L., Cotner, J.B., and McNeill, K. (2008) *Environ. Sci. Technol.*, **42**, 5492–5498.
94. Larson, R.A. and Marley, K.A. (1999) *The Handbook of Environmental Chemistry*, Environmental Photochemistry, Vol. **2L** (ed. P. Boule), Springer, Berlin, pp. 123–137.
95. Leland, J.K. and Bard, A.J. (1987) *J. Phys. Chem.*, **91**, 5076–5083.
96. Faust, B.C. and Hoffmann, M.R. (1986) *Environ. Sci. Technol.*, **20**, 943–948.
97. Faust, B.C., Hoffmann, M.R., and Bahnemann, D.W. (1989) *J. Phys. Chem.*, **93**, 6371–6381.
98. Calza, P., Maurino, V., Minero, C., Pelizzetti, E., Sega, M., and Vincenti, M. (2005) *J. Photochem. Photobiol. A Chem.*, **170**, 61–67.
99. Vione, D., Maurino, V., Minero, C., Vincenti, M., and Pelizzetti, E. (2003) *Environ. Sci. Pollut. Res. Int.*, **10**, 321–324.
100. Hasegawa, K. and Neta, P. (1978) *J. Phys. Chem.*, **82**, 854–857.
101. Saran, M., Beck-Speier, I., Fellerhoff, B., and Bauer, G. (1999) *Free Radical Biol. Med.*, **26**, 482–490.
102. Armbrust, K.L. (1999) *J. Pestic. Sci.*, **24**, 69–73.
103. Bracchini, L., Cozar, A., Dattilo, A.M., Picchi, M.P., Arena, C., Mazzuoli, S., and Loiselle, S.A. (2005) *Ecol. Modell.*, **186**, 43–54.
104. Skoog, D.A. and Leary, J.J. (1992) *Principles of Instrumental Analysis*, 4th edn, Saunders College Publishing.
105. Vione, D., Lauri, V., Minero, C., Maurino, V., Malandrino, M., Carlotti, M.E., Olariu, R.I., and Arsene, C. (2009) *Aquatic. Sci.*, **71**, 34–45 doi: 10.1007/s00027-008-8084-3.
106. Minero, C., Lauri, V., Falletti, G., Maurino, V., Pelizzetti, E., and Vione, D. (2007) *Annal. Chim.*, **97**, 1107–1116.
107. Blough, N.V. and Del Vecchio, R. (2002) in *Biogeochemistry of Marine Dissolved Organic Matter* (eds D.A. Hansell and C.A. Carlson), Academic Press, Amsterdam, pp. 509–546.
108. Del Vecchio, R. and Blough, N.V. (2002) *Mar. Chem.*, **78**, 231–253.
109. Loiselle, S.A., Bracchini, L., Dattilo, A.M., Ricci, M., Tognazzi, A., Cozar, A., and Rossi, C. (2009) *Limnol. Oceanogr.*, **54**, 590–597.
110. Huber, A., Ivey, G.N., Wake, G., and Oldham, C.E. (2008) *J. Hydraulic Eng.*, **134**, 1464–1472.
111. Braslavsky, S.E. (2007) *Pure Appl. Chem.*, **79**, 293–465.
112. Warneck, P. and Wurzinger, C. (1988) *J. Phys. Chem.*, **92**, 6278–6283.
113. Vione, D., Bagnus, D., Maurino, V., and Minero, C. (2009) *Environ. Chem. Lett.*, in press, doi: 10.1007/s10311-009-0208-z.
114. Hejzlar, J., Dubrovsky, M., Buchele, J., and Ruzicka, M. (2003) *Sci. Total Environ.*, **310**, 143–152.
115. Clements, W.H., Brooks, M.L., Kashian, D.R., and Zuellig, R.E. (2008) *Glob. Chang. Biol.*, **14**, 2201–2214.
116. Giralt, S., Julia, R., Leroy, S., and Gasse, F. (2003) *Earth Planet. Sci. Lett.*, **212**, 225–239.
117. Leroy, S., Marret, F., Giralt, S., and Bulatov, S.A. (2006) *Geophys. J. Int.*, **150**, 52–70.

Part IV
Organic Synthesis and Materials

11
Bottom-Up Approaches to Nanographenes through Organic Synthesis[1]

Diego Peña

11.1
Introduction

Graphene, the two-dimensional counterpart of graphite, is a single layer of sp^2-bonded carbon atoms packed into a structure of fused benzene rings (Figure 11.1) [1]. In the last five years, this 2D carbon-based material has attracted a great deal of attention in the physics community because of its unique properties, which include high electron mobility and high elasticity [2]. In particular, graphene can mimic relativistic effects: it is the first material in which electrons within it can behave as massless particles [3]. Evidence for the enormous impact of this material is provided by the fact that the seminal work by Novoselov, Geim, and coworkers, published in 2004 [1a], has been cited more than 1950 times to date. In that paper, the original procedure to prepare graphene follows a *top-down* approach that involves exfoliation of graphite by micromechanical cleavage. In this way some thin layers can be obtained and some of these could be monolayers: graphene flakes. Since then, several methods have been developed to produce graphene and these include vacuum graphitization of silicon carbide [4], reduction of graphite oxide [5], and sonication of exfoliated graphite [6]. All of these methods share a common drawback: each graphene sheet will be different in size and shape.

Structurally, nanosized graphenes (or nanographenes) are nothing but large polycyclic aromatic hydrocarbons (PAHs). The term *graphene* normally refers to a single macromolecule that is several microns in size, while the term *PAH* is usually employed with molecules that are rarely larger than 2 nm. The basic structures of these materials are the same, although the properties can be dramatically different depending on the size and shape (periphery) of structure. During the last century, organic synthesis has provided numerous methods to prepare large PAHs [7]. In the same way, these synthetic methodologies enable the preparation of well-defined nanographenes while avoiding the structural inhomogeneity implicit in the current syntheses of graphenes [8].

1) Dedicated to the memory of Professor Erich Clar for his pioneering achievements.

Ideas in Chemistry and Molecular Sciences: Advances in Synthetic Chemistry. Edited by Bruno Pignataro
Copyright © 2010 WILEY-VCH Verlag GmbH & Co. KGaA, Weinheim
ISBN: 978-3-527-32539-9

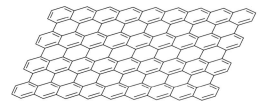

Figure 11.1 Graphene sheet.

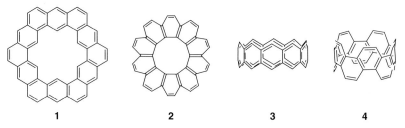

Figure 11.2 Polyarenes characterized by fusion of benzene rings in a macrocyclic arrangement.

For example, intense effort has been focused on the synthesis of large PAHs characterized by fusion of benzene rings in a macrocyclic arrangement (Figure 11.2). The most prominent example of this group that has been successfully prepared is kekulene (**1**) [9]. Other macrocyclic polyarenes such as [10]circulene (**2**) [10] or [6]$_{10}$cyclacene (**3**), a zigzag carbon nanotube subunit [11], have not been reported to date. Notably, the selective degradation of [60]fullerene led to a derivative of cyclo[10]phenacene (**4**) [12], which can be regarded as an armchair carbon nanotube subunit.

In this chapter, we will focus on PAHs formed only by six-membered rings that are not part of a macrocyclic arrangement. In particular, we will discuss alternant PAHs that contain more than 10 fused benzene rings and have been prepared through organic synthesis. As a result, substituted derivatives will not be covered here, although such molecules are frequently soluble compounds that are easier to manipulate and to characterize than unsubstituted ones, and have remarkable properties in materials science and supramolecular chemistry [8, 13]. Furthermore, some other fascinating large polycyclic aromatic compounds obtained through organic synthesis – such as nonalternant polyarenes, sterically congested aromatic hydrocarbons, or bowl-shaped PAHs – are not included in this chapter [14].

The classification of nanographenes in this chapter is based on the number of fused benzene rings. In our opinion, this is a very intuitive approach for readers without a background in chemistry – a significant group within the graphene community. This organization entails grouping together polyarenes with the same number of benzene rings but with different molecular formulae. In alternant PAHs, this formula depends not only on the number of fused benzene rings, but also on the number of internal carbon atoms (i.e., those not located in the periphery): $C_{4n+2-x}H_{2n+4-x}$ (n = number of fused benzene rings, x = number of internal carbon atoms). For example, the molecular formula of a *cata*-condensed

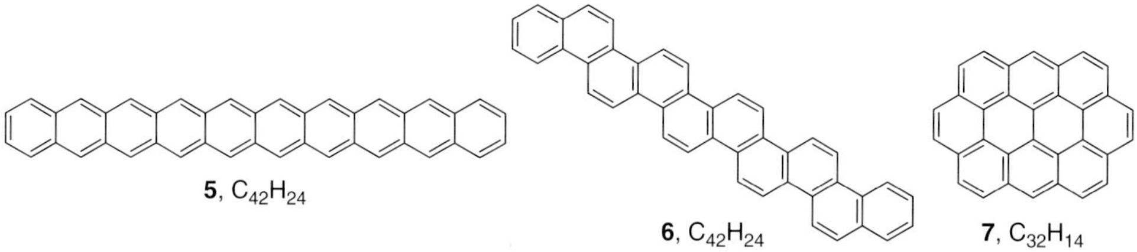

Figure 11.3 Three polyarenes with 10 fused benzene rings: [10]acene (**5**), [10]phenacene (**6**), and ovalene (**7**).

PAH with 10 fused benzene rings, such as [10]acene (**5**) or [10]phenacene (**6**) is $C_{42}H_{24}$ ($n = 10$, $x = 0$), while the formula of ovalene (**7**, Figure 11.3), a polyarene with 10 *peri*-fused benzene rings and 10 internal carbon atoms, is $C_{32}H_{14}$ ($n = 10$, $x = 10$).

Otherwise, nanographenes are normally associated with *peri*-fused polyarenes (e.g., **7**), while *cata*-fused compounds (e.g., **5** or **6**) are often named *nanoribbons*. However, in order to provide a more complete synthetic overview we will comment on the synthesis of both *peri*- and *cata*-condensed PAHs.

11.2
Alternant Polyarenes with 10 Fused Benzene Rings

Among the polyarenes with 10 fused benzene rings that have been synthesized over the last century, the *cata*-condensed structures have the molecular formula $C_{42}H_{24}$. A linearly fused PAH of this group would be [10]acene (**5**, Figure 11.3), a polyarene that has never been detected or obtained, but is predicted theoretically to be extremely reactive [15]. In fact, heptacene is the largest unsubstituted acene isolated to date. In contrast, [10]phenacene (**6**) is expected to be a stable but extremely insoluble molecule [16]. Some remarkable examples of *cata*-condensed polyarenes with 10 fused benzene rings that have been synthesized over the last century are shown in Figure 11.4.

Different methodologies have been employed to prepare the polyarenes displayed in Figure 11.4. Decastarphene (**8**) was prepared by Clar and Mullen from phenanthrene (**19**) following a five-step synthetic route based on Friedel–Crafts acylations and reductions (Scheme 11.1) [17].

By contrast, decahelicene (**9**) was obtained by Martin and coworkers by oxidative photocyclization from the corresponding bis(arylvinyl)benzene precursor **22** (Scheme 11.2) [18]. Similarly, polyarenes **10** [19], **11**–**13** [20], and **14** [21] were also prepared by photochemical methods.

Hexabenzotriphenylene (**15**) was prepared by Pascal and coworkers in 5% yield by vacuum pyrolysis of phenanthrene-9,10-dicarboxylic anhydride (**23**, Scheme 11.3), structural characterization of which by X-ray diffraction showed it to be a D_3-symmetric molecular propeller [22]. Our group described the synthesis of polyarene **15** in 68% yield by palladium-catalyzed cyclotrimerization of 9,10-didehydrophenanthrene (**25**), generated by fluoride-induced decomposition

Figure 11.4 *cata*-Condensed polyarenes with 10 fused benzene rings ($C_{42}H_{24}$).

of *o*-trimethylsilylphenanthryl triflate (**24**) at room temperature [23]. This allowed us to isolate the thermodynamically unstable C_2-symmetric conformer of hexabenzotriphenylene (**15**), which was structurally characterized by X-ray diffraction analysis by Bennett, Wenger, and coworkers [24].

Similarly, palladium-catalyzed [2+2+2]cycloaddition of 4,5-didehydrophenanthrene (**26**, Scheme 11.4) afforded double helicene **16** [25], while the regioisomeric aryne 1,2-didehydrophenanthrene (**27**) led to a mixture of compounds **14** and **17** [26]. Palladium-catalyzed cocyclotrimerization of two molecules of 2,3-didehydrotriphenylene (**28**) and benzyne (**29**) afforded polyarene **18** among other PAHs [27].

An example of a polyarene with 10 fused benzene rings and 2 internal carbon atoms ($C_{40}H_{22}$) is the octacene derivative **30** (Scheme 11.5) [28], which was prepared by Clar *et al.* from hexahydropyrene (**32**) and two molecules of naphthalene-2,3-dicarboxylic anhydride (**31**) following a synthetic route analogous to the one developed for decastarphene (**8**).

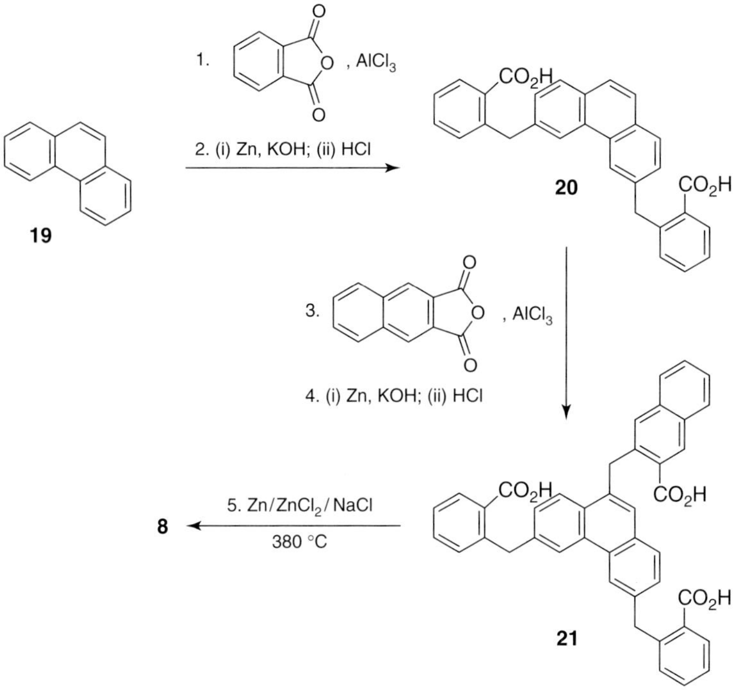

Scheme 11.1 Synthesis of decastarphene (**8**).

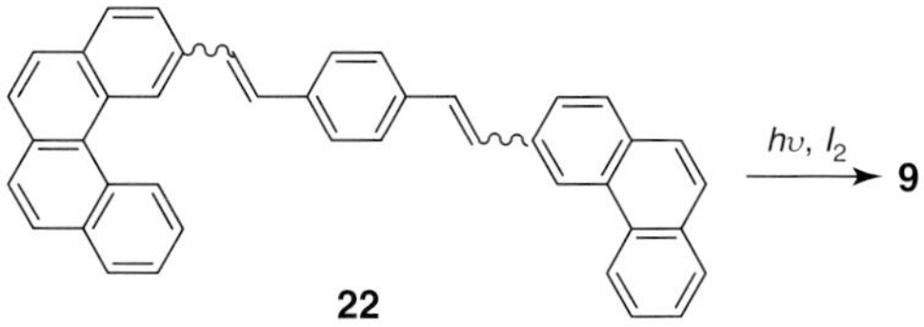

Scheme 11.2 Synthesis of decahelicene (**9**).

Some remarkable examples of polyarenes with 10 fused benzene rings and 4 internal carbon atoms ($C_{38}H_{20}$) were synthesized by Clar and coworkers, and are shown in Figure 11.5.

Dibenzoterrylene **33** was prepared by addition of two molecules of 1-naphthyllithium to 5,12-tetracenequinone (**37**, Scheme 11.6), followed by cyclization/dehydration of **38** with an $AlCl_3/NaCl$ melt [29]. Dibenzoterrylene **34** was prepared in a similar manner [30].

By contrast, polyarenes **35** [31] and **36** [32] were prepared by pyrolysis of the corresponding precursors. In particular, naphthopyrenopyrene **35** was prepared by Elbs pyrolysis of ketone **40** at 430 °C (Scheme 11.7).

Prominent examples of PAHs with 10 fused benzene rings and 6 internal carbon atoms are shown in Figure 11.6. Clar and coworkers synthesized compound **41**, from pyrene and 1-bromonaphthalene [33], and polyarene **42**, by Diels–Alder

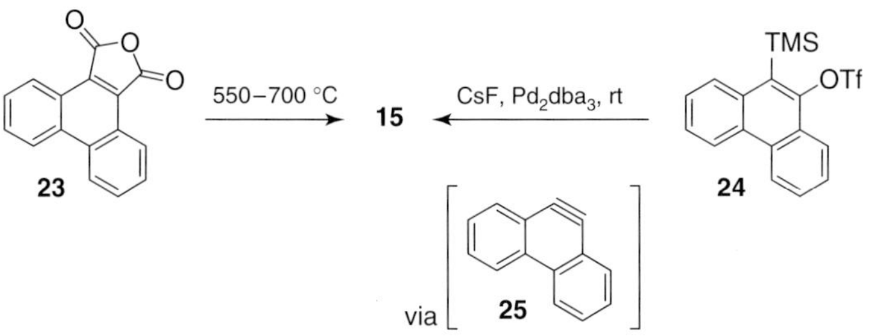

Scheme 11.3 Synthesis of hexabenzotriphenylene (**15**).

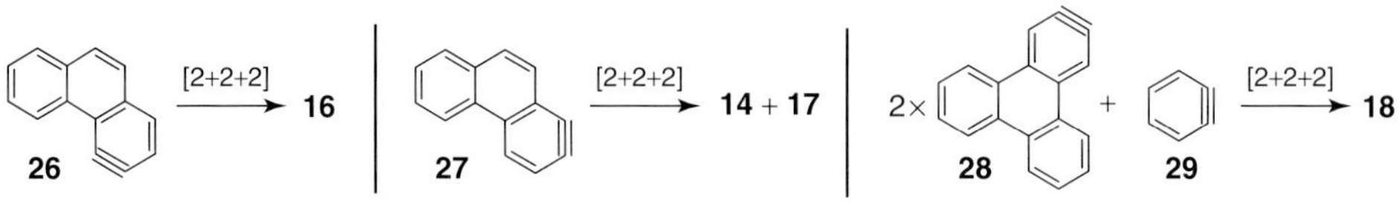

Scheme 11.4 Synthesis of polyarenes by Pd-catalyzed cyclotrimerization of arynes.

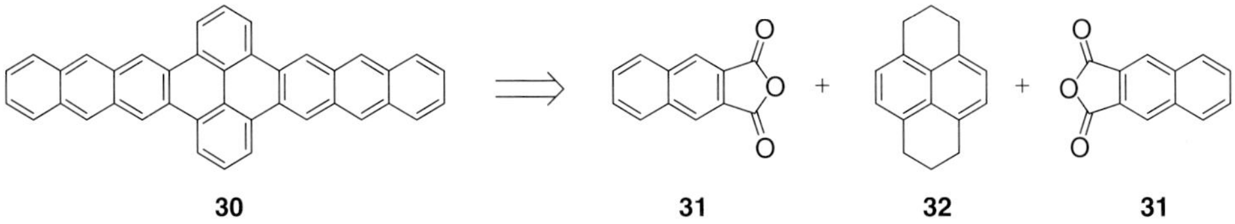

Scheme 11.5 Retrosynthesis of dibenzooctacene **30**.

reaction of dibenzoperylene and phenylacetylene followed by cyclization in an AlCl$_3$/NaCl melt [34a]. In 2000, Müllen and coworkers prepared the same compound by [4+2] cycloaddition of tetraphenylcyclopentadienone with phenylacetylene followed by CO extrusion and cyclodehydrogenation with CuCl$_2$/AlCl$_3$ [34b].

The same group reported the preparation of tribenzocoronene **43** in a procedure initiated by the addition of phenyllithium to a quinone, and tetrabenzanthanthrene **44** by reductive dimerization of a ketone derived from triphenylene with amalgamated aluminum [35]. The synthesis of polyarene **44** as a by-product was also described in the MoCl$_5$-promoted Scholl reaction of o-terphenyl [36]. Anthracenocoronene **45** was prepared by Franck and Zander by reaction of coronene with naphthalene-2,3-dicarboxylic anhydride (**31**) [37]. Umemoto et al. synthesized polyarene **46** [38]. Similar to the synthesis of **43**, compound **47** was prepared in a procedure initiated by the addition of phenyllithium to a quinone [39], while tribenzocoronene **48** was prepared by Fujioka through photochemical cyclization of o,p,o,p,o,p-hexaphenylene [40]. Fetzer and Biggs obtained a mixture of PAHs **49–51** by reductive peri-peri dimerization of phenalenones **53** and **54** (Scheme 11.8) using zinc dust in a melt of NaCl and ZnCl$_2$ at elevated temperature (Clar reaction) [41]. More recently, Fujisawa et al. prepared compound **52** following a procedure initiated by an Ullmann coupling [42].

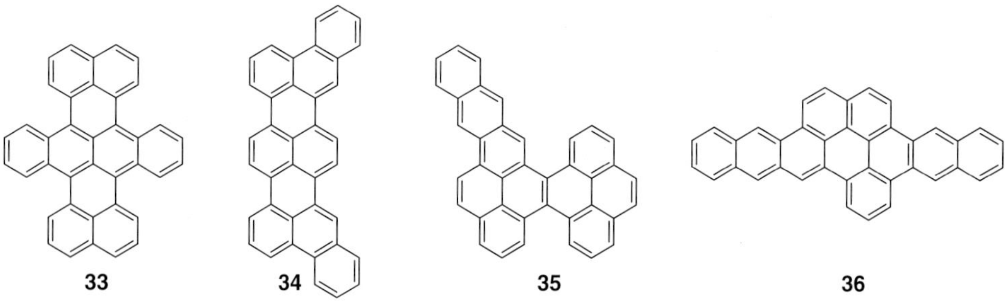

Figure 11.5 Polyarenes with 10 fused benzene rings and 4 internal carbon bonds ($C_{38}H_{20}$).

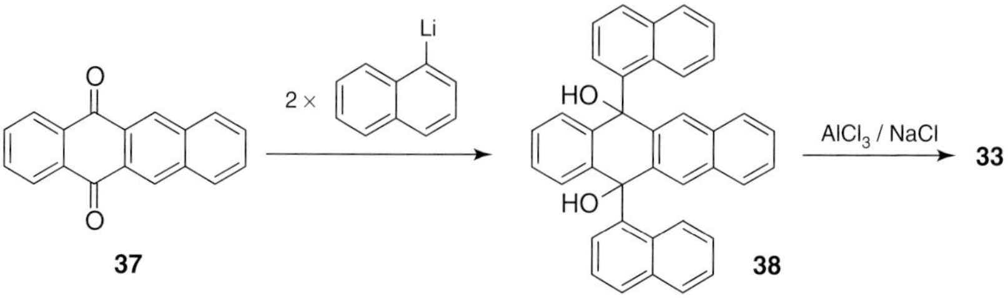

Scheme 11.6 Synthesis of dibenzoterrylene **33**.

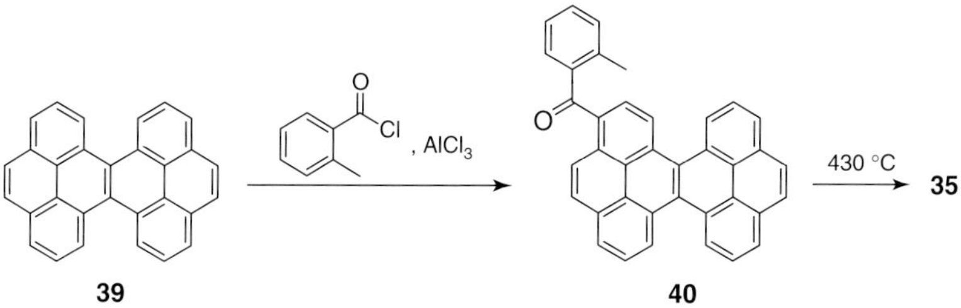

Scheme 11.7 Synthesis of naphthopyrenopyrene **35**.

PAHs **55–57** (Figure 11.7), which have 10 fused benzene rings and 8 internal carbon atoms, were prepared by Clar and coworkers. Dibenzobisanthenes **55** and **56** were originally synthesized from 6- and 8-chlorobenzanthrones [43], while compound **57** was obtained by the synthetic route shown in Scheme 11.9 [33].

Ullmann coupling of 3-bromopyrene (**58**) followed by cyclodehydrogenation of bipyrenil **59** in an $AlCl_3/NaCl$ melt afforded bipyrenilene **60**. Diels–Alder reaction with maleic anhydride in the presence of chloranil as a dehydrogenating agent followed by decarboxylation with soda lime [$Ca(OH)_2$, NaOH, KOH, H_2O] at high temperature led to compound **57** [33]. Remarkably, ovalene (**7**, Figure 11.3), a polyarene with 10 *peri*-fused benzene rings and 10 internal carbon atoms ($C_{32}H_{14}$), was prepared by Clar in a similar manner [44].

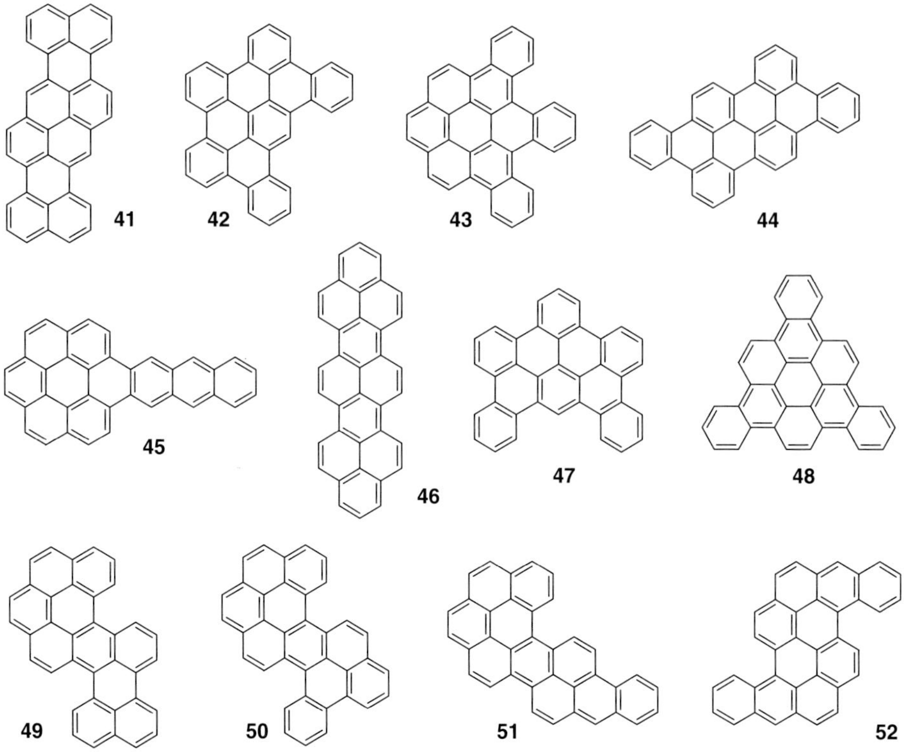

Figure 11.6 Polyarenes with 10 fused benzene rings and 6 internal carbon atoms ($C_{36}H_{18}$).

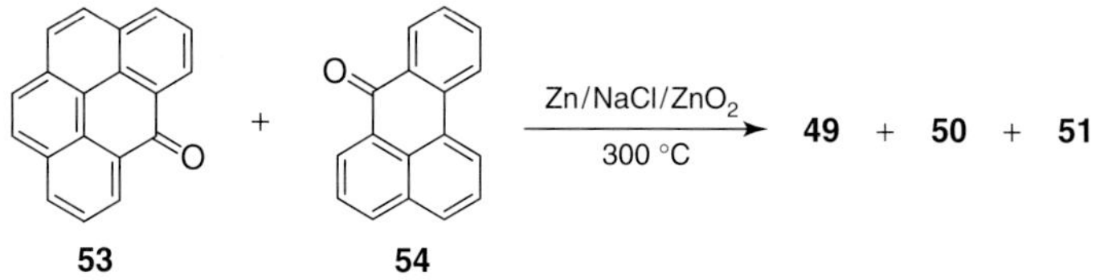

Scheme 11.8 Mixed-ketone Clar reaction.

11.3
Alternant Polyarenes with 11 Fused Benzene Rings

Polyarenes with 11 *cata*-condensed benzene rings have the molecular formula $C_{46}H_{26}$. Photochemical methods were used by Martin and Baes to prepare [11]helicene (**62**, Figure 11.8) [18, 45] and by Laarhoven *et al.* to synthesize double hexahelicene **63** [46]. According to Biermann and Schmidt tetrabenzoheptaphene **64** was prepared by Mackay and Clar [47].

UV-data for polyarene **65** (Figure 11.9), with 11 fused benzene rings and 2 internal carbon atoms ($C_{44}H_{24}$), was reported by Clar and Schmidt [48], although synthetic details were not given.

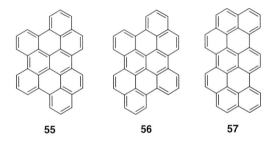

Figure 11.7 Polyarenes with 10 fused benzene rings and 8 internal carbon atoms ($C_{34}H_{16}$).

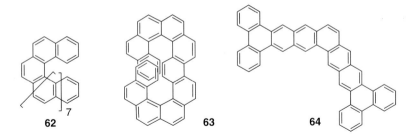

Scheme 11.9 Synthesis of polyarene **57**.

Figure 11.8 *cata*-Condensed polyarenes with 11 fused benzene rings ($C_{46}H_{26}$).

The PAHs prepared with 11 fused benzene rings and 4 internal carbon atoms ($C_{42}H_{22}$) are depicted in Figure 11.10. Tetrabenzoperopyrenes **66** [34, 49] and **67** [42], and dinaphthoperopyrenes **68–69** [50] and **70** [51] were prepared by the reductive *peri-peri* dimerization of phenalenones **74–77** (Scheme 11.10). By contrast, deep green dinaphthoheptacene **71** was obtained by condensation of two molecules of pyrene with pyromellitic dianhydride [52]. Dark-red needle crystals of tetranaphthoheptacene **72** were prepared by Clar and Macpherson and the X-ray diffraction structure was reported by Ferguson and Parvez [53]. Tribenzoterrylene **73** was prepared by Clar and Mullen by addition of two molecules of the Grignard reagent derived from 9-bromophenanthrene to anthraquinone [30].

Examples of PAHs with 11 fused benzene rings and 6 internal carbon atoms are shown in Figure 11.11.

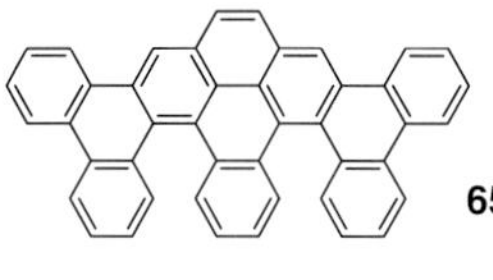

Figure 11.9 Polyarene with 11 fused benzene rings and 2 internal carbon atoms ($C_{44}H_{24}$).

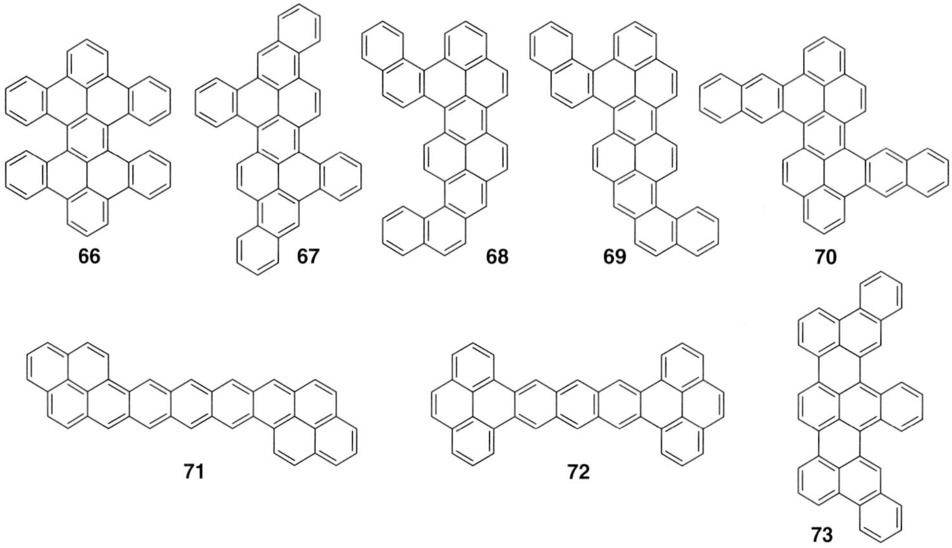

Figure 11.10 Polyarenes with 11 fused benzene rings and 4 internal carbon atoms ($C_{42}H_{22}$).

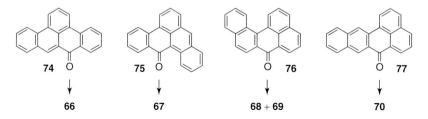

Scheme 11.10 Phenalenones employed to synthesize peropyrene derivatives.

Insoluble quaterrylene (**78**) was prepared by Clar and coworkers through the Scholl reaction of perylene (**81**) in an aluminum chloride/sodium chloride melt (Scheme 11.11) [54a]. Koch and Müllen succeeded in preparing soluble substituted quaterrylenes under milder reaction conditions [54b]. By contrast, pyrolysis of perylene (**81**) at 450 °C led to pentacene derivative **79** [55]. Tetrabenzocoronene **80** was prepared by Clar and McAndrew by reaction of tetrabenzoperylene **82** with maleic anhydride and chloranil, followed by decarboxylation with soda lime at high temperature [56].

Polyarenes with 11 fused benzene rings and 8 internal carbon atoms are shown in Figure 11.12. Orange–red dinaphthoperopyrene **83**, with two cove regions in

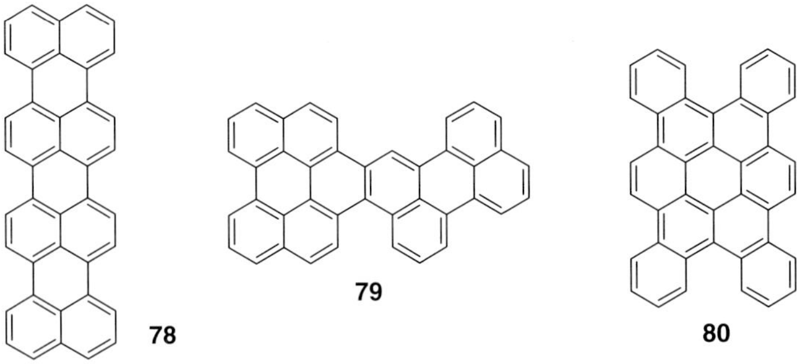

Figure 11.11 Polyarenes with 11 fused benzene rings and 6 internal carbon atoms ($C_{40}H_{20}$).

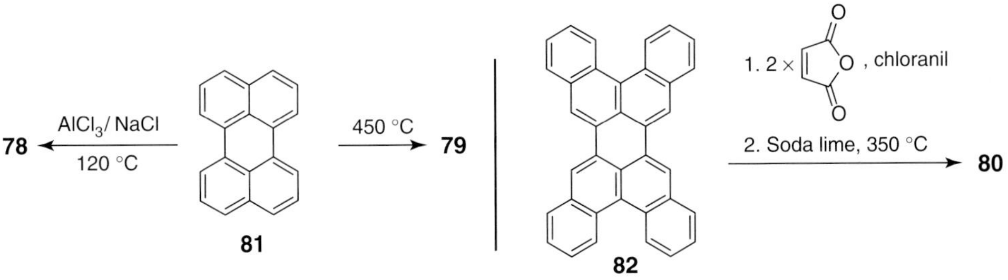

Scheme 11.11 Syntheses of polyarenes **78–80**.

its structure, was originally obtained by Clar *et al.* using the methodology based on Diels–Alder reactions with maleic anhydride [57]. The crystal structure of this compound [58a] and that of its *endo*-peroxide derivative [58b] were reported. In the 1960s, Clar and coworkers described a new method to prepare polyarene **83** through the reductive coupling of phenalenone **53** with zinc dust in a melt of NaCl and $ZnCl_2$ at 300 °C [43a]. Notably, the formation of four additional isomers was detected from the coupling of phenalenone **53**, in particular dinaphthoperopyrenes **84–87** [59]. Recently, polyarene **83** was selectively obtained from **53** by reaction with a low-valent titanium reagent [60].

Orange tetrabenzoperopyrene **88** (Scheme 11.12), a PAH with 11 fused benzene rings and 10 internal carbon atoms, was prepared from 1,5-dichloroanthraquinone (**89**) by reaction with 2-methyl-1-naphthylmagnesium bromide followed by KOH-promoted double cyclization involving the methyl groups [61]. Similarly, dinaphthocoronene **90** was prepared from 1,4-dimethylanthraquinone (**91**) and 1-naphthyllithium [62]. By contrast, polyarene **92** was isolated, together with compounds **49–51**, by reductive dimerization of phenalenones **53** and **54** [41].

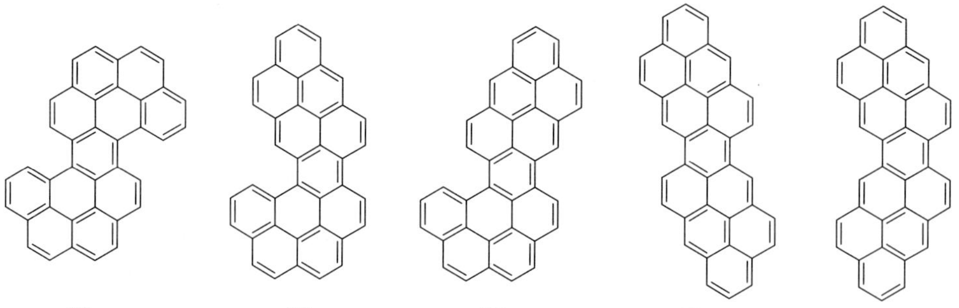

Figure 11.12 Polyarenes with 11 fused benzene rings and 8 internal carbon atoms ($C_{38}H_{18}$).

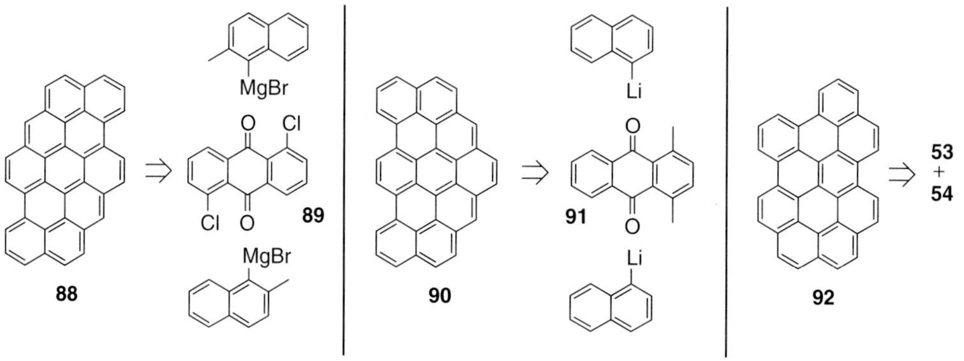

Scheme 11.12 Retrosyntheses of polyarenes with 11 fused benzene rings and 10 internal carbon atoms ($C_{36}H_{16}$).

11.4
Alternant Polyarenes with 12 Fused Benzene Rings

PAHs with 12 *cata*-condensed benzene rings have the molecular formula $C_{50}H_{28}$. Besides [12]helicene (**93**, Figure 11.13) prepared by Martin and Baes [45], double helicenes **94** [63], formed by two fused hexahelicenes, and **95** [64], formed by two heptahelicenes with two rings in common, were prepared by photochemical cyclization.

The syntheses of PAHs with 12 benzene rings and 2, 4, or 6 internal carbon bonds have not been reported to date However, three 12-ring polyarenes with eight internal carbon bonds that have been prepared are shown in Figure 11.14. Tetrabenzoterrylene **96** was prepared from dibenzoterrylene **34** by Diels–Alder reactions with maleic anhydride [30]. Tetrabenzopyranthrene **97** was isolated as a by-product in the reductive dimerization of phenalenone **74** [49a]. Reddish-brown dibenzonaphthopyranthrene **98** was prepared by the same methodology and its crystal structure has been reported [65].

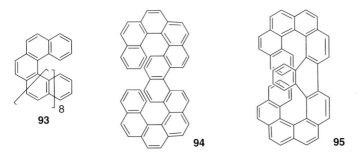

Figure 11.13 *cata*-Condensed polyarenes with 12 fused benzene rings ($C_{50}H_{28}$).

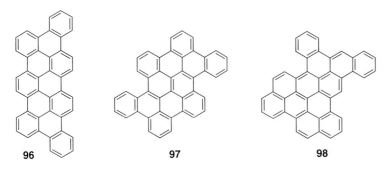

Figure 11.14 Polyarenes with 12 fused benzene rings and 8 internal carbon atoms ($C_{42}H_{20}$).

PAHs formed by 12 fused benzene rings and 10 (**99–101**, $C_{40}H_{18}$) or 12 (**102**, $C_{38}H_{16}$) internal carbon atoms are shown in Figure 11.15. Tetrabenzopyranthrene **99** was prepared by Clar and Kühn from polyarene **41** through Diels–Alder reactions with maleic anhydride followed by decarboxylation [33]. Compound **100** was synthesized by Fetzer through condensation of a mixture of phenalenone-type ketones [66]. Coronene derivative **101** was obtained by the same author by pyrolysis of a coronene–pyrene mixture [67]. Clar and Mackay reported the synthesis of circobiphenyl (**102**), a PAH with 12 fused benzene rings and 12 internal carbon atoms, by a reducing condensation of naphthanthrone (**53**) in a zinc dust melt [59a].

11.5
Alternant Polyarenes with 13 Fused Benzene Rings

Polyarenes with 13 *cata*-condensed benzene rings have the molecular formula $C_{54}H_{30}$. Tridecahelicene (**103**, Figure 11.16) was prepared by Martin *et al.* by a double photocyclization reaction [18, 20b, 68]. By contrast, tris(triphenylene) **104** was synthesized by our group through the palladium-catalyzed cyclotrimerization of 2,3-didehydrotriphenylene (**27**) [27]. Alkoxy-substituted derivatives of *supertriphenylene* **104** prepared by Müllen and coworkers showed interesting mesomorphic

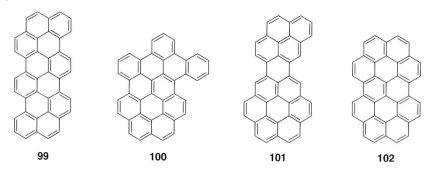

Figure 11.15 Polyarenes with 12 fused benzene rings and 10 (**100–102**, $C_{40}H_{18}$) or 12 (**103**, $C_{38}H_{16}$) internal carbon atoms.

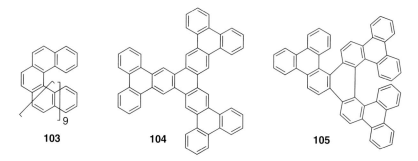

Figure 11.16 *cata*-Condensed polyarenes with 13 fused benzene rings ($C_{54}H_{30}$).

properties [69]. Double helicene **105** was also obtained by Pd-catalyzed [2+2+2] cycloaddition of 1,2-didehydrotriphenylene [70].

Polyarenes with 13 fused benzene rings and 4 (**106**, $C_{50}H_{26}$), 6 (**107**, $C_{48}H_{24}$), or 8 (**108**, **109**, $C_{46}H_{22}$) internal carbon atoms are shown in Figure 11.17. Tetranaphthopentacene **106** was reported by Clar and Schmidt [48]. Hexa-*cata*-hexabenzocoronene (**107**), first prepared by Clar and Stephen by the addition of phenyllithium to a petacenetetrone [71], has recently been prepared by Nuckolls and coworkers through photocyclization [72]. As shown by X-ray diffraction studies, this polyarene is formed by three intersecting pentacene subunits and bends into a zigzag conformation due to the steric congestion. Columnar materials based on this aromatic core have good electrical properties in thin-film transistors. In addition, the behavior of this nonplanar PAH has been investigated by scanning tunneling microscopy (STM) [73]. Tetranaphthopentacene **108** was synthesized from perylene and *p*-benzoquinone [74], while pentabenzoterrylene **109** was prepared from tribenzoterrylene **73** and maleic anhydride [30].

Polyarenes with 13 fused benzene rings and 10 (**110**, $C_{44}H_{20}$), 12 (**111**, $C_{42}H_{18}$) or 14 (**112**, $C_{40}H_{16}$) internal carbon atoms are shown in Figure 11.18. Hexabenzoperopyrene **110** [62b] and hexa-*peri*-hexabenzocoronene **111** [34a] were first prepared by Clar and coworkers. In particular, pale yellow polyarene **111** was obtained by

11.5 Alternant Polyarenes with 13 Fused Benzene Rings

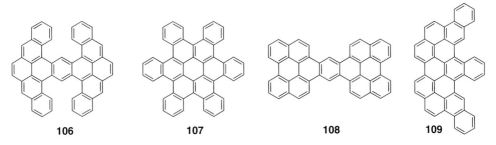

Figure 11.17 Polyarenes with 13 fused benzene rings and 4 (**106**, $C_{50}H_{26}$), 6 (**107**, $C_{48}H_{24}$), or 8 (**108**, **109**, $C_{46}H_{22}$) internal carbon atoms.

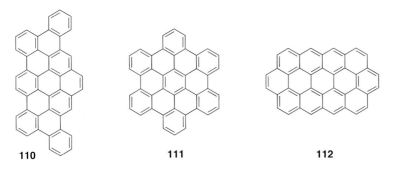

Figure 11.18 Polyarenes with 13 fused benzene rings and 10 (**110**, $C_{44}H_{20}$), 12 (**111**, $C_{42}H_{18}$), or 14 (**112**, $C_{40}H_{16}$) internal carbon atoms.

cyclization of tetrabenzoperopyrene **66** at 480 °C. Halleux et al. synthesized **111** in low yield by cyclodehydrogenation of hexaphenylbenzene in an AlCl₃/NaCl melt and by reductive dimerization of dibenzophenalenone **74** with Zn/ZnCl₂ [49a].

By contrast, Schmidt and coworkers prepared hexabenzocoronene **111** in a procedure initiated by the addition of phenyllithium to a quinone [39]. However, the most efficient method to synthesize compound **111** was developed by the group of Klaus Müllen in a process that involved Co-catalyzed cyclotrimerization of diphenylacetylene (**113**, Scheme 11.13) followed by Cu(II)-promoted Scholl-type oxidative cyclodehydrogenation of hexaphenylbenzene (**114**) catalyzed by AlCl₃ [75]. During the last 15 years this group has reported the syntheses of a vast number of substituted hexabenzocoronenes with promising self-assembly and electronic properties [8].

Circumanthracene (**112**) was prepared by Broene and Diederich in 10 steps starting from 1,4-benzoquinone [76]. Crucial steps in this synthetic route were a fourfold photocyclization of compound **115** to afford **116** in 93% yield (Scheme 11.13) and

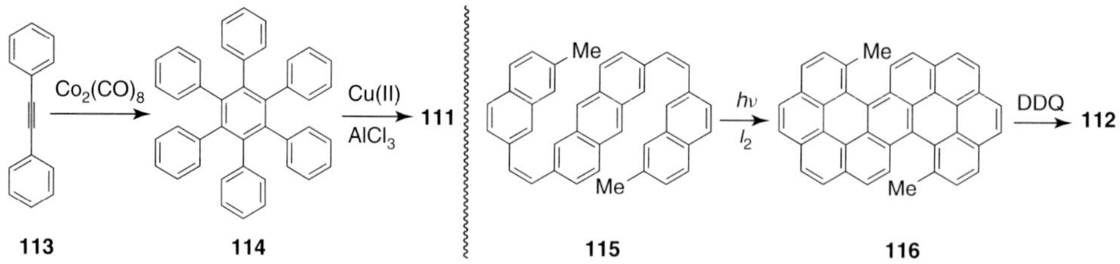

Scheme 11.13 Synthesis of hexa-*peri*-hexabenzocoronene **111** and circumanthracene **112**.

subsequent 2,3-dichloro-5,6-dicyanobenzoquinone (DDQ)-promoted cyclization to obtain **112** in 50% yield.

11.6
Alternant Polyarenes with 14 Fused Benzene Rings

Polyarenes with 14 fused benzene rings are scarce, and those that have been reported are shown in Figure 11.19 [14]. Helicene (**117**) is a *cata*-condensed PAH with molecular formula $C_{58}H_{32}$ and this is the longest helicene prepared to date. The synthesis of this compound was achieved by a double photocyclization reaction [45]. The preparation of penterrylene (**118**), a polyarene with eight internal carbon atoms ($C_{50}H_{24}$), was reported by Naamann *et al.* [77]. A derivative of this polyarene substituted with *tert*-butyl groups was successfully obtained by Koch and Müllen through a cyclization promoted by an $AlCl_3/CuCl_2$ mixture [54b]. Tribenzoquaterrylene **119** (12 internal carbon atoms, $C_{46}H_{20}$) was obtained by pyrolysis of a mixture formed by coronene and benzo[*ghi*]perylene [67]. Among the group of PAHs with 14 fused benzene rings and 14 internal carbon atoms ($C_{44}H_{18}$), cyclodehydrogenation of a tetraphenylbenzopyrene derivative led to the isolation of planar PAH **120**, with a partial zigzag periphery, in 87% yield [78], while peripentacene **121** was obtained during the sublimation of pentacene at temperature greater than 300 °C [79].

11.7
Alternant Polyarenes with More than 15 Fused Benzene Rings

Giant PAHs with more than 15 fused six-membered rings are extremely insoluble compounds, and due to their graphene-like structure and size, they have attracted a huge amount of interest from the materials science community. Undoubtedly, most of the more important synthetic contributions to this group of polyarenes came from the superb work of Klaus Müllen and his coworkers [8].

Polyarenes with 15 fused benzene rings are shown in Figure 11.20. Polybenzoid hydrocarbon **122** was prepared by Müllen and coworkers through double

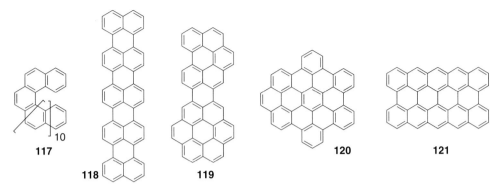

Figure 11.19 Polyarenes with 14 fused benzene rings.

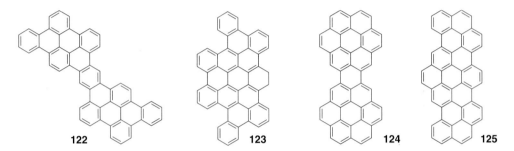

Figure 11.20 Polyarenes with 15 fused benzene rings.

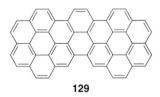

Figure 11.21 Polyarenes with 16 fused benzene rings.

intramolecular [4+2] cycloaddition of **126** followed by dehydrogenation of compound **127** with DDQ and oxidative cyclodehydrogenation of **128** in a mixture of AlCl$_3$/CuCl$_2$ in CS$_2$ (Scheme 11.14) [80].

According to Vacha and Tani [81], benzodiphenanthrobisanthene **123** was synthesized by Hendel and Schmidt, while dicoronene (**124**) was prepared by Zander and Franke through coronene condensation in an AlCl$_3$/NaCl melt [82]. Tetrabenzoquaterrylene **125** is a commercially available compound, although its synthesis has not been reported to date [83].

Compound **129** (Figure 11.21), a polyarene with 16 fused benzene rings, was prepared by Zander and Friedrichsen through reaction of dicoronene (**124**) with maleic anhydride [84].

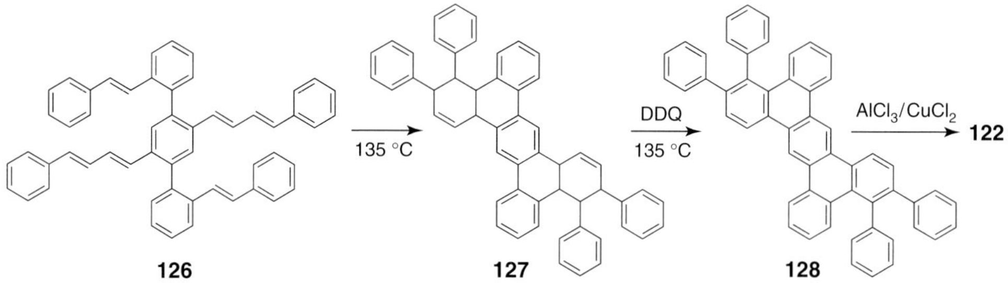

126 127 128

Scheme 11.14 Synthesis of polyarene **122**.

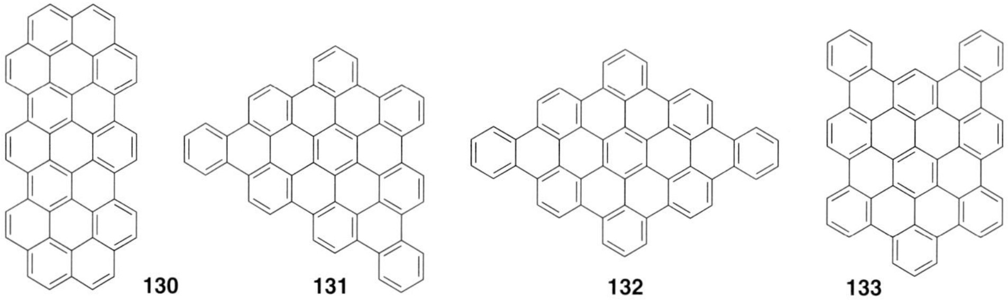

130 131 132 133

Figure 11.22 Polyarenes with 17 fused benzene rings.

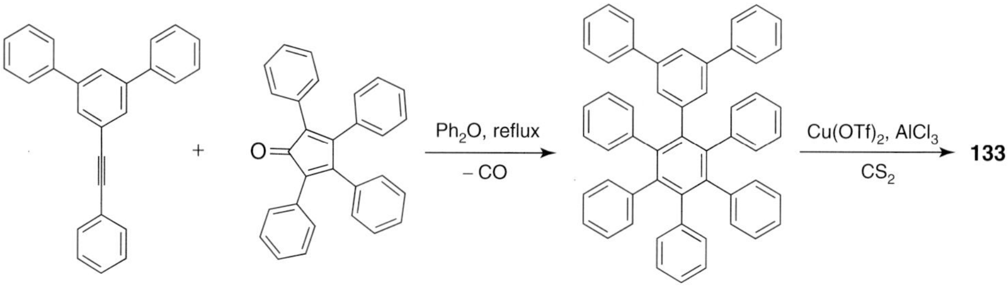

134 135 136

Scheme 11.15 Synthesis of polyarene **133**.

Polyarenes with 17 fused benzene rings are shown in Figure 11.22. Circumparaterphenyl **130** was prepared by Zander and Friedrichsen by reaction of **129** with maleic anhydride [84]. The synthesis of compound **131** by a cycloaddition–cyclodehydrogenation sequence from a stilbenoid was reported by Müllen and coworkers, although conclusive structural proof was not presented [85]. The same group successfully prepared the rhombus polyarene **132** using the same methodology [80] and the arrowlike PAH **133** was synthesized following the route depicted in Scheme 11.15 [86].

Diels–Alder cycloaddition of alkyne **134** with tetraphenylcyclopenadienone (**135**) followed by CO extrusion led to oligophenylene **136** in 77% yield.

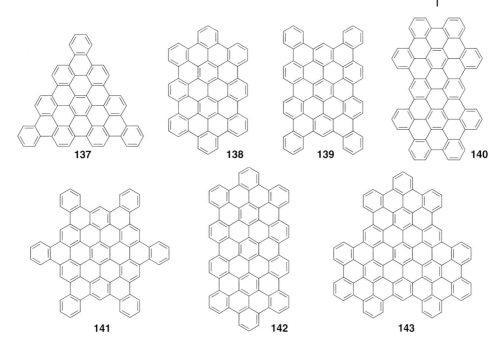

Figure 11.23 Nanographenes with 19–34 fused benzene rings.

Cyclodehydrogenation of this compound with copper(II) trifluoromethanesulfonate/aluminum(III) chloride afforded the desired polyarene **133** as a brownish orange powder in 75% yield. A tetrasubstituted derivative of compound **133** with four dodecyl chains was prepared by the same route, and the presence of these alkyl chains provided sufficient solubility for characterization [86].

The polyarenes shown in Figure 11.23 were prepared by Müllen and coworkers in excellent yields through a final oxidative cyclodehydrogenation step from the corresponding oligophenylene with $Cu(OTf)_2/AlCl_3$ or $FeCl_3$. In particular, they reported the synthesis of triangle-shaped **137** (19 rings) [87], decagram quantities of the C60 hydrocarbon **138** (20 rings) [88], cove-type edge polyarene **139** (21 rings) [86], supernaphthalene **140** (24 rings) [89], starlike PAH **141** (25 rings) [86], C78 hydrocarbon **142** (27 rings) [90], and C96 superphenalene **143** (34 rings) [89a].

The nanographenes depicted in Figure 11.24 were also synthesized by Müllen and coworkers using the powerful cyclodehydrogenation methodology: graphitic molecule with partial zigzag periphery **144** (37 rings) [78], armchair-like periphery PAH **145** (43 rings) [86], supertriphenylene **146** (46 rings) [89a], C132 hydrocarbon **147** (50 rings) [91], threefold symmetric C150 nanographene **148** (55 rings) [92], C156 polyarene **149** (59 rings) [93], and C222 nanographene disk **150** (91 rings) [89a], [94], which is the largest PAH synthesized to date. This giant polyarene has a diameter of 3.1 nm and was prepared by planarizing the corresponding oligophenylene precursor through the removal of 108 hydrogens. The material was characterized by isotopically resolved MALDI-TOF mass spectroscopy.

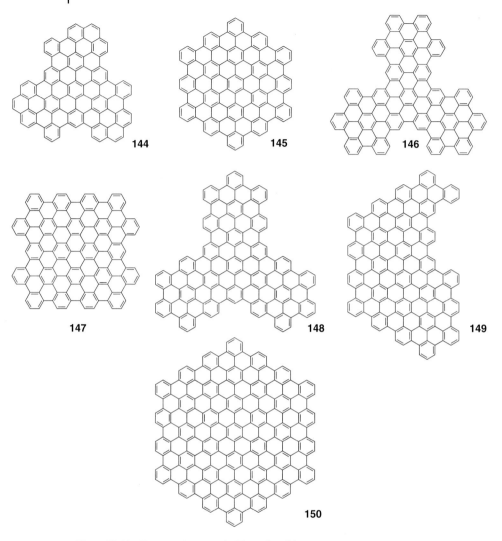

Figure 11.24 Nanographenes with 37–91 fused benzene rings.

11.8
Conclusions and Outlook

Over the past six decades, from Clar's synthesis of a polyarene with 10 fused benzene rings (ovalene) to Müllen's preparation of a graphene disk formed by 91 fused rings, organic synthesis has provided numerous methods to obtain large PAHs. The chemistry of these compounds underwent a remarkable revival when graphene, the two-dimensional counterpart of graphite with unique properties, burst onto the scientific scene. Synthetic methods that had been developed to synthesize large PAHs can now be used to prepare graphenes through bottom-up approaches.

The main advantage of using these chemical methodologies based on organic synthesis is that they avoid the structural inhomogeneity implicit in the current approaches to prepare graphenes. In addition, these chemical methods provide access to nanographenes with different peripheries and substitution, features that are crucial to develop new functional materials. On the other hand, as recently shown by Müllen and coworkers [93], chemical synthesis in solution, in particular, planarization by cyclodehydrogenation, faces limitations for graphenes that exceed 4 nm in size. New synthetic methods will be required to prepare nanographenes beyond this size, probably assisted by surface chemistry, polymer technology, and scanning probe microscopy.

Acknowledgments

Financial support from the Xunta de Galicia, FEDER, and the Spanish Ministry of Education and Science is gratefully acknowledged. I am deeply grateful to Enrique Guitián, Dolores Pérez, and all of our coworkers.

References

1. (a) Novoselov, K.S., Geim, A.K., Morozov, S.V., Jiang, D., Zhang, Y., Dubonos, S.V., Grigorieva, I.V., and Firsov, A.A. (2004) *Science*, **306**, 666; (b) Novoselov, K.S., Geim, A.K., Morozov, S.V., Jiang, D., Katsnelson, M.I., Grigorieva, I.V., Dubonos, S.V., and Firsov, A.A. (2005) *Nature*, **438**, 197; (c) Meyer, J.C., Geim, A.K., Katsnelson, M.I., Novoselov, K.S., Booth, T.J., and Roth, S. (2007) *Nature*, **446**, 60; (d) Geim, A.K. and Novoselov, K.S. (2007) *Nat. Mater.*, **6**, 183.
2. (a) Lee, C., Wei, X., Kysar, J.W., and Hone, J. (2008) *Science*, **321**, 385; (b) Kim, K.S., Zhao, Y., Jang, H., Lee, S.Y., Kim, J.M., Kim, K.S., Ahn, J.-H., Kim, P., Choi, J.-Y., and Hong, B.H. (2009) *Nature*, **457**, 706.
3. Heersche, H.B., Jarillo-Herrero, P., Oostinga, J.B., Vandersypen, L.M.K., and Morpurgo, A.F. (2007) *Nature*, **446**, 56.
4. Berger, C., Shong, Z., Li, X., Wu, X., Brown, N., Naud, C., Mayou, D., Li, T., Hass, J., Marchenkov, A.N., Conrad, E.H., First, P.N., and de Heer, W.A. (2006) *Science*, **312**, 1191.
5. (a) Dikin, D.A., Stankovich, S., Zimney, E.J., Piner, R.D., Dommett, G.H.B., Evmenenko, G., Nguyen, S.T., and Ruoff, R.S. (2007) *Nature*, **448**, 457; (b) Xu, Y., Bai, H., Lu, G., Li, C., and Shi, G. (2008) *J. Am. Chem. Soc.*, **130**, 5856.
6. Li, X., Wang, X., Zhang, L., Lee, S., and Dai, H. (2008) *Science*, **319**, 1129.
7. (a) Scholl, R., Seer, C., and Weitzenböck, R. (1910) *Chem. Ber.*, **43**, 2202; (b) Clar, E. (1964) *Polycyclic Hydrocarbons*, vol. I/II, Academic Press, New York; (c) Harvey, R.G. (1997) *Polycyclic Aromatic Hydrocarbons*, Wiley-VCH Verlag GmbH, New York; (d) Fetzer, J.C. (2000) *Large (C>=24) Polycyclic Aromatic Hydrocarbons: Chemistry and Analysis*, Wiley-VCH Verlag GmbH, New York; (e) Harvey, R.G. (2004) *Curr. Org. Chem.*, **8**, 303.
8. (a) Wu, J., Pisula, W., and Müllen, K. (2007) *Chem. Rev.*, **107**, 718; (b) Müllen, K. and Rabe, J.P. (2008) *Acc. Chem. Res.*, **41**, 511; (c) Zhi, L. and Müllen, K. (2008) *J. Mater. Chem.*, **18**, 1472.
9. Staab, H.A. and Diederich, F. (1983) *Chem. Ber.*, **116**, 3487.
10. Christoph, H., Grunenberg, J., Hopf, H., Dix, I., Jones, P.G., Scholtissek, M., and Maier, G. (2008) *Chem. Eur. J.*, **14**, 5604.

11. (a) Girreser, U., Giuffrida, D., Kohnke, F.H., Mathias, J.P., Philp, D., and Stoddart, J.F. (1993) *Pure Appl. Chem.*, **65**, 119; (b) Choi, H.S. and Kim, K.S. (1999) *Angew. Chem. Int. Ed.*, **38**, 2256; (c) Chen, Z., Jiang, D., Lu, X., Bettinger, H.F., Dai, S., Schleyer, P.R., and Houk, K.N. (2007) *Org. Lett.*, **9**, 5449.

12. Matsuo, Y., Tahara, K., Sawamura, M., and Nakamura, E. (2004) *J. Am. Chem. Soc.*, **126**, 8725.

13. (a) Hill, J.P., Jin, W., Kosaka, A., Fukushima, T., Ichihara, H., Shimomura, T., Ito, K., Hashizume, T., Ishii, N., and Aida, T. (2004) *Science*, **304**, 1481; (b) Grimsdale, A.C. and Müllen, K. (2005) *Angew. Chem. Int. Ed.*, **44**, 5592; (c) Basu, C., Barthes, C., Sadhukhan, S.K., Girdhar, N.K., and Gourdon, A. (2007) *Eur. J. Org. Chem.*, 136; (d) Wang, X., Zhi, L., Tsao, N., Tomović, Z., Li, J., and Müllen, K. (2008) *Angew. Chem. Int. Ed.*, **47**, 2990.

14. (a) Yamamoto, K., Saitho, Y., Iwaki, D., and Ooka, T. (1991) *Angew. Chem. Int. Ed. Engl.*, **30**, 1173; (b) Boorum, M.M., Vasil'ev, Y.V., Drewello, T., and Scott, L.T. (2001) *Science*, **294**, 828; (c) Scott, L.T., Boorum, M.M., McMahon, B.J., Hagen, S., Mack, J., Blank, J., Wegner, H., and de Meijere, A. (2002) *Science*, **295**, 1500; (d) Rosei, F., Schunack, M., Jiang, P., Gourdon, A., Laegsgaard, E., Stensgaard, I., Joachim, C., and Besenbacher, F. (2002) *Science*, **296**, 328; (e) Elliott, E.L., Orita, A., Hasegawa, D., Gantzel, P., Otera, J., and Siegel, J.S. (2004) *Org. Biomol. Chem.*, **3**, 581; (f) Gómez-Lor, B., González-Cantalapiedra, E., Ruiz, M., de Frutos, Ó., Cárdenas, D.J., Santos, A., and Echavarren, A.M. (2004) *Chem. Eur. J.*, **10**, 2601; (g) Lu, J., Ho, D.M., Vogelaar, N.J., Kraml, C.M., and Pascal, R.A. Jr. (2004) *J. Am. Chem. Soc.*, **126**, 11168; (h) Pascal, R.A. Jr. (2006) *Chem. Rev.*, **106**, 4809; (i) Wu, Y.-T. and Siegel, J.S. (2006) *Chem. Rev.*, **106**, 4843; (j) Tsefrikas, V.M. and Scott, L.T. (2006) *Chem. Rev.*, **106**, 4868; (k) Rim, K.T., Siaj, M., Xiao, S., Myers, M., Carpentier, V.D., Liu, L., Su, C., Steigerwald, M.L., Hybertsen, M.S., McBreen, P.H., Flynn, G.W., and Nuckolls, C. (2007) *Angew. Chem. Int. Ed.*, **46**, 7891; (l) Sygula, A., Fronczek, F.R., Sygula, R., Rabideau, P.W., and Olmstead, M.M. (2007) *J. Am. Chem. Soc.*, **129**, 3842; (m) Otero, G., Biddau, G., Sánchez-Sánchez, C., Caillard, R., López, M.F., Rogero, C., Palomares, F.J., Cabello, N., Basanta, M.A., Ortega, J., Méndez, J., Echavarren, A.M., Pérez, R., Gómez-Lor, B., and Martín-Gago, J.A. (2008) *Nature*, **454**, 865.

15. (a) Aihara, H. (1999) *Phys. Chem. Chem. Phys.*, **1**, 3193; (b) Bendikov, M., Duong, H.M., Starkey, K., Houk, K.N., Carter, E.A., and Wudl, F. (2004) *J. Am. Chem. Soc.*, **126**, 7416; (c) Mondal, R., Shah, B.K., and Neckers, D.C. (2006) *J. Am. Chem. Soc.*, **128**, 9612; (d) Anthony, J.E. (2008) *Angew. Chem. Int. Ed.*, **47**, 452.

16. Mallory, F.B., Butler, K.E., Evans, A.C., Butler, K.E., Evans, A.C., Brondyke, E.J., Mallory, C.W., Yang, C., and Ellenstein, A. (1997) *J. Am. Chem. Soc.*, **119**, 2119.

17. Clar, E. and Mullen, A. (1968) *Tetrahedron*, **24**, 6719.

18. Moradpour, A., Kagan, H., Baes, M., Morren, G., and Martin, R.H. (1975) *Tetrahedron*, **31**, 2139.

19. Laarhoven, W.H. and Nivard, R.J.F. (1976) *Tetrahedron*, **32**, 2445.

20. (a) Laarhoven, W.H. and Cuppen, H.J.M. (1971) *Tetrahedron Lett.*, **12**, 163; (b) Laarhoven, W.H. and Nivard, R.J.F. (1972) *Tetrahedron*, **28**, 1803; (c) Laarhoven, W.H. and Cuppen, H.J.M. (1973) *Recl. Trav. Chim. Pays-Bas*, **92**, 553.

21. Mehta, G., Panda, G., Shah, S.R., and Kunwar, A.C. (1997) *J. Chem. Soc., Perkin Trans. 1*, 2269.

22. Barnett, L., Ho, D.M., Baldridge, K.K., and Pascal, R.A. Jr. (1999) *J. Am. Chem. Soc.*, **121**, 727.

23. Peña, D., Cobas, A., Pérez, D., Guitián, E., and Castedo, L. (2000) *Org. Lett.*, **2**, 1629.

24. Bennett, M.A., Kopp, M.R., Wenger, E., and Willis, A.C. (2003) *J. Organomet. Chem.*, **667**, 8.

25. Peña, D., Cobas, A., Pérez, D., Guitián, E., and Castedo, L. (2003) *Org. Lett.*, **5**, 1863.

26. Guitián, E., Pérez, D., and Peña, D. (2005) *Top. Organomet. Chem.*, **14**, 109.

27. Romero, C., Peña, D., Pérez, D., and Guitián, E. (2006) *Chem. Eur. J.*, **12**, 5677.
28. Clar, E., Guye-Vuillème, J.F., McCallum, S., and Macpherson, I.A. (1963) *Tetrahedron*, **19**, 2185.
29. Clar, E. and Willicks, W. (1955) *Chem. Ber.*, **88**, 1205.
30. Clar, E. and Mullen, A. (1971) *Tetrahedron*, **27**, 5239.
31. (a) Clar, E., Guye-Vuillème, J.F., and Stephen, J.F. (1964) *Tetrahedron*, **20**, 2107; (b) Clar, E., Robertson, J.M., Schlögl, R., and Schmidt, W. (1981) *J. Am. Chem. Soc.*, **103**, 1320.
32. Clar, E., Ironside, C.T., and Zander, M. (1966) *Tetrahedron*, **22**, 3527.
33. Clar, E. and Kühn, O. (1956) *Liebigs Ann.*, **601**, 181.
34. (a) Clar, E., Ironside, C.T., and Zander, M. (1959) *J. Chem. Soc.*, 142; (b) Kübel, C., Eckhardt, K., Enkelmann, V., Wegner, G., and Müllen, K. (2000) *J. Mater. Chem.*, **10**, 979.
35. Clar, E. and McCallum, A. (1964) *Tetrahedron*, **20**, 507.
36. King, B.T., Kroulík, J., Robertson, C.R., Rempala, P., Hilton, C.L., Korinek, J.D., and Gortari, L.M. (2007) *J. Org. Chem.*, **72**, 2279.
37. Franck, H.-G. and Zander, M. (1966) *Chem. Ber.*, **99**, 1272.
38. Umemoto, S., Kawashima, T., Sakata, Y., and Misumi, S. (1975) *Tetrahedron Lett.*, 1005.
39. Hendel, W., Khan, Z.H., and Schmidt, W. (1986) *Tetrahedron*, **42**, 1127.
40. Fujioka, Y. (1985) *Bull. Chem. Soc. Jpn.*, **58**, 481.
41. Fetzer, J.C. and Biggs, W.R. (1988) *Org. Prep. Proced. Int.*, **20**, 223.
42. Fujisawa, S., Takekawa, M., Nakamura, Y., Uchida, A., Ohshima, S., and Oonishi, I. (1999) *Polycyclic Aromat. Compd.*, **14-15**, 99.
43. (a) Clar, E., Fell, G.S., Ironside, C.T., and Balsillie, A. (1960) *Tetrahedron*, **10**, 26; (b) Aoki, J. (1964) *Bull. Chem. Soc. Jpn.*, **37**, 1079.
44. Clar, E. (1948) *Nature*, **161**, 238.
45. Martin, R.H. and Baes, M. (1975) *Tetrahedron*, **31**, 2135.
46. Laarhoven, W.H., Cuppen, J.H.M., and Nivard, R.J.F. (1974) *Tetrahedron*, **30**, 3343.
47. Biermann, D. and Schmidt, W. (1980) *J. Am. Chem. Soc.*, **102**, 3173.
48. Clar, E. and Schmidt, W. (1979) *Tetrahedron*, **35**, 1027.
49. (a) Halleux, A., Martin, R.H., and King, G.S.D. (1958) *Helv. Chim. Acta*, **41**, 1177; (b) Fujimaki, Y., Takekawa, M., Fujisawa, S., Ohshima, S., and Sakamoto, Y. (2004) *Polycyclic Aromat. Compd.*, **24**, 107.
50. Stephenson, M. and Sutcliffe, F.K. (1962) *J. Chem. Soc.*, 3516.
51. Fujimaki, Y., Suga, A., Takekawa, M., Ohshima, S., and Sakamoto, Y. (2004) *Polycyclic Aromat. Compd.*, **24**, 279.
52. Boggiano, B. and Clar, E. (1957) *J. Chem. Soc.*, 2681.
53. Ferguson, G. and Parvez, M. (1979) *Acta Crystallogr.*, **B35**, 2419.
54. (a) Clar, E., Kelly, W., and Laird, R.M. (1956) *Monatsh. Chem.*, **87**, 391; (b) Koch, K.-H. and Müllen, K. (1991) *Chem. Ber.*, **124**, 2091.
55. (a) Zinke, A., Nudmueller, H., and Ott, R. (1955) *Monatsh. Chem.*, **86**, 853; (b) Ott, R., Zeschko, E., and Zinke, A. (1963) *Monatsh. Chem.*, **94**, 51.
56. Clar, E. and McAndrew, B.A. (1972) *Tetrahedron*, **28**, 1137.
57. Clar, E., Kelly, W., Robertson, J.M., and Rossmann, M.G. (1956) *J. Chem. Soc.*, 3878.
58. (a) Robertson, J.M. and Trotter, J. (1959) *J. Chem. Soc.*, 2614; (b) Ohshima, S., Uchida, A., Horiguchi, S., Suzuki, A., Fujisawa, S., and Oonishi, I. (1994) *Bull. Chem. Soc. Jpn.*, **67**, 924.
59. (a) Clar, E. and Mackay, C.C. (1972) *Tetrahedron*, **28**, 6041; (b) Fujisawa, S., Aoki, J., Takekawa, M., and Iwashima, S. (1979) *Bull. Chem. Soc. Jpn.*, **52**, 2159; (c) Fujisawa, S., Oonishi, I., Aoki, J., Ohashi, Y., and Sasada, Y. (1982) *Bull. Chem. Soc. Jpn.*, **55**, 3424; (d) Fetzer, J.C. (1999) *Polycyclic Aromat. Compd.*, **14-15**, 1; (e) Nakamura, Y., Uchida, A., Ohshima, S., Oonishi, I., and Fujisawa, S. (1999) *Polycyclic Aromat. Compd.*, **14-15**, 265.
60. Pogodin, S. and Agranat, I. (1999) *Org. Lett.*, **1**, 1387.

61. Clar, E. and Kelly, W. (1956) *J. Chem. Soc.*, 3875.
62. (a) Clar, E., Schmidt, W., and Zauder, T. (1979) German Patent Application P2914994.8; (b) Clar, E. and Schmidt, W. (1978) *Tetrahedron*, **34**, 3219.
63. Laarhoven, W.H. and de Jong, M.H. (1973) *Recl. Trav. Chim. Pays-Bas*, **92**, 651.
64. Martin, R.H., Eyndels, Ch., and Defay, N. (1974) *Tetrahedron*, **30**, 3339.
65. Fujisawa, S., Oonishi, I., Aoki, J., and Ohashi, Y. (1986) *Acta Cryst.*, **C42**, 1390.
66. Fetzer, J.C. (1988) *Adv. Chem. Ser.*, **217**, 309.
67. Fetzer, J.C. (1996) *Polycyclic Aromat. Compd.*, **11**, 317.
68. Martin, R.H., Morren, G., and Schurter, J.J. (1969) *Tetrahedron Lett.*, **10**, 3683.
69. Yatabe, T., Harbison, M.A., Brand, J.D., Wagner, M., Müllen, K., Samori, P., and Rabe, J.P. (2000) *J. Mater. Chem.*, **10**, 1519.
70. Romero, C., Peña, D., Pérez, D., and Guitián, E. (2008) *J. Org. Chem.*, **73**, 7996.
71. Clar, E. and Stephen, J.F. (1965) *Tetrahedron*, **21**, 467.
72. Xiao, S., Myers, M., Miao, Q., Sanaur, S., Pang, K., Steigerwald, M.L., and Nuckolls, C. (2005) *Angew. Chem. Int. Ed.*, **44**, 7390.
73. (a) Rim, K.T., Siaj, M., Xiao, S., Myers, M., Carpentier, V.D., Liu, L., Su, C., Steigerwald, M.L., Hybertsen, M.S., McBreen, P.H., Flynn, G.W., and Nuckolls, C. (2007) *Angew. Chem. Int. Ed.*, **46**, 7891; (b) Treier, M., Ruffieux, P., Gröning, P., Xiao, S., Nuckolls, C., and Fasel, R. (2008) *Chem. Commun.*, 4555.
74. Ott, R., Wiedemann, F., and Zinke, A. (1968) *Monatsh. Chem.*, **99**, 2032.
75. (a) Stabel, A., Herwig, P., Müllen, K., and Rape, J.P. (1995) *Angew. Chem. Int. Ed.*, **34**, 1609; (b) Müller, M., Kübel, C., and Müllen, K. (1998) *Chem. Eur. J.*, **4**, 2099; (c) Rempala, P., Kroulík, J., and King, B.T. (2006) *J. Org. Chem.*, **71**, 5067.
76. Broene, R.D. and Diederich, F. (1991) *Tetrahedron Lett.*, **32**, 5227.
77. Naamann, H., Neumann, P., and Muench, V. (1983) German Patent 3226780.
78. Wang, Z., Tomović, Ž., Kastler, M., Pretsch, R., Negri, F., Enkelmann, V., and Müllen, K. (2004) *J. Am. Chem. Soc.*, **126**, 7794.
79. (a) Roberson, L.B., Kowalik, J., Tolbert, L.M., Kloc, C., Zeis, R., Chi, X., Fleming, R., and Wilkins, C. (2005) *J. Am. Chem. Soc.*, **127**, 3069; (b) Northrop, B.H., Norton, J.E., and Houk, K.N. (2007) *J. Am. Chem. Soc.*, **129**, 6536.
80. Müller, M., Petersen, J., Strohmaier, R., Günther, C., Karl, N., and Müllen, K. (1996) *Angew. Chem. Int. Ed.*, **35**, 886.
81. Vacha, M. and Tani, T. (1997) *J. Phys. Chem. A*, **101**, 5027.
82. (a) Zander, M. and Franke, W. (1958) *Chem. Ber.*, **91**, 2794; (b) Goddard, R., Haenel, M.W., Herndon, W.C., Krüger, C., and Zander, M. (1995) *J. Am. Chem. Soc.*, **117**, 30.
83. The Biomarker Catalogue, Chiron AS. (2008).
84. Zander, M. and Friedrichsen, W. (1991) *Chem.-Ztg.*, **115**, 360.
85. Müller, M., Mauermann-Düll, H., Wagner, M., Enkelmann, V., and Müllen, K. (1995) *Angew. Chem. Int. Ed.*, **34**, 1583.
86. Dötz, F., Brand, J.D., Ito, S., Gherghel, L., and Müllen, K. (2000) *J. Am. Chem. Soc.*, **122**, 7707.
87. Feng, X., Wu, J., Ai, M., Pisula, W., Zhi, L., Rabe, J.P., and Müllen, K. (2007) *Angew. Chem. Int. Ed.*, **46**, 3033.
88. Iyer, V.S., Yoshimura, K., Enkelmann, V., Epsch, R., Rabe, J.P., and Müllen, K. (1998) *Angew. Chem. Int. Ed.*, **37**, 2696.
89. (a) Iyer, V.S., Wehmeier, M., Brand, J.D., Keegstra, M.A., and Müllen, K. (1997) *Angew. Chem. Int. Ed.*, **36**, 1604; (b) Wasserfallen, D., Kastler, M., Pisula, W., Hofer, W.A., Fogel, Y., Wang, Z., and Müllen, K. (2006) *J. Am. Chem. Soc.*, **128**, 1334.
90. (a) Müller, M., Iyer, V.S., Kübel, C., Enkelmann, V., and Müllen, K. (1997) *Angew. Chem. Int. Ed.*, **36**, 1607; (b) Böhme, T., Simpson, C.D., Müllen, K., and Rabe, J. (2007) *Chem. Eur. J.*, **13**, 7349.

91. Samori, P., Severin, N., Simpson, C.D., Müllen, K., and Rabe, J.P. (2002) *J. Am. Chem. Soc.*, **124**, 9454.
92. Wu, J., Tomović, Z, Enkelmann, V., and Müllen, K. (2004) *J. Org. Chem.*, **69**, 5179.
93. Simpson, C.D., Mattersteig, G., Martin, K., Gherghel, L., Bauer, R.E., Räder, H.J., and Müllen, K. (2004) *J. Am. Chem. Soc.*, **126**, 3139.
94. Simpson, C.D., Brand, J.D., Berresheim, A.J., Przybilla, L., Räder, H.J., and Müllen, K. (2002) *Chem. Eur. J.*, **8**, 1424.

12
Differentiated Ligands for the Sequential Construction of Crystalline Heterometallic Assemblies

Stéphane A. Baudron

12.1
Introduction

Coordination networks and discrete assemblies have received much interest over the past few years as a class of compounds with appealing physical properties resulting from both their structure and their metallo-organic nature [1–6]. In particular, recent focus has been on so-called metal-organic frameworks (MOFs) for their potential applications as gas storage materials or in catalysis [7–12]. Such systems result from the assembly of at least one multitopic ligand and a metal center. The vast majority of these compounds are homometallic systems. On the other hand, heterometallic architectures, incorporating at least two different types of metal centers, are particularly fascinating owing to their very heterometallic nature that provides avenues for structural diversity and modulation of their physical properties. As a prototype to illustrate these features, the family of Prussian blue analogs, one of the oldest classes of heterometallic systems, is especially worth mentioning [13, 14]. These compounds result from the reaction of a hexacyanometallate with another metal center leading to three-dimensional crystalline architectures (Figure 12.1). Depending on the nature of the metal centers, the magnetic properties of these systems are modulated [13, 14], as well as their hydrogen gas sorption properties, since the enthalpy of gas adsorption is metal dependent [15, 16].

When considering the elaboration of a heteronuclear metallo-organic architecture in the crystalline state, one is faced with a synthetic challenge. Indeed, the preparation of homometallic systems involves the direct combination of an organic ligand with one metal center (Figure 12.2). The transposition to heterometallic systems is illustrated in Figure 12.2, in the case of the formation of one-dimensional bimetallic networks. Reaction of an organic ligand with two transition metal ions, concurrently and in the absence of any differentiation at both the organic and metallic levels, can lead to a mixture of systems among which are the undesired homometallic architectures and a variety of heterometallic ones resulting from a statistical distribution of the metal centers.

264 *12 Differentiated Ligands for the Sequential Construction of Crystalline Heterometallic Assemblies*

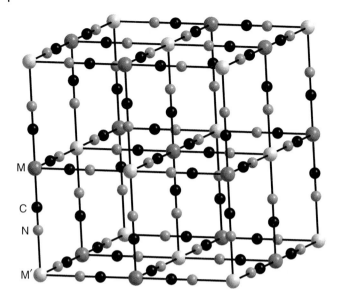

Figure 12.1 A fragment of the crystal structure of an idealized Prussian blue analog of formula [M(CN)$_6$]M′.

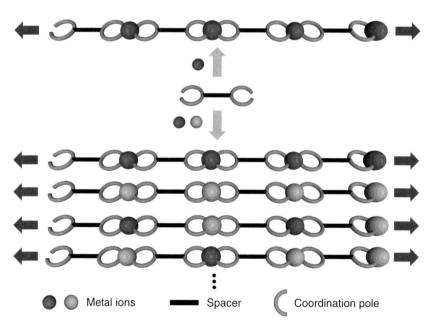

Figure 12.2 Formation of a 1D homometallic network (top) and a mixture of networks in the presence of two metal centers and an undifferentiated ligand (bottom).

A precise control of the relative position of the metal ions therefore calls for another approach [17]. Two aspects can be pointed out here to circumvent this problem: the differentiation and the sequential synthesis of these systems. Considering again the case of a heterometallic one-dimensional network, we can envision the use of an organic ligand bearing two different coordination poles, a primary and a secondary one (Figure 12.3). This differentiation can be realized by modifying the nature and/or the charge of these poles and gives rise to a hierarchy of coordination. Therefore, in the presence of a first metal ion, a discrete metal complex can be isolated bearing at its periphery, secondary coordination poles available for further ligation to another metal ion. This latter unit can then be regarded either as a metalloligand for the elaboration of discrete heterometallic complexes or as a metallatecton [18–20] for the generation of infinite molecular networks in the solid state by periodic repetition of the coordination event to a second metal ion. Using this approach, the precise relative arrangement of metal ions could be tuned.

As the ligands are synthesized using the tools of organic chemistry, a broad range of derivatives can be envisioned for the application of this strategy. We will not make a comprehensive list of such differentiated ligands here, but rather illustrate the approach, again for the formation of a one-dimensional network, with three examples selected from the literature. As a first class of ligands, we present a system based on functionalized terpyridine derivatives [21]. The 4-pyridyl appended derivative, **1**, comprises a tridentate and a monodentate coordination pole and forms with Ru^{2+} cations, a discrete complex $[Ru(1)_2]^{2+}$ (Figure 12.4) [22]. The latter bears at its periphery, two available pyridine groups and can therefore be regarded as a metallo-organic expanded analog of 4,4'-bipyridine. Upon reaction with a silver salt, a one-dimensional network is obtained in the crystalline state as a result of the coordination of the peripheral pyridine groups to the silver cation (Figure 12.4) [23].

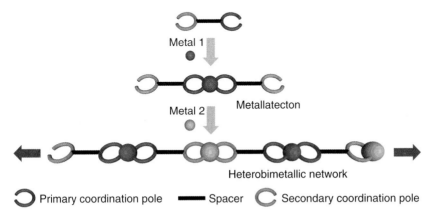

Figure 12.3 Sequential construction of a heterometallic coordination network using a differentiated ligand bearing two coordination poles.

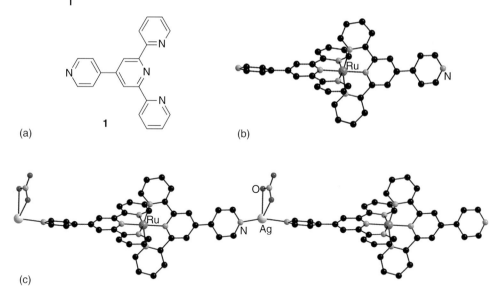

Figure 12.4 The differentiated ligand **1** (a) forms with Ru^{2+} the metallatecton [Ru(**1**)$_2$]$^{2+}$ (b) which self assembles with AgNO$_3$ to afford a crystalline heterometallic Ru/Ag network (c) [22, 23]. Hydrogen atoms have been omitted for clarity.

The second example is based on ligands of the oxamato type which have been successfully used by Kahn and coworkers for the elaboration of magnetic heterobimetallic architectures [24, 25]. Indeed, these ligands such as H$_4$pba (1,3-propylenebisoxamato), **2**, shown in Figure 12.5, form complexes with copper, for example, featuring four peripheral oxygen atoms. Interestingly, these oxygen atoms can not only act as two bidentate poles, but also carry about 10% of the spin density. Therefore, upon reaction with a second metallic spin carrier, magnetic chains can be obtained. The crystal structure of such a one-dimensional Mn/Cu chain is depicted in Figure 12.5 [26, 27]. The latter is a ferrimagnetic chain; unfortunately these chains are antiferromagnetically coupled to one another at low temperature.

Another class of potential ligands for the above-mentioned approach consists of macrocyclic derivatives such as porphyrins. These molecules have been extensively used for the construction of coordination networks [28–30]. The macrocycle can host a first metal center and can also be readily functionalized with other peripheral coordination sites at the *meso* positions. Interest in these derivatives stems from the many possible functionalized isomers that can be isolated, as well as the inherent catalytic activity and optical properties associated with these molecules. The example of a pyridine functionalized porphyrin, **3**, is illustrated in Figure 12.6. The copper derivative [Cu(**3**)] assembles with Cd(NO$_3$)$_2$ to afford again a one-dimensional network upon coordination of the peripheral pyridine groups to the Cd^{2+} ions [30]. It is worth emphasizing here that although the metallatecton bears two peripheral coordination sites in a nonlinear fashion, the resulting heterobimetallic network is

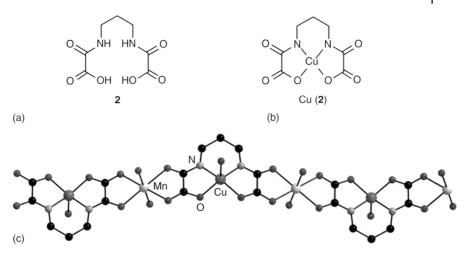

Figure 12.5 Ligand **2** (a) forms square-planar complexes with Cu(II) (b) bearing coordination peripheral oxygen atoms leading to ferromagnetic chain of the type Mn[Cu(**2**)](H$_2$O)$_3$·2H$_2$O (c) [26, 27]. Hydrogen atoms have been omitted for clarity.

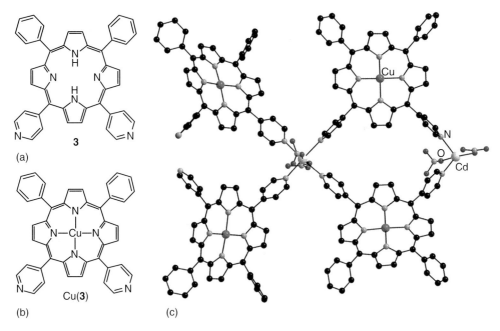

Figure 12.6 A one-dimensional heterometallic network (c) based on a pyridine functionalized porphyrin (a) as its Cu complex (b) [30]. Hydrogen atoms have been omitted for clarity.

one-dimensional as a result of the coordination preferences of the second metal center, Cd^{2+}.

The three examples presented above have been focusing on the elaboration of one-dimensional networks. This was done for illustrative purposes; we should note, however, that architectures of higher dimensionality or discrete species can be obtained by tuning the organic ligands (the number and relative arrangement of the coordination poles, their denticity, ...) and the metal centers (coordination preferences, number of available coordination sites, the relative affinities for the different poles ....). Note also, that there are numerous examples of other ligands that can be used. In addition to the ones presented above, we can mention here, 2,2′-bipyridine derivatives [31, 32], Schiff bases [33], dithiolenes [34], for example.

Over the past four years, we have developed and applied this strategy using mainly two families of ligands, 4,5-diazafluorene appended dithiolates and dipyrrin derivatives. We will describe in the following our results based on both types of ligands.

12.2
Dithiolate Ligands

Metal dithiolate complexes have been shown to be versatile building blocks for the construction of molecular materials [35]. In particular, 1,1- and 1,2-dithiolates, such as dithiolenes, have been studied owing to their ability to bind a wide range of metals and to stabilize several of their oxidation states [35–37]. Interested by this versatility, we have developed ligands **4** and **5**, comprising a 1,1- or 1,2-dithiolate fragment conjugated to a 4,5-diazafluorene group (Scheme 12.1) [38, 39]. The two chelating coordination poles are differentiated here by their nature and their charge. It was therefore envisioned that, in a first step, coordination at the sulfur atoms would allow the formation of a metal complex bearing peripheral neutral nitrogen-based chelates.

12.2.1
Metallatectons

Following this strategy, several mononuclear metal complexes of the $[ML_2]^{2-}$ type were prepared, isolated, characterized, and their crystal structure determined by single-crystal X-ray diffraction (Figure 12.7) [17, 38, 39]. As expected, the relative orientation of the two 4,5-diazafluorene groups can be tuned depending on the metal center. For example, the Ni(II) complexes are square planar and the Hg(II) complexes show a tetrahedral coordination geometry around the metal center. Note here that when comparing complexes $[Hg(4)_2]^{2-}$ and $[Hg(5)_2]^{2-}$, the difference in bite angle between the two ligands is apparent with a tendency to dimerization for the former complex with the 1,1-enedithiolate derivative. In addition, beyond the purely structural differences, the electrochemical properties also differ. Whereas $[Ni(5)_2]^{2-}$ features two reversible oxidation waves in DMF, the other complexes show only irreversible oxidation processes under the same conditions.

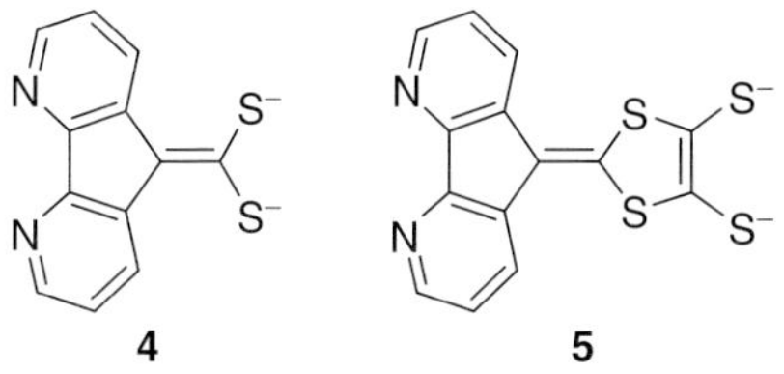

Scheme 12.1 The 4,5-diazafluorene appended 1,1- and 1,2-enedithiolate ligands **4** and **5**.

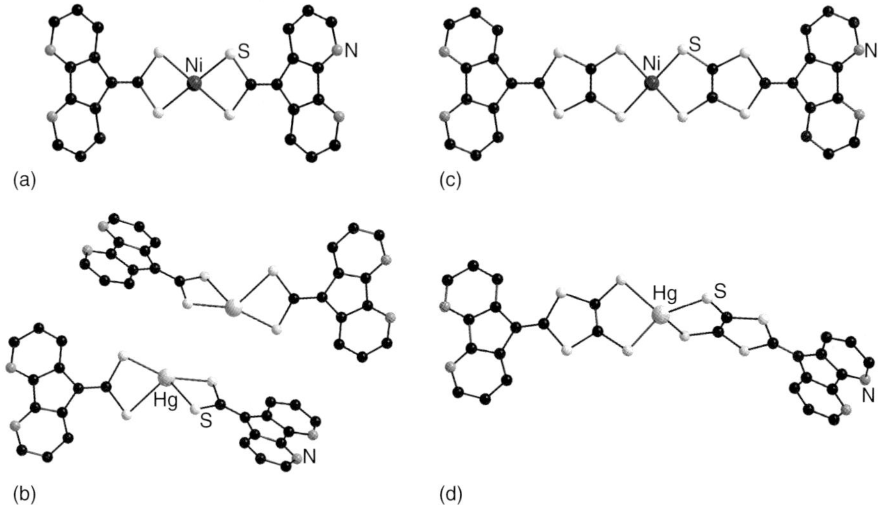

Figure 12.7 Crystal structure of complexes [Ni(**4**)$_2$]$^{2-}$ (a), [Hg(**4**)$_2$]$^{2-}$ (b), [Ni(**5**)$_2$]$^{2-}$ (c) and [Hg(**5**)$_2$]$^{2-}$ (d) [17, 38]. The hydrogen atoms are not shown for clarity.

Complexes with Pd(II), Zn(II), Ni(II), Hg(II), and Cu(II) have been prepared. However, the case of copper-based species with 1,1-enedithiolate ligand is rather interesting. It has been reported with the prototypical *i*-mnt (*iso*-maleonitriledithiolate) ligand, **6**, that depending on the oxidation state of copper, either a mononuclear [Cu^{II}(**6**)$_2$]$^{2-}$ complex or an octanuclear cubic complex [Cu$^{I}_8$(**6**)$_6$]$^{4-}$ can be obtained [40, 41]. Similarly, while with Cu(II), a [Cu^{II}(**4**)$_2$]$^{2-}$ species is isolated, reaction with Cu(I) affords a tetraanionic octanuclear complex [Cu$^{I}_8$(**4**)$_6$]$^{4-}$ [39]. The latter was characterized in solution by ^{1}H and ^{13}C NMR, UV–Visible spectroscopy, and in the solid state by single-crystal X-ray diffraction. The crystal structures of these species are depicted in Figure 12.8. The polynuclear compound (Figure 12.8b) features six peripheral 4,5-diazafluorene chelate groups available for further coordination to a second metal center. Unfortunately, its limited solubility in common organic solvents has hindered, so far, its use as a metallatecton. It is nonetheless interesting to note that by simply modifying the oxidation state of the metal source, two metallatectons of different nuclearity, geometry, and charge can be isolated.

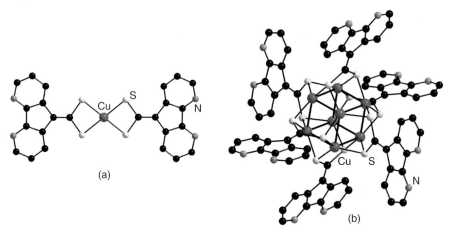

Figure 12.8 Crystal structure of the mononuclear [Cu^{II}(4)$_2$]$^{2-}$ (a) and the octanuclear [Cu$_8^I$(4)$_6$]$^{4-}$ (b) complexes [39]. Hydrogen atoms have been omitted for clarity.

12.2.2
Infinite Chains

The dianionic metal complexes [ML$_2$]$^{2-}$ have been isolated as tetraalkylammonium salts. A simple substitution of Na$^+$ cations for the organic cations has allowed the isolation of crystalline one-dimensional networks. Compounds [Cu(4)$_2$]$^{2-}$, [Ni(4)$_2$]$^{2-}$, [Pd(4)$_2$]$^{2-}$ lead to isostructural one-dimensional chains where the metal complexes are bridged by [Na$_2$(DMSO)$_5$]$^{2+}$ units coordinated to the 4,5-diazafluorene moieties, as shown Figure 12.9 [17]. Rather interestingly, in the

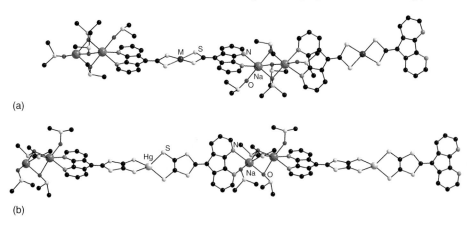

Figure 12.9 Crystal structures of the one-dimensional networks [M(4)$_2$][Na$_2$(DMSO)$_5$] (M = Ni, Pd, Cu) (a) and [Hg(5)$_2$][Na$_2$(DMSO)$_5$] (b). Hydrogen atoms and solvent molecules have been omitted for clarity [17, 38].

case of the mercury complex incorporating ligand **5**, a one-dimensional chain is also obtained resulting from the same type of connectivity [38]. The major difference between the two cases lies in the periodicity of the chain as well as its twist, owing to the more extended sulfur system and the nature of the first metal center – mercury, here – that forms a tetrahedral metal complex while the former three metallatectons are square planar.

While Na$^+$ cations cannot be considered as transition metal ions, these examples can be seen as proof of principles of the strategy illustrated in Figure 12.3, for the preparation of heterobimetallic systems. Our attempts to prepare heterometallic systems incorporating at least two different transition metals have been hindered not by the coordination event but by the difficulty of isolating crystalline materials. This perhaps can be circumvented when targeting discrete species.

12.2.3
Discrete Assemblies

To form discrete assemblies using the strategy elaborated above (Figure 12.3), the periodic repetition of the coordination event has to be prevented at some stage. This can be easily accomplished by introducing a blocking ligand either on the second metal center (Figure 12.10a) or on the first one (Figure 12.10b). Using both approaches, a series of heteronuclear metal complexes has been obtained in the crystalline state.

To demonstrate the strategy illustrated Figure 12.10a, the tetraazamacrocycle 1,4,7,10-tetraazacyclododecane (cyclen), was employed as a capping ligand for Ni(II) cations leaving two available coordination sites in the *cis* position. Reaction of [Pd(**4**)$_2$]$^{2-}$ with 2 equiv. of [(cyclen)Ni]$^{2+}$ leads to the formation of the heterobimetallic trinuclear species [(cyclen)Ni(Pd(**4**)$_2$)Ni(cyclen)]$^{2+}$ [17]. Its crystal structure is shown Figure 12.11. In this compound, the two octahedral Ni(II) centers carry a spin S = 1, but are only weakly magnetically coupled owing to their long separation (18.5 Å) and the diamagnetic nature of the bridging complex.

Regarding the strategy illustrated in Figure 12.10b, a series of heteroleptic complexes [(dppp)ML] (dppp = bis(diphenylphosphino)propane, M = Pd(II), Pt(II), L = **4**, **5**) has been prepared. These species were first self-assembled with Cu(I)

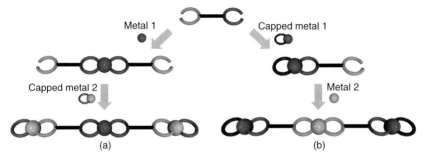

Figure 12.10 Strategies for the sequential elaboration of discrete heterometallic assemblies.

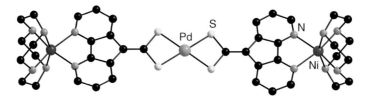

Figure 12.11 Crystal structure of the trinuclear hetero-bimetallic discrete complex [(cyclen)Ni(Pd(**4**)$_2$)Ni(cyclen)]$^{2+}$ [17]. Hydrogen atoms have been omitted for clarity.

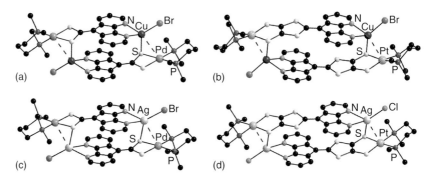

Figure 12.12 Crystal structures of a series of heteronuclear metallamacrocycles [(dppp)MLM'X]$_2$ (M = Pd, Pt, L = **4**, **5**, M' = Ag, Cu, X = Cl, Br). Hydrogen atoms as well as the phenyl groups of the dppp ligands have been omitted for clarity [42].

salts to form heteronuclear metallamacrocycles [(dppp)MLCuBr]$_2$ (Figure 12.12) [42]. These assemblies result from the coordination of the 4,5-diazafluorene moiety to Cu(I), as expected, and from an additional Cu–S interaction with another metal complex. Note that for both ligands **4** and **5**, analogous metallamacrocycles are obtained (Figure 12.12 a and b). The only difference lies in the length of these species, a direct result of the ligand design. It is also worth emphasizing here that a Cu–M distance shorter than the sum of the van der Waals radii is observed in all instances suggesting d^8–d^{10} interactions [43]. To further probe the robustness of this approach, Ag(I), another d^{10} metal, was used. Reaction of [(dppp)ML] complexes with Ag(OTf) in a CH$_2$Cl$_2$/CH$_3$CN mixture leads also to heteronuclear metallamacrocycles [(dppp)MLAgCl]$_2$ built upon coordination of Ag to both a diazafluorene group and a sulfur atom of another complex. Surprisingly, an AgCl unit is observed in these complexes. It results from the decomposition of the solvent, as the reaction in CH$_2$Br$_2$ leads to the [(dppp)MLAgBr]$_2$ analog (Figure 12.12c). Here again, a short Ag–M distance is observed. In this series, it has been possible to modulate the bridging ligand as well as the d^8 and d^{10} metal centers.

The family of 4,5-diazafluorene appended ene-dithiolate ligands has proved its efficiency in demonstrating the sequential strategy (Figures 12.3 and 12.10) for the construction of heterometallic coordination networks and discrete architectures. Further work will be centered on the characterization of the optical properties of these species as well as the synthesis of derivatives bearing other secondary coordination poles.

12.3
Dipyrrin Ligands

We have also applied the strategy elaborated above using another class of ligands: dipyrrins [44]. Known since the 1920s [45], these molecules have received a revived interest over the past few years, with the synthesis of α, β-unsubstituted derivatives by Dolphin [46] and Lindsey [47]. Indeed, they can be readily functionalized in their position 5, equivalent to the *meso* position of porphyrins, and their conjugate base acts as a monoanionic chelate and forms neutral complexes with a large variety of metal centers. This allows the straightforward synthesis of derivatives bearing a secondary coordinating site in addition to the bispyrrolic chelate. This feature has been advantageously used by the group of Cohen [48–51] as well as ours [52, 53] for the elaboration of heterometallic architectures. For example, a Co(III) neutral complex incorporating the 4-pyridyl-dipyrrin, **7**, can be prepared. This neutral complex, [Co(**7**)$_3$], forms a three-dimensional heterometallic network upon reaction with Ag(OTf) (Figure 12.13) [48].

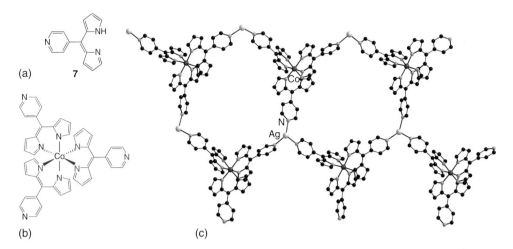

Figure 12.13 Sequential construction of a heterometallic network (c) by Cohen and coworkers based on the assembly of a silver salt with a neutral Co(III) complex (b) incorporating ligand **7** (a) [48].

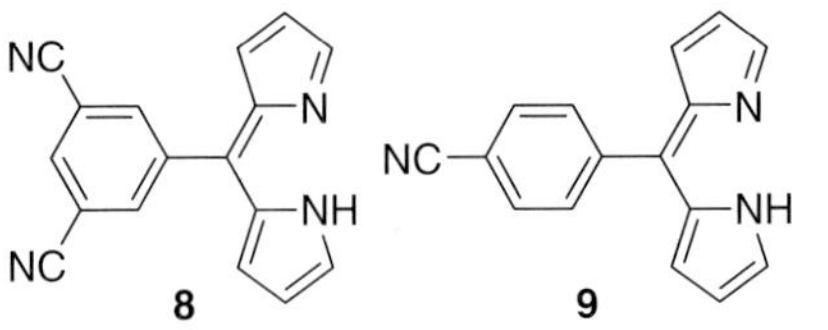

Scheme 12.2 The dipyrrin-based ligands **8** and **9**.

We have been interested in using the 5-(3,5-dicyanophenyl)dipyrrin, **8**, and 5-(4-cyanophenyl)dipyrrin, **9** (Scheme 12.2) as ligands where the dipyrrin moiety may act as the primary coordination pole and the neutral nitrile group(s) as secondary coordination pole(s) [52, 53].

12.3.1
Metallatectons

Starting from these ligands, under basic conditions, a series of homo- and heteroleptic complexes have been prepared, isolated, and characterized. The crystal structures of two of these compounds are presented Figure 12.14 [52, 53]. Playing with the presence of a capping ligand and the number of nitrile groups on the dipyrrin-based ligand, diverse secondary peripheral coordination sets can be obtained.

12.3.2
Heterometallic Architectures Built with the Assistance of an Ag–π Interaction

Upon reaction of complex [Zn(**8**)$_2$] with Ag(OTf) in a benzene/CH$_3$CN mixture, crystals of a two-dimensional heterometallic network are obtained [52]. Interestingly, crystal structure determination revealed that the silver cations are coordinated to triflate anions and the peripheral nitrile groups, but also seem to sit in an unsymmetrical fashion on top of a C=C bond of a pyrrole ring of the dipyrrin ligand (called *C2*=*C3* hereafter) with Ag–C distances ranging from 2.397(4) to 2.583(4) Å (Figure 12.15). These geometrical characteristics suggest the presence of an Ag–π interaction. The latter has been long known with aromatics. Indeed, in

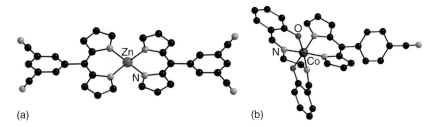

Figure 12.14 Crystal structures of [Zn(**8**)$_2$], [(salen)Co(**9**)]. Salen = N,N'-bis(salicylidene)ethylediamine. Hydrogen atoms have been omitted for clarity [52, 53].

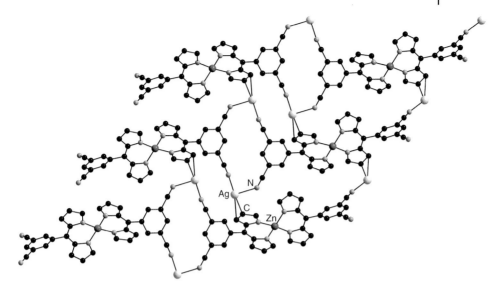

Figure 12.15 A two-dimensional network based on Ag–NC coordination and an Ag–π interaction in [Zn(**8**)$_2$][Ag(OTf)]$_2$(C$_6$H$_6$)$_{2.5}$ [52]. The hydrogen atoms as well as the benzene solvate molecules and the triflate anions are not shown for clarity.

the course of his determination of the molecular weight of Ag(ClO$_4$) in benzene, Hill hypothesized that such interactions existed and were leading to a polymeric network [54]. This was later confirmed by the crystal structure determination of AgClO$_4$·C$_6$H$_6$ [55]. In the case of polypyrrolic systems, this interaction has been characterized and exploited in solution [56, 57], but not described in the solid state. This surprising result is nonetheless interesting, particularly in light of the work of Cohen and coworkers who have reported heterobimetallic networks incorporating silver cations via a "classical" coordination to the peripheral coordinating group [48–50] (Figure 12.13). This called for a further investigation as to whether or not this interaction could be observed in the crystalline phase in other instances. Several questions are then being raised, such as, can this be observed with complexes of metals other than Zn, of different coordination geometry, or incorporating a ligand other than **8**? This study was undertaken starting first by the reaction of [(hfac)Cu(**8**)] (hfac, hexafluoroacetylacetonate) with Ag(OTf) under the same conditions as for the zinc derivative. Although the crystals were not of high quality, the preliminary X-ray analysis clearly indicated the presence of Ag–π interaction also in this architecture. This suggests that the nature of the primary metal center is not at stake in the presence of this mode of complexation of silver cations.

In order to assess whether the position of the nitrile groups was involved in this interaction, derivative **9** was used instead of **8**. Interestingly both the heteroleptic [(hfac)Cu(**9**)] and homoleptic [Cu(**9**)$_2$] complexes lead, in the presence of a silver salt, to crystalline heterometallic networks where the Ag–π interaction

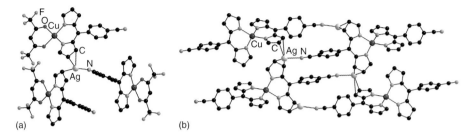

Figure 12.16 Silver–π interaction in [(hfac)Cu(**9**)][Ag(OTf)]$_2$(H$_2$O) (a) and [Cu(**9**)$_2$](AgOTf)(C$_6$H$_6$)$_{1.5}$ (b) [52]. Hydrogen atoms, solvate molecules and triflate anions are not shown for clarity.

is present (Figure 12.16). The Ag–C distances are in the range observed for this type of interaction in the solid state with aromatics [58]. The Ag–π interaction can therefore be observed with Zn(II) and Cu(II) complexes of ligands **8** and **9**. Interestingly, a coordination network based on the homoleptic octahedral complex [Co(**9**)$_3$] and Ag(OTf) built solely on coordination of the dipyrrin to the cobalt centers and of the nitrile groups to the silver cations has been reported [49]. This prompted us to investigate other octahedral metallatectons such as [(salen)Co(**9**)].

Reaction of [(salen)Co(**9**)] with Ag(SbF$_6$) in benzene/CH$_3$CN afforded crystalline materials of a discrete [2+2] metallamacrocycle [(salen)Co(**9**)Ag(H$_2$O)(NCCH$_3$)]$_2$ (SbF$_6$)$_2$ (Figure 12.17a) [53]. In this compound, the silver ion is coordinated to a water and acetonitrile molecule, as well as to the nitrile group of the dipyrrin-based ligand and is again complexed by the C2=C3 bond of a pyrrolic moiety of another Co complex. It is worth emphasizing here that this compound is obtained with the noncoordinating SbF$_6^-$ anion, while all the former species were isolated starting from silver triflate salts. In spite of the octahedral coordination geometry around the Co center, the Ag–π interaction is observed, at variance with what has been reported with the homoleptic analog [49]. This difference might be due to the fact that the C2=C3 bond is more accessible in the case of the heteroleptic complex than with the homoleptic one.

Moreover, there is a striking analogy between this macrocycle and the series of compounds described in 12.2.3 based on the 4,5-diazafluorene appended ene-dithiolate ligands. Built upon the application of the same construction strategy, although with different types of interactions, the architectures are structurally similar, underlining the generality of this approach.

This type of interaction seems to be rather general, as it has been observed with homo- and heteroleptic complexes of Zn(II), Cu(II), Co(III) incorporating ligands **8** and **9**, with diverse coordination geometries. A final question remains regarding the existence of this interaction when a silver salt is combined with one of these ligands. Interestingly, reaction of **9** with Ag(BF$_4$) leads again to a [2+2] metallamacrocycle, but built on coordination of Ag(I) ions to the nitrile group and to the nitrogen atom of a pyrrolic ring of the dipyrrin ligand (Figure 12.17b) [53].

This study shows that in the presence of silver ions, nitrile functionalized dipyrrins can be regarded as ligands actually incorporating three coordination poles, namely the dipyrrin chelate, the CN groups, as well as the C2=C3 bond of the pyrrolic ring. These ligands are viable for the application of the strategy elaborated in Figure 12.3, but with an added degree of complexity owing to the Ag–π interaction. It would be interesting to investigate the presence of this interaction with other dipyrrin-based ligands as well as other polypyrrolic derivatives such as tripyrrins or porphyrins.

12.4
Conclusion and Outlook

The synthetic challenge represented by the elaboration of heterometallic architectures can be faced using a sequential construction strategy (Figures 12.3 and 12.10). One of the strengths of this approach stems from the organic nature of the ligands used. Indeed, with the tools of organic chemistry, diverse ligands can be imagined and synthesized. These ligands can be tuned with respect to the nature and relative arrangement of the coordination poles, having therefore a direct influence on the structure of the architectures. We should note here that although the derivatives presented here have been designed with only two coordination poles (dithiolate, 4,5-diazafluorene, dipyrrin, nitrile), many more could be introduced leading to systems incorporating more than two types of metal centers. Beyond the structural beauty of these systems, their physical properties are of prime importance. For example, a further development of the dipyrrin-based architectures relies on the luminescence associated with some of their complexes [59–61], which could lead to luminescent heterometallic architectures. The redox properties of the metal dithiolene complexes and the inherent electronic conductivity of some these species in the solid state [35] could be further exploited within heterometallic systems.

While our main focus here has been on single-crystalline systems that can be analyzed by X-ray diffraction, future directions will be centered on working at different scales. Recent work has shown that surfaces can be patterned with coordination

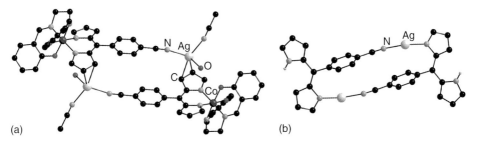

Figure 12.17 Crystal structures of the [2+2] metallamacrocycles [(salen)Co(**9**)Ag(H$_2$O)(NCCH$_3$)]$^+_2$ (a) and [(**9**)Ag]$_2$$^{2+}$ (b) [53]. Hydrogen atoms and anions have been omitted for clarity.

networks [62–64] or discrete assemblies [65, 66]. The stepwise construction strategy could be also applied at this scale and characterized by scanning tunneling microscopy. Composite crystals represent also an interesting class of species [67–69]. They are built upon the epitaxial growth of isomorphous crystals onto other crystals. This way a crystal having a particular physical property could be embedded in a crystal with different properties, and so on. Another approach is inspired by the work done on solid-phase synthesis, which consists in attaching a ligand to a polymer, construct a metal complex, then attach another ligand, and to continue this process until cleavage of the heterometallic complex off the polymer support [70]. These avenues of research are promising for the further development of the sequential construction strategy of heterometallic architectures.

Acknowledgments

This work has been developed in the laboratory of Prof. Mir Wais Hosseini, whose permanent support and mentorship are gratefully acknowledged. The contribution of Dr Domingo Salazar-Mendoza is gratefully acknowledged. Financial support from the Université de Strasbourg, the CNRS, the French Ministry of Education and Research, the Institut Universitaire de France, and CONACYT is gratefully acknowledged.

References

1. Lehn, J.M. (1995) *Supramolecular Chemistry, Concepts and Perspectives*, Wiley-VCH Verlag GmbH, Weinheim.
2. Constable, E.C. (1994) Higher oligopyridines as a structural motif in metallosupramolecular chemistry, in *Progress in Inorganic Chemistry*, vol. 42 (ed. K.D. Karlin), pp. 67–138.
3. Sauvage, J.-P. (ed.) (1999) *Transition Metals in Supramolecular Chemistry*, John Wiley & Sons, Inc., New-York.
4. Lenninger, S., Olenuyk, B., and Stang, P.J. (2000) Self-assembly of discrete cyclic nanostructures mediated by transition metals. *Chem. Rev.*, **100**, 853–908.
5. Swiegers, G.F. and Malefetse, T.J. (2000) New self-assembled structural motifs in coordination chemistry. *Chem. Rev.*, **100**, 3483–3537.
6. Fujita, M., Tominaga, M., Hori, A., and Therrien, B. (2000) Coordination assemblies from a Pd(II)-cornered square complex. *Acc. Chem. Res.*, **38**, 369–378.
7. Batten, S.R. and Robson, R. (1998) Interpenetrated nets: ordered, periodic entanglement. *Angew. Chem. Int. Ed.*, **37**, 1460–1494.
8. Eddaoudi, M., Moler, D.B., Li, H., Chen, B., Reineke, T.M., O'Keefe, M., and Yaghi, O.M. (2001) Modular chemistry: secondary building units as a basis for the design of highly porous and robust metal-organic carboxylate frameworks. *Acc. Chem. Res.*, **34**, 319–330.
9. Kitagawa, S., Kitaura, R., and Noro, S.I. (2004) Functional porous coordination polymers. *Angew. Chem. Int. Ed.*, **43**, 2334–2375.
10. Férey, G., Mellot-Draznieks, C., Serre, C., and Millange, F. (2005) Crystallized frameworks with giant pores: are there limits to the possible? *Acc. Chem. Res.*, **38**, 217–225.
11. Dinca, M. and Long, J.R. (2008) Hydrogen storage in microporous metal-organic frameworks with exposed

metal sites. *Angew. Chem. Int. Ed.*, **47**, 6766–6779.

12. Maspoch, D., Ruiz-Molina, D., and Veciana, J. (2007) Old materials with new tricks: multifunctional open-framework materials. *Chem. Soc. Rev.*, **36**, 770–818.

13. Dunbar, K.R. and Heintz, R.A. (1997) Chemistry of transition metal cyanide compounds: modern perspectives, in *Progress in Inorganic Chemistry*, vol. 45 (ed. K.D.Karlin), pp. 283–391.

14. Verdaguer, M., Bleuzen, A., Marvaud, V., Vaissermann, J., Seuleiman, M., Desplanches, C., Scuiller, A., Train, C., Garde, R., Gelly, G., Lomenech, C., Rosenman, I., Veillet, P., Cartier, C., and Villain, F. (1999) Molecules to build solids: high Tc molecule based magnets by design and recent revival of cyano complexes chemistry. *Coord. Chem. Rev.*, **190–192**, 1023–1047.

15. Kaye, S.S. and Long, J.R. (2005) Hydrogen storage in the dehydrated Prussian blue analogues M3[Co(CN)6]2. *J. Am. Chem. Soc.*, **127**, 6505–6507.

16. Chapman, K.W., Southon, P.D., Weeks, C.L., and Kepert, C.J. (2005) Reversible hydrogen gas uptake in nanoporous Prussian blue analogues. *Chem. Commun.*, 3322–3324.

17. Baudron, S.A., Hosseini, M.W., Kyritsakas, N., and Kurmoo, M. (2007) A stepwise approach to the formation of heterometallic discrete complexes and infinite architectures. *Dalton Trans.*, 1129–1139.

18. Simard, M., Su, D., and Wuest, J.D. (1991) Use of hydrogen bonds to control molecular aggregation – self-assembly of 3-dimensional networks with large chambers. *J. Am. Chem. Soc.*, **113**, 4696–4698.

19. Mann, S. (1993) Molecular tectonics in biomineralization and biomimetic materials. *Nature*, **365**, 499–505.

20. Hosseini, M.W. (2005) Molecular tectonics: from simple tectons to complex molecular networks. *Acc. Chem. Res.*, **38**, 313–323.

21. Constable, E.C. (2008) Expanded ligands – an assembly principle for supramolecular chemistry. *Coord. Chem. Rev.*, **252**, 842–855.

22. Constable, E.C. and Cargill Thompson, A.M.W.J. (1994) Pendant-functionalised ligands for metallosupramolecular assemblies; ruthenium(II) and osmium(II) complexes of 4′-(4-pyridy1)-2,2′: 6′,2′′-terpyridine. *J. Chem. Soc., Dalton Trans.*, 1409–1418.

23. Beeves, J.E., Constable, E.C., Housecroft, C.E., Kepert, C.J., and Price, D.J. (2007) The first example of a coordination polymer from the expanded 4,4′-bipyridine ligand [Ru(pytpy) 2]2+ (pytyp = 4′-(4-pyridyl)-2,2′:6′, 2′′-terpyridine). *CrystEngComm*, **9**, 456–459.

24. Kahn, O. (2000) Chemistry and physics of supramolecular materials. *Acc. Chem. Res.*, **33**, 647–657.

25. For a recent review: Pardo, E., Ruiz-Garcia, R., Cano, J., Ottenwaelder, X., Lescouëzec, R., Journaux, Y., Lloret, F., and Julve, M. (2008) Ligand design for multidimensional magnetic materials: a metallosupramolecular perspective. *Dalton Trans.*, 2780–2805.

26. Pei, Y., Verdaguer, M., Kahn, O., Sletten, J., and Renard, J.P. (1987) Magnetism of manganese(II)copper(II) and nickel(II)copper(II) ordered bimetallic chains. Crystal structure of MnCu(pba)(H2O)3.2H2O (pba = 1,3-propylenebis(oxamato)). *Inorg. Chem.*, **26**, 138–143.

27. Baron, V., Gillon, B., Cousson, A., Mathonière, C., Kahn, O., Grand, A., Öhrström, L., Delley, B., Bonnet, M., and Boucherle, J.X. (1997) Spin density maps for the ferrimagnetic chain compound MnCu(pba) (H2O)3•2H2O (pba = 1,2-Propylenebis(oxamato)): polarized neutron diffraction and theoretical studies. *J. Am. Chem. Soc.*, **119**, 3500–3506.

28. Suslick, K.S., Bhyrappa, P., Chou, J.P., Kosal, M.E., Nakagaki, S., Smithenry, D.W., and Wilson, S.R. (2005) Microporous porphyrin solids. *Acc. Chem. Res.*, **38**, 283–291.

29. Goldberg, I. (2008) Crystal engineering of nanoporous architectures and chiral porphyrin assemblies. *CrystEngComm*, **10**, 637–645.

30. Deiters, E., Bulach, V., and Hosseini, M.W. (2008) Molecular tectonics: ribbons type coordination networks

based on porphyrins bearing two pyridine and two pyridine N-oxide units. *New. J. Chem.*, **32**, 99–104.

31. Balzani, V., Juris, A., Venturi, M., Campagna, S., and Serroni, S. (1996) Luminescent and redox-active polynuclear transition metal complexes. *Chem. Rev.*, **96**, 759–834.

32. Kaes, C., Katz, A., and Hosseini, M.W. (2000) Bipyridine: the most widely used ligand. A review of molecules comprising at least two 2,2'-bipyridine units. *Chem. Rev.*, **100**, 3553–3590.

33. Kitaura, R., Onoyama, G., Sakamoto, H., Matsuda, R., Noro, S., and Kitagawa, S. (2004) Immobilization of a metallo schiff base into a microporous coordination polymer. *Angew. Chem. Int. Ed.*, **43**, 2684–2687.

34. Dawe, L.N., Miglioi, J., Turnbow, L., Taliaferro, M.L., Shum, W.W., Bagnato, J.D., Zakharov, L.N., Rheingold, A.L., Arif, A.M., Fourmigué, M., and Miller, J.S. (2005) Structure and magnetic properties of (meso-tetraporphyrinato)manganese(III) bis(dithiolato)nickelates. *Inorg. Chem.*, **44**, 7530–7539.

35. Karlin, K.D. and Stiefel, E. I. (eds) (2004) *Dithiolene Chemistry, Synthesis, Properties and Applications, Progress in Inorganic Chemistry*, vol. 52, John Wiley & Sons, Inc., Hoboken.

36. Coucouvanis, D. (1970) Chemistry of the dithioicacid and 1,1-dithiolate complexes, in *Progress in Inorganic Chemistry*, vol. 11 (ed. S.J.Lippard), pp. 233–3791.

37. Coucouvanis, D. (1979) The chemistry of the dithioicacid and 1,1-dithiolate complexes, 1968–1977, in *Progress in Inorganic Chemistry*, vol. 26 (ed. S.J.Lippard), pp. 301–469.

38. Baudron, S.A. and Hosseini, M.W. (2006) Sequential generation of 1-D networks based on a differentiated bischelate-type ligand bearing both 4,5-diazafluorene and dithiolene units. *Inorg. Chem.*, **45**, 5260–5262.

39. Baudron, S.A., Hosseini, M.W., and Kyritsakas, N. (2006) Octanuclear cubic Cu(I) complex decorated with six peripheral chelates. *New J. Chem.*, **30**, 1083–1086.

40. Gompper, R. and Töpfl, W. (1962) Substituierte dithiocarbonsaüren und ketenmercaptale. *Chem. Ber.*, **95**, 2861–2870.

41. McCandlish, L.E., Bissell, E.C., Coucouvanis, D., Fackler, J.P., and Knox, K. (1968) A new metal cluster system containing a cube of metal atoms. *J. Am. Chem. Soc.*, **90**, 7357–7359.

42. Baudron, S.A. and Hosseini, M.W. (2008) Modular construction of a series of heteronuclear metallamacrocycles. *Chem. Commun.*, 4558–4560.

43. Balch, A.L., Catalano, V.J., and Olmstead, M.M. (1990) Structure and photoluminescence of a heterodinuclear d8-d10 complex, [AuIr(CO)Cl(μ-Ph2PCH2PPh2)2][PF6]. *Inorg. Chem.*, **29**, 585–586.

44. Wood, T.E. and Thompson, A. (2007) Advances in the chemistry of dipyrrins and their complexes. *Chem. Rev.*, **107**, 1831–1861.

45. Fischer, H. and Schubert, M. (1924) Synthetische Versuche mit Blutfarbstoff-Spaltprodukten und Komplexsalz-Bildung bei Dipyrryl-methenen. *Ber. Dtsch. Chem. Ges.*, **57**, 610–617.

46. Brückner, C., Karunaratne, V., Rettig, S.J., and Dolphin, D. (1996) Synthesis of meso-pehnyl-4,6-dipyrrins, preparation of their Cu(II), Ni(II), and Zn(II) chelates, and their structural characterization of bis[meso-pehnyl-4,6-dipyrrinato]Ni(II). *Can. J. Chem.*, **74**, 2182–2193.

47. Wagner, R.W. and Lindsey, J.S. (1996) Boron-dipyrromethene dyes for incorporation in synthetic multi-pigment light harvesting arrays. *Pure Appl. Chem.*, **68**, 1373–1380.

48. Halper, S.R. and Cohen, S.M. (2005) Heterometallic metal-organic frameworks based on tris(dipyrrinato) coordination complexes. *Inorg. Chem.*, **44**, 486–488.

49. Murphy, D.L., Malachowski, M.R., Campana, C.F., and Cohen, S.M. (2005) A chiral, heterometallic metal-organic framework derived from a tris(chelate) coordination complex. *Chem. Commun.*, 5506–5508.

50. Halper, S.R., Do, L., Stork, J.R., and Cohen, S.M. (2006) Topological control in heterometallic metal-organic frameworks by anion templating and metalloligand design. *J. Am. Chem. Soc.*, **128**, 15255–15268.
51. Garibay, S.J., Stork, J.R., Wang, Z., Cohen, S.M., and Telfer, S.G. (2007) Enantiopure vs. racemic metalloligands: impact on metal-organic framework structure and synthesis. *Chem. Commun.*, 4881–4883.
52. Salazar-Mendoza, D., Baudron, S.A., and Hosseini, M.W. (2007) Beyond classical coordination: silver-π interactions in metal dipyrrin complexes. *Chem. Commun.*, 4558–4560.
53. Salazar-Mendoza, D., Baudron, S.A., and Hosseini, M.W. (2008) Many faces of dipyrrins: from H-bonded networks to homo- and hetero-nuclear metallamacrocycles. *Inorg. Chem.*, **47**, 766–768.
54. Hill, A.E. (1921) The distribution of a strong electrolyte between benzene and water. *J. Am. Chem. Soc.*, **43**, 254–268.
55. Smith, H.G. and Rundle, R.E. (1958) The silver perchlorate-benzene complex $C_6H_6 \cdot ClO_4$, crystal structure and charge transfer energy. *J. Am. Chem. Soc.*, **80**, 5075–5080.
56. Thompson, A. and Dolphin, D. (2000) Nuclear magnetic resonance studies of helical dipyrromethene-zinc complexes. *Org. Lett.*, **2**, 1315–1318.
57. Ikeda, M., Takeuchi, M., Shinkai, S., Tani, F., Naruta, Y., Sakamoto, S., and Yamaguchi, K. (2002) Allosteric binding of an Ag+ ion to cerium(IV) bis-porphyrinates enhances the rotational activity of porphyrin ligands. *Chem. Eur. J.*, **8**, 5542–5550.
58. Lindeman, S.V., Rathore, R., and Kochi, J.K. (2000) Silver(I) complexation of (poly)aromatic ligands. Structural criteria for depth penetration in cis-stibenoid cavities. *Inorg. Chem.*, **39**, 5707–5716.
59. Sazanovich, I.V., Kirmaier, K., Hindin, E., Yu, L., Bocian, D.F., Lindsey, J.S., and Holten, D. (2004) Structural control of the excited-state dynamics of bis(dipyrrinato)zinc complexes: self-assembling chromophores for light-harvesting architectures. *J. Am. Chem. Soc.*, **126**, 2664–2665.
60. Thoi, V.S., Stork, J.R., Magde, D., and Cohen, S.M. (2006) Luminescent dipyrrinato complexes of trivalent group 13 metal ions. *Inorg. Chem.*, **45**, 10688–10697.
61. Smalley, S.J., Waterland, M.R., and Telfer, S.G. (2009) Heteroleptic dipyrrin/bipyridine complexes of ruthenium(II). *Inorg. Chem.*, **48**, 13–15.
62. De Feyter, S., Abdel-Mottaleb, M.M.S., Schuurmans, N., Berkuijl, B.J.V., van Esch, J.V., Feringa, B.L., and De Schryver, F.C. (2002) Metal ion complexation: a route to 2D templates? *Chem. Eur. J.*, **10**, 1124–1132.
63. Stepanow, S., Lingenfelder, M., Dmitriev, A., Spillmann, H., Delvigne, E., Lin, N., Deng, X., Cai, C., Barth, J.V., and Kern, K. (2004) Steering molecular organization and host-guest interactions using two-dimensional nanoporous coordination systems. *Nat. Mater.*, **3**, 229–233.
64. Surin, M., Samori, P., Jouaiti, A., Kyritsakas, N., and Hosseini, M.W. (2007) Molecular tectonics on surfaces: bottom-up fabrication of 1D coordination networks that form 1D and 2D arrays on graphite. *Angew. Chem. Int. Ed.*, **46**, 245–249.
65. Newkome, G.R., Wang, P., Moorefield, C.N., Cho, T.J., Mohapatra, P.P., Li, S., Hwang, S.H., Lukoyanova, O., Echegoyen, L., Palagatto, J.A., Iancu, V., and Hla, S.W. (2006) Nanoassembly of a fractal polymer: a molecular "Sierpinski hexagonal gasket. *Science*, **312**, 1782–1785.
66. Li, S.-S., Northrop, B., Yuan, Q.-H., Wan, L.-J., and Stang, P.J. (2009) Surface confined metallosupramolecular architectures: formation and scanning tunnelling microscopy characterization. *Acc. Chem. Res.*, **42**, 249–259.
67. MacDonald, J.C., Dorrestein, P.C., Pilley, M.M., Foote, M.M., Lundburg, J.L., Henning, R.W., Schultz, A.J., and Manson, J.L. (2000) design of layered crystalline materials using coordination chemistry and hydrogen bonds. *J. Am. Chem. Soc.*, **122**, 11692–11702.
68. Noveron, J.C., Lah, M.S., Del Sesto, R.E., Arif, A.M., Miller, J.S., and Stang, P.J. (2002) Engineering the

structure and magnetic properties of crystalline solids via the metal-directed self-assembly of a versatile molecular building unit. *J. Am. Chem. Soc.*, **124**, 6613–6625.

69. Dechambenoit, P., Ferlay, S., and Hosseini, M.W. (2005) From tectons to composite crystals. *Cryst. Growth Des.*, **5**, 2310–2312.

70. Heinze, K., Beckmann, M., and Hempel, K. (2008) Solid-phase synthesis of transition metal complexes. *Chem. Eur. J.*, **14**, 9468–9480.

13
Water-Soluble Perylene Dyes

Cordula D. Schmidt and Andreas Hirsch

13.1
History and Functionalization of Perylene Dyes

Perylene is a polycyclic aromatic molecule that consists of five rings and can be naturally found in coal tar. It has been discovered and synthesized for the first time in 1910 by Roland Scholl, who obtained perylene from oxidative coupling of two naphthalene units in *peri*-positions with anhydrous aluminum trichloride [1]. This condensation of aromatic rings has been named after its inventor as the "Scholl reaction" and perylene is the short form of *peri*-dinaphthalene [1].

The ring system is planar and symmetrical to inversion. It belongs to the point group D_{2h}. Similar to polymers, the perylene skeletal structure could be elongated to poly(*peri*-naphthalenes), which represent the so-called rylene dyes (Figure 13.1b). The smallest but most important homolog of this class is perylene.

Perylene derivatives have only been used as vat dyes until the late 1950s due to their low solubility. As soon as soluble perylene compounds were developed their fluorescence and their commercial use has been realized. Nowadays, many different perylene pigments for coloring yellow, orange, red, pink, purple, or black are commercially available [2].

The starting material for the synthesis of perylene-3,4,9,10-tetracarboxylic acid diimides (PDIs, also called perylene-3,4:9,10-bis(dicarboximides), PBIs) is perylene-3,4,9,10-tetracarboxylic acid dianhydride (PTCDA). The anhydride can be easily transformed into the imide by condensation reactions. The PDIs can be modified in two different ways: either the imide or the bay positions or both can be functionalized (Figure 13.1c). Substituents in both positions affect the solubility and aggregation behavior; also, the photostability and the color of the perylene dyes can be controlled via the rests R and R' [3].

In the following, possible transformations of PTCDA into soluble perylene compounds are presented and general properties and applications of perylene dyes are summarized. After an overview of water-soluble perylene dyes and their applications, chiral water-soluble perylene dyes are presented in the last part.

Ideas in Chemistry and Molecular Sciences: Advances in Synthetic Chemistry. Edited by Bruno Pignataro
Copyright © 2010 WILEY-VCH Verlag GmbH & Co. KGaA, Weinheim
ISBN: 978-3-527-32539-9

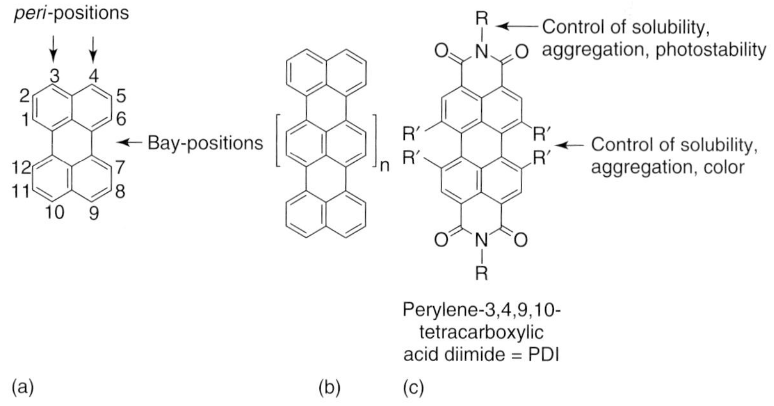

Figure 13.1 (a) Nomenclature of perylene, (b) general structure of rylenes, and (c) functional sites of perylene diimides (PDIs).

13.2
Derivatization of Perylene Dyes

PTCDA is a cheap commercially available substance, and therefore serves as starting material for a great variety of perylene compounds and higher aromatic systems (**A–V**, Scheme 13.1). PTCDA can be directly transformed to either perylene tetraesters (**A**) [4], perylene diimides (PDIs, **B**) or to bay-substituted derivatives (**K–V**).

Formation of PDIs (**B**) can be achieved under different conditions. Langhals has developed the strategy with imidazole as base and zinc acetate as catalyst [5]. Under basic alcoholic conditions, he and his coworkers have observed the rearrangement of PDIs (**B**) to the lactame ring contracted by-product **C** [6]. Unsymmetric PDIs (**F**) can be obtained via the perylene monoanhydride monoimide (**E**). **E** can be synthesized in two different ways: either one substituent of the symmetric PDI (**B**) is removed under basic conditions and the anhydride is reclosed by addition of acid [7] or the perylene monoanhydride monopotassium salt (**D**) is first formed from PTCDA, which is then condensed with an amine followed by acidification [7–9]. Intermediate (**E**) can also be transformed to perylene monoimide diesters (**G**) [10]. Perylene monoimide (**H**) is accessible by autoclave reactions either from PDI (**B**) or from PTCDA followed by a condensation reaction. Compound **H** is the starting material for extended rylene dyes; terrylene diimides **I** are obtained by coupling of **H** with naphtylimide, quaterrylene diimides **I** by coupling of two units of perylene monoimide (**H**). Coupling can be achieved by Suzuki cross-coupling [11], Stille heterocoupling, or Yamamoto homocoupling conditions with subsequent cyclodehydrogenation [12] or just under basic conditions with, for example, potassium *tert*-butylate and DBN (1,5-Diazabicydo[4.3.0] non-5-ewe) [13–15].

Introduction of substituents at the carboxylic scaffold in the so-called bay area are possible via bromination (**K, L, R**) or chlorination (**S**) reactions. Bromination

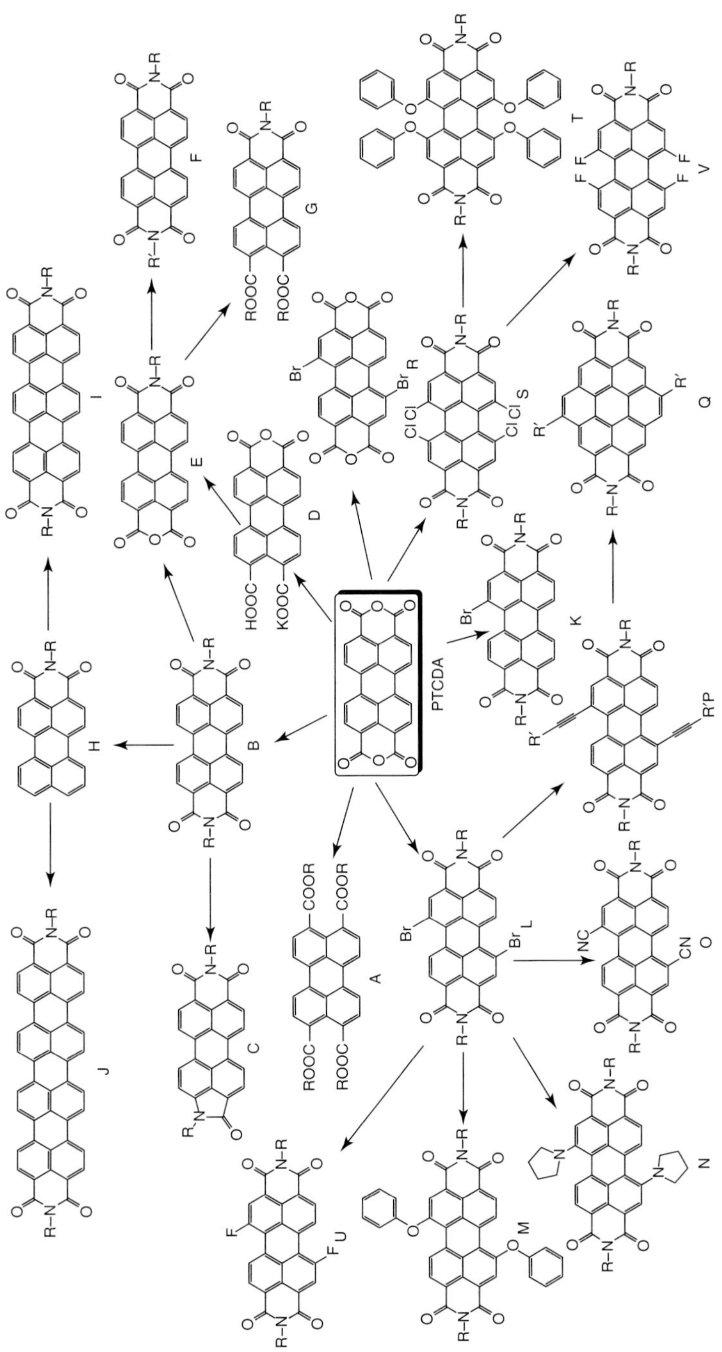

Scheme 13.1 Possible reactions of perylene-3,4,9,10-tetracarboxylic acid dianhydride (PTCDA) [3].

of PTCDA or PDIs (**B**) leads to disubstituted products (**R** and **L**) [16–18]. Under mild conditions also, the monobrominated derivative (**K**) can be obtained [19]. The versatility of this reaction is based on the easily achievable exchange of the bromine substituents with a variety of nucleophiles, for example, alcohols [16] (**M**), amines [20, 21] (**N**), cyano [22] (**O**), or carbon [23] derivatives (**P**). Under nonnucleophilic basic conditions an isomeric mixture of a corene bisimide **Q** is available from the acetylenic derivative **P** [11, 17, 18]. Tetrafold bay-substituted PDIs (**S, T, V**) are accessible by chlorination. The chlorine substituents of the tetrafold chlorinated perylene diimide (**S**) can be displaced by phenoxy groups to yield **T** [24, 25]. Furthermore, core-fluorinated PDIs (**U, V**) can be achieved by nucleophilic substitution of chlorinated (**S**) or brominated (**L**) PDIs via the Halex reaction [26].

13.3
Properties and Applications of Perylene Dyes

Perylene dyes are characterized not only by their color intensities and enormous fluorescence quantum yields (up to 99% [27]), but also by their outstanding chemical, thermal, and photochemical stability [3] based on the electron-deficient character of PDIs, which impedes oxidation reactions [3]. Therefore, applications of perylenes are manifold: they are used as pigments, vat dyes, dyes for ink-jet prints [28, 29], varnishes, and polymers, as laser dyes, [29–31], or as fluorescence dyes. As the latter, they could serve as labels or tracers for drugs or biological molecules [33–37] with detection limits of 4×10^{-18} M [38].

Nowadays, the electronic properties of perylenes have become of increasing interest. It has been found that PTCDA crystallizes in tightly packed one-dimensional $\pi-\pi$ stacks. The small distance of only 3.2 Å facilitates π-orbital, overlap and therefore it is regarded as the "archetype molecular semiconductor" [39]. As in inorganic semiconductors, photoluminescent charge transfer is observed. In the meantime, analog columnar aggregates of PDI derivatives have also been found. The n-type semiconductivity is based on the high electron affinity of rylenes. As PDIs are considered to be the best organic n-type semiconductors available to date, they have been used in electrophotography as xerographic photoreceptors [40] and in organic solar cells [41–43] for more than one decade. They could be employed also for data storage in CDs or DVDs [44]. Other applications include scintillation counters [43], organic light emitting diodes (OLEDs) [45], or organic field effect transistors [40–42, 46, 47].

13.4
Advantages and Applications of Water-Soluble Perylene Derivatives

As water is a cheap, nonpolluting, easily obtainable solvent, it should be aspired to create water-soluble building blocks for industrial applications. However, although

13.4 Advantages and Applications of Water-Soluble Perylene Derivatives

a huge variety of perylene dyes exist nowadays, there are only very few known structures of perylene dyes, which are soluble in aqueous media.

Most of these examples are functionalized in the bay positions to prevent aggregation and achieve good fluorescence quantum yields. Such hydrophilic groups could be, for example, sulfonic acid groups (Figure 13.2a, **1**), sulfonate salts (Figure 13.2a, **2**), or quaternized amines (Figure 13.2a, **3**) as published by Müllen and coworkers [33, 48]. Owing to their fluorescence properties and their water-solubility these perylene dyes can be used as labels for enzymes, virions, or drugs [49, 50], which can then be investigated by single-molecule spectroscopy to obtain information of biological processes in the living cell. To study the enzyme activity of phospholipase, Müllen synthesized perylene and terrylene dyes with four sulfonic acid groups in the bay region for water-solubility, and one N-hydroxysuccinimide ester or maleimide group as the imide substituent for covalent linkage to the enzyme [33]. As described before, perylenes are good n-type material candidates for molecular electronics [51–54] and also water-soluble perylene dyes can be used as building blocks. Webber was the first to show that from the water-soluble anionic and cationic bay-functionalized PDIs **1–3** perylene diimide films can be fabricated without loss of subsequent polyelectrolyte extraction by layer-by-layer (LBL) technology (for LBL assemblies see also Section 13.6.2) [45]. In these films an efficient energy transfer to a trapping layer has been observed [43, 55].

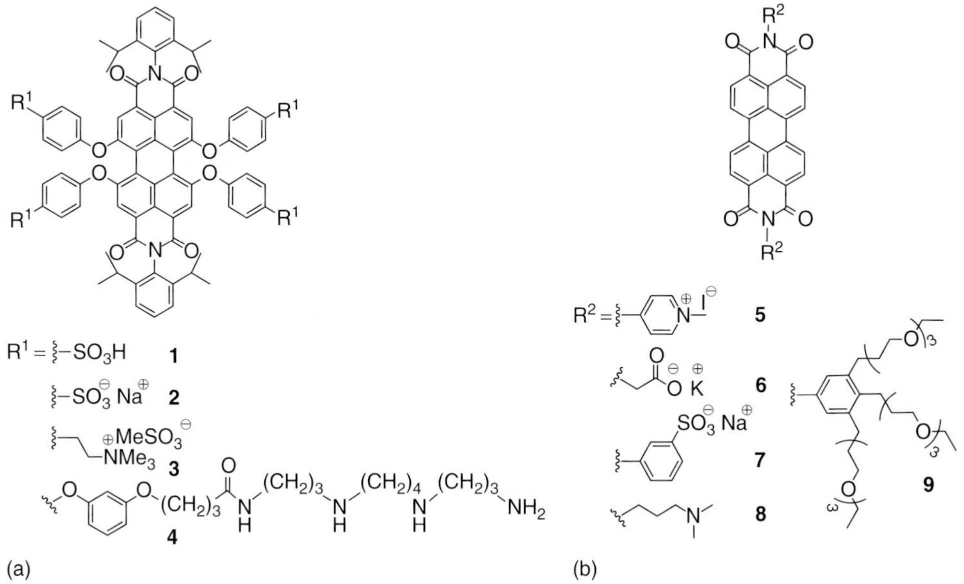

Figure 13.2 Examples of water-soluble PDIs with hydrophilic groups (a) in the bay region and (b) in the imide region.

In other approaches, the imide nitrogens of the PDI are connected to substituents carrying quaternized amine salts [48, 56] (Figure 13.2b, **5**), hydrophilic groups like carboxylic or sulfonic acids, and their corresponding salts (Figure 13.2b, **6, 7**) [48, 57] or poly(ethylene oxide) (PEO) chains (Figure 13.2b, **9**) [58]. Also, aliphatic substituted PDIs with free amine groups that can react to the homolog hydrochlorides have been published [35, 36]. As Müllen has shown, perylene dyes with one PEO chain in the imide region reflect the polarity of their environment because they show strong solvatochromic behavior: in nonpolar solvents they exhibit strong fluorescence, whereas in polar solvents their fluorescence is very weak due to aggregation [33]. Therefore, such dyes can be used for staining cells [33].

Water-soluble perylene dyes like N,N'-bis[3,3'-(dimethylamino)-propylamine]-3,-4,9,10-perylene tetra carboxylic diimide (DAPER, Figure 13.2, **8**) with or without additional tertiary amine functions in the bay region can also act as telomerase inhibitors and therefore as antitumor agents [35, 36]. Telomerase expression can be inhibited by triple helix formation with antisense oligonucleotides. With DAPER (**8**), the triple helix stability and therefore telomerase inhibition is increased [35]. Also, noncovalent interactions between PDI molecules and DNA have been studied with a PDI tetrafold bay-functionalized with spermine (**4**) [59]. Moreover, PDIs can be covalently linked to DNA; these "artificial nucleosides" can be either at terminal or at internal positions of the DNA duplex [60–64, 70], and lead to an enhancement of the stability of the duplex [61]. With PDIs as DNA building blocks the transfer of holes and electrons through DNA can be investigated [62]. Besides DNA-duplexes, other nanostructures like triplex DNA [63] or hairpins [61, 62] can also be designed. When several PDI units are incorporated into a DNA strand, thermophilic foldable polymers can be obtained [64].

Aggregation phenomena of perylene dyes have been extensively studied in organic solvents: The formation of linear nanobelts, spherical nanoparticles [65], double-helical [3], and columnar nanostructures [39, 66] has been reported with examples of linear J-type $\pi-\pi$ stacking of core-twisted perylene dyes [67] as well as the self-organization of linear polymeric H-type $\pi-\pi$ stacks [68]. The aggregation constants strongly depend on the substituents of the PDIs as well as on the solvent. The lowest aggregation constants are found for bay-substituted PDIs because of their twisted π-system (30–700 M^{-1} for tetrachlorinated PDIs **S**) [69]. For PDIs without substituents in the bay area, and electron-rich imide substituents in little polarizable aliphatic solvents like methylcyclohexane, the highest aggregations constants with values between 1×10^4 M^{-1} (for trialkoxybenzyl substituents) and $>10^8$ M^{-1} (for oligo(p-phenylene-vinylene) units) were obtained [69].

In contrast to organic solvents, which in general, strongly reduce the Gibbs free energy for $\pi-\pi$ aggregates compared to the gas phase, water is a special solvent and can strongly increase the free energies for aggregation due to hydrophobic effects [69]. Considering this fact, it is surprising that up to now, only a rather few number of water-soluble perylene dyes have been investigated. Micelles, vesicles, and rod aggregates of amphiphilic PDIs in water containing 2% THF have been reported by Zhang et al. [58]. Ford found an aggregation constant of 1.0×10^7 M^{-1} for the glycine-substituted PDI **6** in basic aqueous solution [57]. Würthner

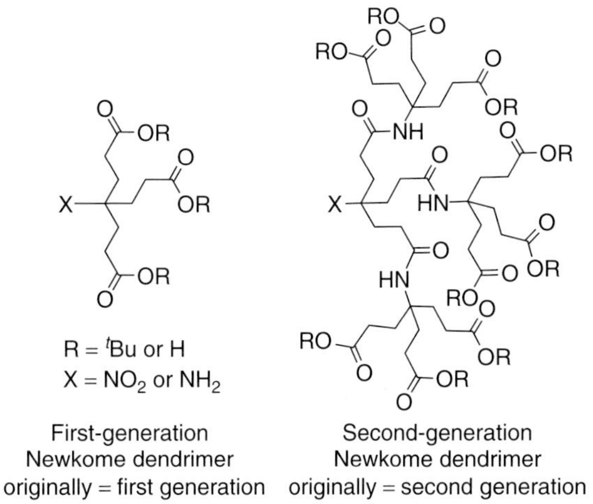

Figure 13.3 Newkome dendrimers of first and second generation.

and coworkers even determined for PDI **9**, with several PEO chains in the imide position in water, an aggregation constant $>10^8$ M^{-1}. These values are among the highest reported for PDIs [69].

In our group, water-solubility of nonpolar molecules like fullerenes or porphyrins is achieved by attaching polyacid Newkome dendrons of the first and second generation (Figure 13.3) [71, 72]. The Newkome dendrons [73–75] are versatile building blocks due to the high solubility of the polyamide derivatives in organic solvents and the hydrophilicity of the polyacid compounds. It is also possible to create well-defined amphiphiles with the help of these dendrimers. Examples in our group include amphiphilic fullerenes and calixarenes [76, 77]. These dendrons have also been used to create symmetric as well as amphiphilic PDI dyes, which are soluble in water as shown below.

13.5
Symmetric Water-Soluble PDIs

Symmetric PDIs are easily obtainable by condensation of PTCDA with primary amines at high temperatures with quinoline or molten imidazole as solvent, and zinc acetate or dicyclohexylcarbodiimide (DCC) as catalytic dehydrating agents [47].

Newkome dendrons have been used as hydrophilic groups, but direct condensation of the amines of the first or second generation failed due to steric hindrance. Therefore, the dendrons have been elongated with ε-amino caproic acid as C6-spacer [78]. Two equivalents of the obtained elongated dendritic amines were then condensed with 1 equiv. of PTCDA, by melting in imidazole and with zinc acetate as catalyst to yield the symmetric PDI **10**. Acidic hydrolysis with formic acid or trifluoroacetic acid resulted in the formation of the water-soluble compounds **11** and **12**, respectively (Scheme 13.2) [78].

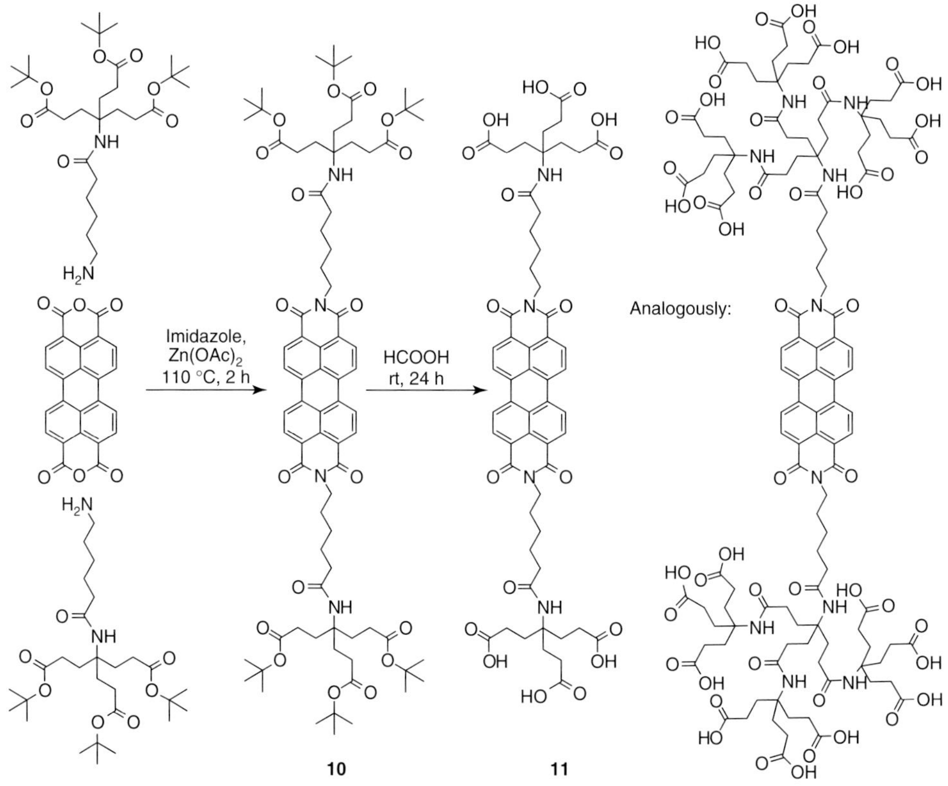

Scheme 13.2 Synthesis of symmetric water-soluble PDIs **11** and **12** [78].

The aggregation behavior of the symmetric water-soluble compounds **11** and **12** in buffered water at pH 7.2 has been investigated by absorption and fluorescence spectroscopy. For the first-generation compound **11**, a typical absorption spectrum for aggregated perylene dyes was obtained (Figure 13.4a). The absorption maximum is centered at 501 nm, a smaller maximum is at 545 nm, and a shoulder is detected at around 476 nm. From the absorption ratio between the $0 \to 0$ (around 550 nm) and $0 \to 1$ (around 500 nm) transition, direct conclusion to the degree of aggregation can be drawn; while monomeric PDIs have a normal Franck–Condon progression with $A^{0\to0}/A^{0\to1} \approx 1.6$, aggregated PDIs have inversed intensity distribution among their vibrionic states with $A^{0\to0}/A^{0\to1} \leqslant 0.7$ [57, 79]. The first-generation bolaamphiphile **11** shows an absorption ratio of $A^{0\to0}/A^{0\to1} = 0.51$, and therefore, it can be concluded that it is strongly aggregated. In contrast, the second-generation derivative **12** shows a completely different absorption spectrum (Figure 13.4b). The three "fingers" are increasing with higher wavelengths, with maxima located at 472, 499, and 534 nm and the band ratio $A^{0\to0}/A^{0\to1} = 1.15$, which indicates a very low degree of aggregated dye molecules.

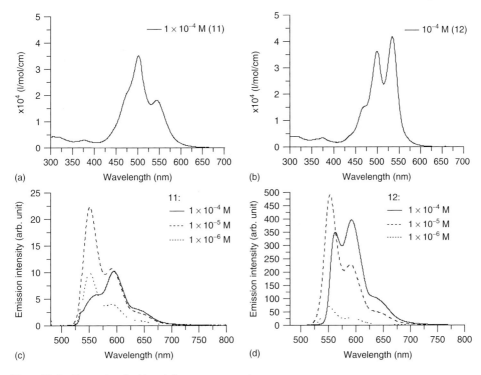

Figure 13.4 Absorption (a, b) and fluorescence spectra (c, d) of the symmetric water-soluble perylenes **11** (a, c) and **12** (b, d).

Also, the fact that the second-generation bolaamphiphile **12** has higher extinction coefficients than the analog, the first-generation compound **11** is indicative for less aggregation.

In Figure 13.4c and d also, the fluorescence spectra of **11** and **12** are shown at different concentrations. For comparability, all fluorescence spectra were recorded under the same conditions (same cuvette, same slot width) and were excited at 534 nm. It is obvious that for the first-generation compound **11**, the fluorescence intensity is much lower than for the second-generation compound. This is another hint for aggregation of **11**, because upon aggregation, fluorescence quenching takes place. Also, the shape of the fluorescence spectra of **11** and **12** are different at a concentration of 1×10^{-4} M: the first-generation bolaamphiphile **11** shows an emission maximum at 594 nm and two shoulders around 565 and 640 nm, while the second-generation derivative **12** shows more hypsochromic emission maxima at 562 and 592 nm and a shoulder around 635 nm.

All these findings indicate that the first-generation compound **11** is strongly aggregated and the second-generation compound **12** is only weakly aggregated at a concentration of 1×10^{-4} M. This can be explained on one hand by steric reasons: the large second-generation Newkome dendritic substituents are much more sterically demanding than the first-generation dendritic substituents, and therefore mask the aromatic perylene area so that aggregation by $\pi-\pi$ stacking between the perylene units is hindered. On the other hand, the number of hydrophilic groups plays an important role: the second-generation bolaamphiphile **12** has 18 carboxylic acid groups, which are mostly deprotonated at pH 7.2, and therefore, the electrostatic repulsion between the negative charges also disfavors aggregation, while the first-generation bolaamphiphile **11** has only six carboxylic acids groups, and therefore, less electrostatic repulsion. Additionally the water-solubility of the second-generation derivative **12** is better due to the more hydrophilic groups, and therefore, the hydrophobic forces are much weaker.

During dilution of the first-generation compound **11** to 1×10^{-5} M, the maximum at 594 nm decreases and shifts to 590 nm, and the shoulder at 565 nm increases and appears as new emission maximum at 551 nm. At a concentration of 1×10^{-6} M, reduced emission intensity according to dilution effects but no shift of the emission maxima are observed (Figure 13.4c). For the second-generation bolaamphiphile **12**, the following spectral changes are obtained: at 1×10^{-4} M, the emission maxima are at 562 and 592 nm, upon dilution to 1×10^{-5} M, the maximum at 592 nm decreases and shifts to 590 nm, while the maximum at 592 nm increases and shifts to 553 nm (Figure 13.4d).

13.6
Unsymmetric PDIs

Unsymmetric PDIs can be derived from perylene monoanhydride monoimides. As already mentioned in Section 13.2, there are two different routes to obtain the monoimides starting from PTCDA: either, first the symmetric PDI **B** is formed and then one imide substituent is removed under basic conditions (this way is chosen for branched aliphatic substituents) [7] or, first the perylene tetracarboxylic acid monoanhydride monopotassium salt **D** is formed, which can then be condensed with 1 equiv. of amine [7–9]. In both cases, acidification leads to reclosure of the cyclic anhydride function and the perylene monoanhydride monoimide derivative **E** is obtained. This anhydride function can then be condensed with another amine to give the unsymmetric perylene diimide **F** (Scheme 13.3). For the water-soluble second-generation amphiphile **13** shown in Scheme 13.3, the synthetic route via the perylene monopotassium salt **D** was chosen. Condensation with dodecylamine led to the perylene monoanhydride monoimide **E**, which was then condensed with the *tert*-butyl protected second-generation Newkome dendron elongated with ε-amino caproic acid to avoid steric hindrance. Acidic hydrolysis of the *tert*-butyl esters with formic acid or trifluoro acetic acid finally yields the water-soluble amphiphile **13** [78].

Investigations with absorption spectroscopy have revealed that this amphiphile is aggregated in phosphate buffered solution at pH 7.2 at very low concentrations (from 1×10^{-4} to 1.3×10^{-6} M) [78]. Transmission electron microscopy (TEM) shows that amphiphile **13** forms regular micelles with an average diameter of 16 nm in phosphate buffer solution at pH 7.2 (Figure 13.5) [78].

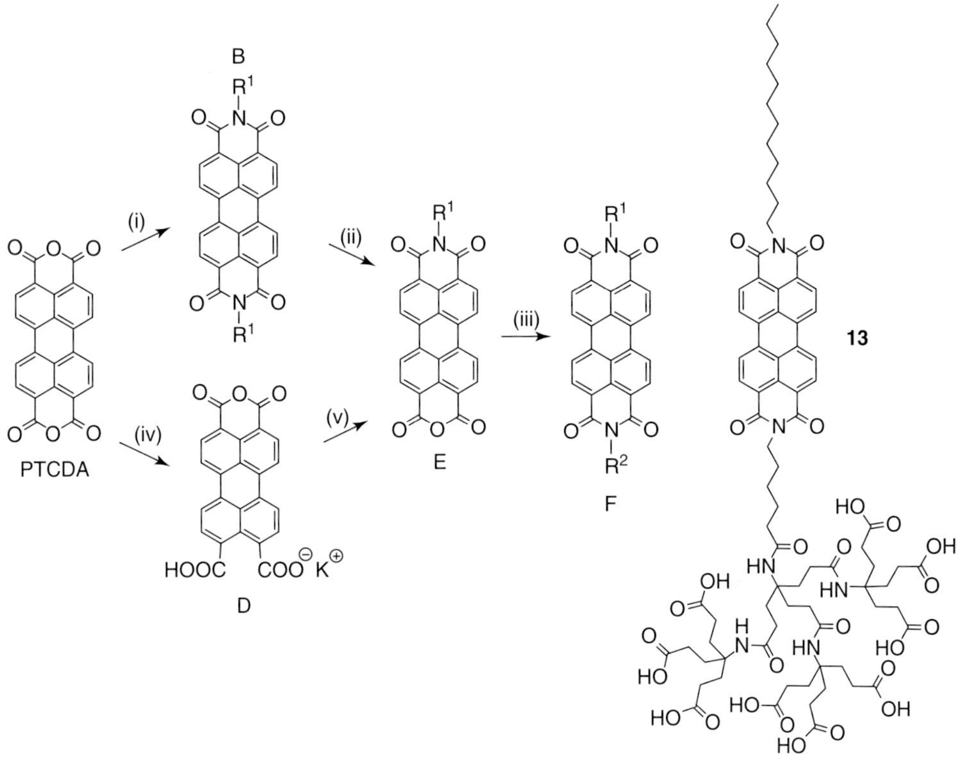

Scheme 13.3 General synthetic pathways to unsymmetric PDIs; (i) 2 equiv. R¹-NH₂, imidazole, Zn(OAc)₂, 90–160 °C, 2–4 hours; (ii) 1. KOH, *tert*-BuOH, 80 °C, 2. AcOH, 10% HCl; (iii) 1 equiv. R²-NH₂, imidazole, Zn(OAc)₂, 90–160 °C, 2–4 hours; (iv) 1. KOH, H₂O, Δ, 2. AcOH; (v) 1. R¹-NH₂, H₂O:EtOH 1 : 1, rt, 4 hours → 90 °C, 2 hours, 2. 10% HCl. For PDI **13** synthetic route (iv) → (v) → (iii) was chosen with R¹ = dodecyl and R² = C₆ – second-generation Newkome dendrimer.

The amphiphilic compound **13** can be used for several applications, for example, as surfactant for carbon nanotubes (see Section 13.6.1) or as reporter electrolyte (see Section 13.6.2).

13.6.1
Noncovalent Functionalization of Carbon Nanotubes

One application for water-soluble PDIs is the noncovalent functionalization of single-wall carbon nanotubes (SWNTs) (Figure 13.6).

Despite their outstanding mechanical and electronic properties, applications of SWNTs have been hampered by their intrinsic insolubility in water and/or organic solvents due to their tendency to form bundles. Another problem is that SWNTs are not homogeneous, well-defined compounds, but are always produced as a mixture of different types of SWNTs: there are semiconducting and metallic

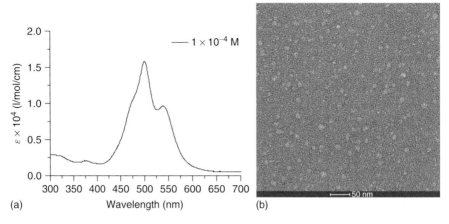

Figure 13.5 (a) Absorption spectra and (b) TEM image of amphiphile **13** [78].

SWNTs with different lengths and diameters and different chirality. To overcome these problems, it is necessary to individualize and separate SWNTs before well-defined functionalization is possible. One solution to reach these aims can be the noncovalent functionalization with polycyclic aromatic compounds like PDIs. The noncovalent strategy has the advantage of implementing multifunctional groups without compromising the main properties of SWNTs as opposed to

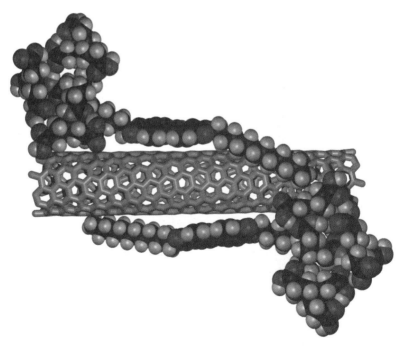

Figure 13.6 Carbon nanotube noncovalently functionalized via $\pi-\pi$ stacked PDIs **13**.

covalent functionalization; the sp^2 carbon backbone is not altered. Perylene dyes are adequate polycyclic aromatic compounds due to their reasonably sized π-system; furthermore, perylenes are excellent electron acceptors, and therefore, it is possible to create SWNT-based electron-donor-acceptor hybrids with SWNTs as electron donors [80].

The amphiphilic derivative **13**, as well as the symmetric first and second-generation compounds **11** and **12** enable the dispersion of SWNTs in buffered water, and also afford a high degree of nanotube exfoliation according to microscopic investigations by TEM and AFM(atomic force microscopy) [80, 81]. Figure 13.7a shows a typical TEM image of the SWNT-perylene **13** composite displaying very long and straight nanotubes. Furthermore, it has been shown by statistical AFM height analysis that the bundle size is strongly reduced compared to SWNTs dispersed in the commercial surfactant SDBS (sodium dodecyl benzene sulfonate). (Figure 13.7b) [81]

Besides suspension and debundling of SWNTs, separation of the above-mentioned different kinds of SWNTs is an important aspect; especially, the separation of metallic from semiconducting SWNTs is very interesting with regard to industrial applications. One separation approach is based on density gradient ultracentrifugation, which is capable of separating objects according to their buoyant density. When samples of highly individualized nanotubes are subjected to this treatment, for example, when dispersed in perylene **13**, [82], they are fractioned according to diameter or type.

The principle of noncovalent functionalization with water-soluble perylenes can also be extended from SWNT to graphene sheets. The interactions between perylene dyes and graphene should be even stronger, because grapheme, in contrast to SWNTs, is a flat aromatic compound and therefore the π-orbital overlap is maximized. First experiments with the symmetric second- generation PDI **12** have been promising [83].

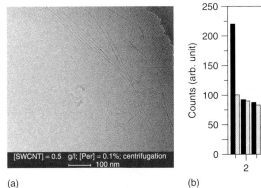
(a)

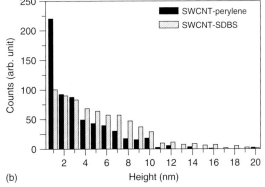
(b)

Figure 13.7 (a) TEM image of SWNTs individualized by perylene **13** and (b) statistical AFM analysis of individualized SWNTs by perylene **13** compared to the commercially available surfactant SDBS [81].

Not only are the $\pi-\pi$ interactions of the perylene dyes interesting to investigate, there are also applications for ionic interactions between the water-soluble dyes and other charged molecules as described in the next chapter.

13.6.2
PDIs as Reporter Electrolytes for Implantation Technology

To achieve faster and better adhesion of orthopedic or dental implants to the organism cells, coatings with polymeric electrolytes [84–86] have already been tested. However, the use of monodisperse oligoelectrolytes (like amphiphile **13**) instead of polymers, leads to well-defined and therefore better manageable multilayer systems [87]. So-called "reporter electrolytes" can be used to investigate the formation as well as the degradation of, for example, implant coatings. Owing to its color and fluorescence properties on the one hand and its structural similarity to natural building blocks for cell membranes, for example, phospholipids (Figure 13.8) on the other hand, the amphiphilic perylene diimide **13** is perfectly suited as a reporter electrolyte.

For this purpose, titanium implants are modified using the so-called LBL self-assembly introduced by Decher and coworkers [88]. This technique is based on attractive electrostatic forces between oppositely charged polyelectrolytes. First, the negatively charged implant is coated with a monolayer of a cationic molecule (in our case, cationic dendrimer) [87]. By dipping the titanium sample into an aqueous solution of the charged molecule, these molecules bind electrostatically to the surface. After a washing step, the procedure is repeated with a solution of a negatively charged oligoelectrolyte (in our case, negatively charged PDI **13**). By following this protocol repeatedly, the titanium samples have been coated with several layers of the above-mentioned molecular oligoelectrolytes. After steam sterilization, human fetal osteoblasts (hFOB cells) were cultured on these modified samples for several days to investigate the *in vitro* behavior of the coating as well as the influence of components of the coating on osteoblastic cells. Cell proliferation and attachment of hFOB cells have been observed by means of cell number and fluorescence microscopy; fluorescence spectroscopy has been used to examine the degradation behavior of the coating. Detection of the fluorescent signal in the cell culture medium proofs the release of the PDI **13** out of the LBL-coating to the cell medium. Increased number of dye layers on the sample leads to higher fluorescence emission in the corresponding cell medium. The coherence between the number of layers and fluorescence emission is strictly linear. In addition to the cell medium, the applied perylene dye **13** has also been found integrated in the hFOB cells, which could be visualized by fluorescence microscopy. Owing to the similarity of the used reporter dye **13** to phospholipids that are a major component of all biological membranes (Figure 13.8), it is assumed that dye **13** was incorporated into the cell membrane of the hFOB cells. The LBL-coated titanium samples turned out to be an efficient transporting and delivery system for film components into cells [89].

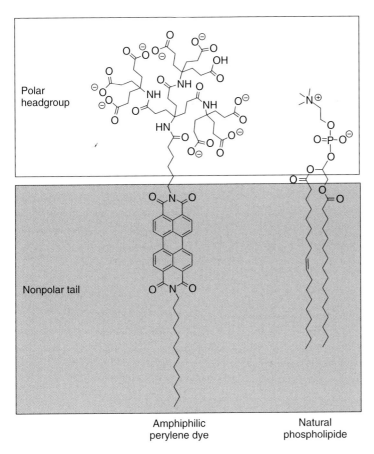

Figure 13.8 Structure of the amphiphilic perylene dye **13** compared to the natural phospholipid lecithin.

In future studies also, charged bioactive substances like growth factors or drugs should be incorporated in these LBL assemblies to positively affect the response of bone cells and thereby, the performance of biomaterials should be enhanced.

13.7
Chiral Water-Soluble PDIs

PDIs without bay-substituents easily form aggregates with other molecules like nanotubes [80–82] or with themselves via $\pi-\pi$ stacking. The control of this self-aggregation process is aspired for constructing electronic devices like dye sensitized solar cells or organic field effect transistors. For some bay-unsubstituted PDIs, it is experimentally proven by crystal structures and optical polarization microscopy that they prefer to form columnar π-stacks due to rotational displacements of neighboring molecules, and this type of aggregation was also calculated

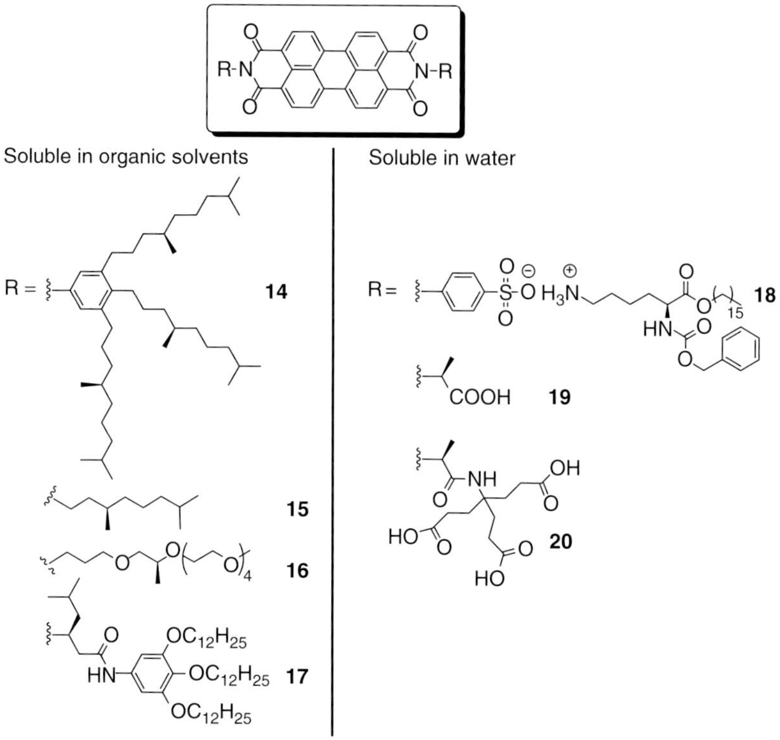

Figure 13.9 Examples of chiral non–water-soluble PDIs (**14–17**) and chiral water-soluble PDIs (**18–20**).

as the energy minimum [66, 90–93]. As with achiral PDIs there is no control of self-aggregation, because racemic mixtures of helical assemblies are assumed; PDIs with chiral substituents show a preference for a single helical formation. Meijer [92] and Würthner [93] evidenced helical aggregation in organic solvents for compounds **14–17** (Figure 13.9) and obtained *M*- and *P*-helices (Figure 13.10) dependent on the absolute configuration of the chiral side chains. In water, the self-aggregation of PDIs should be strongly increased compared to organic solvents due to hydrophobic effects. However, only very few examples of chiral water-soluble PDIs exist in literature: Faul and coworkers [94] described the formation of chiral complexes between an achiral perylene sulfonic acid diimide salt and chiral cationic ligands by ionic self-assembly (Figure 13.9, **18**), while Sun *et al.* [95] created PDIs with amino acids as substituents (Figure 13.9, **19**), but the obtained chiral PDIs showed only very low solubility in water. Recently in our group, we have published the synthesis of chiral water-soluble PDIs with Newkome dendritic amino acids as side chains (Figure 13.9, **20**) [96].

To the best of our knowledge, this is the first example of very water-soluble covalently linked chiral PDIs. As chiral subunits, we have used the enantiopure amino acids alanine and lysine and Newkome dendrimers of the first and second

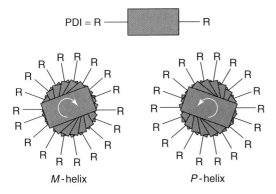

Figure 13.10 M- and P-helical arrangements of PDIs.

generation as polar head groups. In Scheme 13.4 the synthesis of the first-generation alanine compound **20** is schematically depicted.

First, the enantiopure Cbz-protected alanine is condensed with the first-generation dendritic Newkome amine under DCC coupling conditions to give **21**. After removing the Cbz-protection group by hydrogenation, 2 equiv. of the obtained chiral dendritic amine **22** are condensed with PTCDA to give the chiral *tert*-butyl protected PDI **23**. After acidic cleavage of the *tert*-butyl esters the chiral water-soluble PDI **20** is obtained. The water-soluble (L)- and (D)-target molecules have been investigated by absorption and circular dichroism (CD) spectroscopy. The concentration dependent UV/Vis spectra of **20** (Figure 13.11a) measured in phosphate buffer solution at pH 7.2 show that at a concentration of 1×10^{-4} M the molecules are aggregated, the band ratio $A^{0 \to 0}/A^{0 \to 1} = 0.7$, and the absorption maxima are at 502 and 537 nm. By diluting, disaggregation takes place: the absorption maximum at 537 nm strongly increases. Additionally, both maxima are hypsochromically shifted to 534 and 498 nm, respectively. At a concentration of 1×10^{-6} M mostly monomeric chiral dyes are observed ($A^{0 \to 0}/A^{0 \to 1} = 1.4$).

The recorded CD spectra of the two enantiomers **20** at a concentration of 1×10^{-4} M and a pH of 7.2 are depicted in Figure 13.11b. With CD spectroscopy, circular polarized light is sent through a colored chiral sample. Left and right circular polarized light passes the sample with different velocity and is differently strong absorbed; hence the difference of the extinction coefficients ($\Delta \varepsilon$) is measured. For achiral molecules or racemic mixtures, $\Delta \varepsilon$ is zero whereas chiral compounds give CD effects in their absorption region. A pair of enantiomers must show mirror-inverted spectra with the zero line as a mirror plane. This is the case for the water-soluble first-generation alanine compounds: the (L)-enantiomer gives a minimum at 500 nm ($\Delta \varepsilon = -36$) and a maximum at 563 nm ($\Delta \varepsilon = +32$), whereas the (D)-enantiomer gives a maximum at 500 nm ($\Delta \varepsilon = +37$) and minimum at 563 nm ($\Delta \varepsilon = -34$). The strong bisignate Cotton effects have a zero-crossing at 523 nm, which is equivalent to the minimum at the absorption spectrum at 1×10^{-4} M, and the peak positions of the CD couplet almost coincides with the absorption maxima of the aggregated species at a concentration of 1×10^{-4} M.

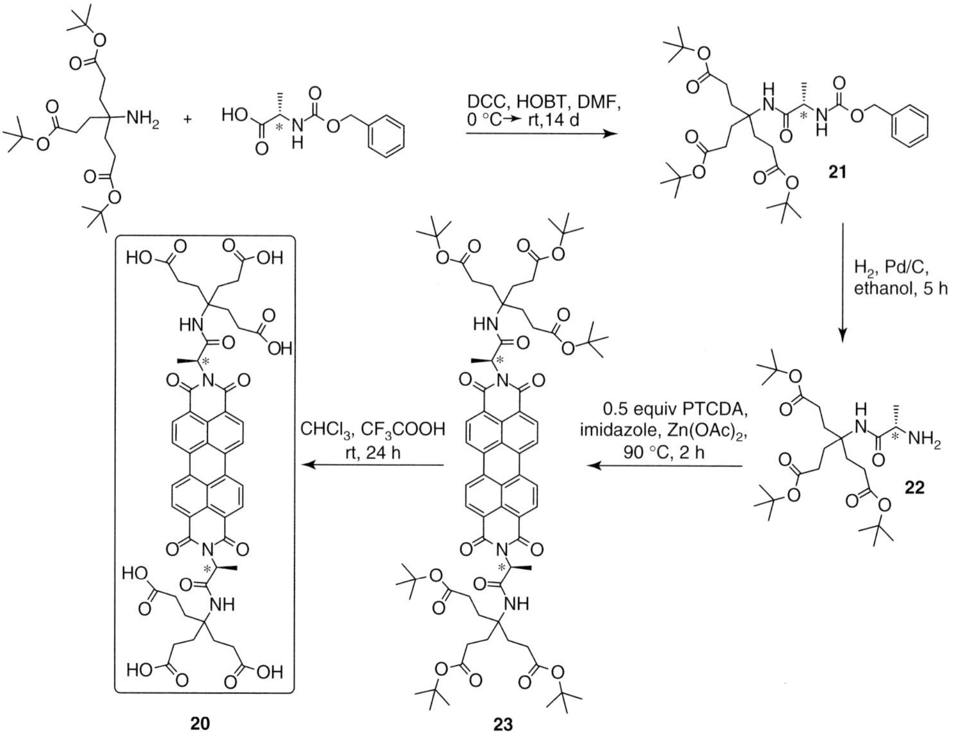

Scheme 13.4 Synthesis of the first-generation-(L)-alanine-PDI **20**, the synthesis of the (D)-compounds **20–23** starting from the Cbz-protected (D)-alanine were carried out analogously.

Upon dilution, the CD effects diminish without shift of the maxima or minima (not shown). From these observations a helical arrangement of two or more dyes can be assumed [97]. Furthermore, direct conclusions from the sign of the CD couplet can be drawn: a positive sign of the couplet (change from negative to positive $\Delta\varepsilon$ values with increasing wavelength) observed for the (L)-enantiomer indicates a right-handed, clockwise (*P*)-helical arrangement whereas the mirror-imaged negative CD couplet obtained for the (D)-enantiomer indicates a left-handed, counterclockwise (*M*)-helicity for the PDI aggregates [97].

13.8
Conclusion and Outlook

In this chapter, manifold reactions starting from PTCDA to yield different types of perylene dyes have been presented (Section 13.2). The optical properties as well as the stability of perylene dyes makes this substance class appropriate for a huge variety of applications ranging from vat dyes to electronic devices (Section 13.3).

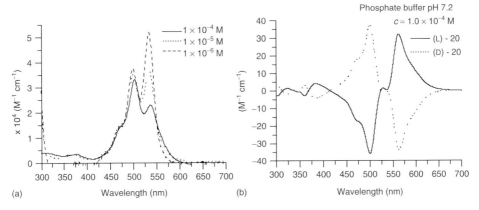

Figure 13.11 (a) Absorption spectra of **20** at different concentrations and (b) CD spectra of (L)- and (D)-**20**.

Although perylenes have been known for one century, the field of water-soluble perylene dyes is still relatively narrow (Section 13.4). Besides sulfonic acid or quaternary amine groups as polar subunits, hydrophilicity of perylene dyes can be achieved with the use of Newkome dendrons as building blocks (Sections 13.4–13.6). With the help of these dendrimers it was possible for the first time to also create fully water-soluble chiral PDIs (Section 13.7). Possible applications for water-soluble PDIs as solubilizing detergents for SWNTs (Section 13.6.1) or as reporter electrolytes for LBL-film coated implants (Section 13.6.2) have been described. In future also, SWNTs individualized by chiral water-soluble compounds should be investigated to elucidate whether there is any chirality preference of the perylene dye for SWNTs.

Acknowledgments

We thank Claudia Backes and Christian Ehli for the investigations of PDIs **11–13** with SWNTs, Jan Englert for studies of PDI **12** with graphene, Christoph Böttcher for recording TEM image 13.5b, Karin Rosenlehner for assembling the LBL-films, Sabine Ponader for the cell studies and Frank Hauke for modeling Figure 13.6. Furthermore, we want to thank the students Tugce Akdas and Kirsten Heußner, who partly synthesized the chiral water-soluble PDI **20**.

References

1. Scholl, R., Seer, C., and Weitzenböck, R. (1910) *Ber. Dtsch. Chem. Ges.*, **43**, 2202–2209.
2. Greene, M. (2002) in *High Performance Pigments* (ed. H.M.Smith), Wiley-VCH Verlag GmbH, Weinheim, pp. 249–261.
3. Würthner, F. (2004) *Chem. Commun.*, 1564–1579.
4. Eilingsfeld, H. and Patsch, M. (1975) Perylene-3,4,9,10-tetracarboxylates. Ger. Offen., DE 2512516, filed March 21, 1975 and issued Sept. 30, 1976.

5. Langhals, H. (1995) *Heterocycles*, **40**, 477–500.
6. Langhals, H. and von Unold, P. (1995) *Angew. Chem. Int. Ed.*, **34** (20), 2234–2236.
7. Kaiser, H., Lindner, J., and Langhals, H. (1991) *Chem. Ber.*, **124** (3), 529–535.
8. Tröster, H. (1983) *Dyes Pigm.*, **4** (3), 171–177.
9. Nagao, Y., Naito, T., Abe, Y., and Misono, T. (1996) *Dyes Pigm.*, **32** (2), 71–83.
10. Yang, L., Shi, M., Wang, M., and Chen, H. (2008) *Tetrahedron*, **64** (22), 5404–5409.
11. Nolde, F., Pisula, W., Müller, S., Kohl, C., and Müllen, K. (2006) *Chem. Mater.*, **18** (16), 3715–3725.
12. Geerts, Y., Quante, H., Platz, H., Mahrt, R., Hopmeier, M., Böhm, A., and Müllen, K. (1998) *J. Mater. Chem.*, **8** (11), 2357–2369.
13. Langhals, H., Büttner, J., and Blanke, P. (2005) *Synthesis*, 364–366.
14. Langhals, H. and Poxleitner, S. (2008) Fluorescent bi-chromophoric dyes comprising perylene imide and naphthaleneimide chromophores. Ger. Offen., DE 102007004016, filed Jan. 26, 2007, issued July 31, 2008.
15. Sakamoto, T. and Pac, C. (2001) *J. Org. Chem.*, **66** (1), 94–98.
16. Böhm, A., Arms, H., Henning, G., and Blaschka, P. (1997), Aromatic derivatives of 3,4,9,10-perylenetetracarboxylic acids, dianhydrides, and diimides, their preparation and their use. Ger. Offen. DE 19547209, filed Dec. 18, 1995, issued June 19, 1977.
17. Rohr, U., Kohl, C., Müllen, K., van de Craats, A., and Warman, J. (2001) *J. Mater. Chem.*, **11** (7), 1789–1799.
18. Rohr, U., Böhm, A., Groß, M., Meerholz, K., Bräuchle, C., and Müllen, K. (1998) *Ang. Chem. Int. Ed.*, **37**, 1434–1437.
19. Rajasingh, P., Cohen, R., Shirman, E., Shimon, L.J.W., and Rybtchinski, B. (2007) *J. Org. Chem.*, **72** (16), 5973–5979.
20. Zhao, Y. and Wasielewski, M.R. (1999) *Tetrahedron Lett.*, **40** (39), 7047–7050.
21. Lukas, A.S., Zhao, Y., Miller, S.E., and Wasielewski, M.R. (2002) *J. Phys. Chem. B*, **106**, 1299–1306.
22. Ahrens, M.J., Fuller, M.J., and Wasielewski, M.R. (2003) *Chem. Mater.*, **15**, 2684–2686.
23. Böhm, A., Arms, H., Henning, G., and Blaschka, P. (1997) Derivatives of 3,4,9,10-perylenetetracarboxylic acids, dianhydrides, and diimides, their preparation and their use, Ger. Offen. DE 19547210, filed Dec. 18, 1995, issued June 19, 1977.
24. Seybold, G. and Wagenblast, G. (1989) *Dyes Pigments*, **11**, 303–317.
25. Seybold, G. and Stange, A. (1987), Fluorescent aryloxy-substituted 3,4,9,10-perylenetetracarboxylic diimide dyes for plastics for surface-modified concentration of light. Ger. Offen. DE 3545004, filed Dec. 19, 1985, issued June 25, 1977.
26. Würthner, F., Osswald, P., Schmidt, R., Kaiser, T.E., Mansikkamäki, H., and Könemann, M. (2006) *Org. Lett.*, **8**, 3765–3768.
27. Langhals, H. (1985) *Chem. Ber.*, **118**, 4641–4645.
28. Mattersteig, G., Petermann, R., Fanghänel, E., and Marx, J. (2006) Water-soluble perylene dyes for ink-jet inks. Ger. Offen., DE 102005019743, filed April 28, 2005, issued Nov. 2, 2007.
29. Langhals, H. (1988) Water-soluble perylenetetracarboxylic acid bisimide fluorescent dyes. Ger. Offen., DE 3703513, filed Feb. 5, 1987, issued Aug. 18, 1988.
30. Löhmannsröben, H.-G. and Langhals, H. (1989) *Appl. Phys.*, **48** (Part B) 449–452.
31. Ebeid, E.Z., El-Daly, S.A., and Langhals, H. (1988) *J. Phys. Chem.*, **92** (15), 4565–4568.
32. Sadrai, M. and Bird, G.R. (1984) *Opt. Commun.*, **51** (1), 62–64.
33. Weil, T., Abdalla, M.A., Jatzke, C., Hengstler, J., and Müllen, K. (2005) *Biomacromolecules*, **6** (1), 68–79.
34. Peneva, K., Mihov, G., Nolde, F., Rocha, S., Hotta, J., Braeckmans, K., Hofkens, J., Uji-i, H., Herrmann, A., and Müllen, K. (2008) *Angew. Chem. Int. Ed.*, **47** (18), 3372–3375.

35. Rossetti, L., D'Isa, G., Mauriello, C., Varra, M., Santis, P.D., Mayol, L., and Savino, M. (2007) *Biophys. Chem.*, **129** (1), 70–81.
36. Alvino, A., Franceschin, M., Cefaro, C., Borioni, S., Ortaggi, G., and Bianco, A. (2007) *Tetrahedron*, **63** (33), 7858–7865.
37. Ma, T., Li, C., and Shi, G. (2008) *Langmuir*, **24** (1), 43–48.
38. Aubert, C., Fünfschilling, J., Zschokke-Gränacher, I., and Langhals, H. (1985) *Z. Anal. Chem.*, **320** (1), 361–364.
39. Würthner, F., Chen, Z., Dehm, V., and Stepanenko, V. (2006) *Chem. Commun.*, **7** (11), 1188–1190.
40. Law, K.-Y. (1993) *Chem. Rev.*, **93**, 449–486.
41. Schmidt-Mende, L., Fechtenkötter, A., Müllen, K., Moons, E., Friend, R.H., and MacKenzie, J.D. (2001) *Science*, **293**, 1119–1122.
42. Yakimov, A. and Forrest, S.R. (2002) *Appl. Phys. Lett.*, **80**, 1667–1669.
43. Tang, T., Herrmann, A., Peneva, K., Müllen, K., and Webber, S.E. (2007) *Langmuir*, **23** (8), 4623–4628.
44. Langhals, H., Ring, U., and Kratz, O. (2007) Perylene fluorescence dyes with chiral terminal groups, Ger. Offen., DE 102006017000, filed April 11, 2006, issued Oct. 18, 2007.
45. Tang, T., Qu, J., Müllen, K., and Webber, S.E. (2006) *Langmuir*, **22** (18), 7610–7616.
46. Würthner, F. (2001) *Angew. Chem. Int. Ed.*, **40**, 1037–1039.
47. Rademacher, A., Märkle, S., and Langhals, H. (1982) *Chem. Ber.*, **115** (8), 2927–2934.
48. Kohl, C., Weil, T., Qu, J., and Müllen, K. (2004) *Chem. Eur. J.*, **10** (21), 5297–5310.
49. Bräuchle, C., Seisenberger, G., Endreß, T., Ried, M.U., Buning, H., and Hallek, M. (2002) *ChemPhysChem*, **3**, 299–303.
50. Seisenberger, G., Ried, M.U., Endreß, T., Buning, H., Hallek, M., and Bräuchle, C. (2001) *Science*, **294**, 1929–1932.
51. Tauber, M.J., Kelley, R.F., Giaimo, J.M., Rybtchinski, B., and Wasielewski, M.R. (2006) *J. Am. Chem. Soc.*, **128** (6), 1782–1783.
52. Ishi-I, T., Murakami, K., Imai, K., and Mataka, S. (2005) *Org. Lett.*, **7**, 3175–3178.
53. An, Z., Yu, J., Jones, S.C., Barlow, S., Yoo, S., Domercq, B., Prins, P., Siebbeles, L.D.A., Kippelen, B., and Marder, S.R. (2005) *Adv. Mater.*, **17** (21), 2580–2583.
54. Weiss, E.A., Ahrens, M.J., Sinks, L.E., Gusev, A.V., Ratner, M.A., and Wasielewski, M.R. (2004) *J. Am. Chem. Soc.*, **126** (17), 5577–5584.
55. Tang, T., Qu, J., Müllen, K., and Webber, S.E. (2006) *Langmuir*, **22** (1), 26–28.
56. Schnurpfeil, G., Stark, J., and Wöhrle, D. (1995) *Dyes Pigm.*, **27** (4), 339–350.
57. Ford, W.E. (1987) *J. Photochem.*, **37** (1), 189–204.
58. Zhang, X., Chen, Z., and Würthner, F. (2007) *J. Am. Chem. Soc.*, **129** (16), 4886–4887.
59. Krauß, S., Lysetska, M., and Würthner, F. (2005) *Lett. Org. Chem.*, **4**, 349–353.
60. (a) Wagner, C. and Wagenknecht, H.-A. (2006) *Org. Lett.*, **8** (19), 4191–4194; (b) Baumstark, D. and Wagenknecht, H.-A. (2008) *Angew. Chem. Int. Ed.*, **47** (14), 2612–2614; (c) Baumstark, D. and Wagenknecht, H.-A. (2008) *Chem.-Eur. J.*, **14** (22), 6640–6645.
61. Zheng, Y., Long, H., Schatz, G.C., and Lewis, F.D. (2005) *Chem. Commun.*, **38**, 4795–4797.
62. Rahe, N., Rinn, C., and Carell, T. (2003) *Chem. Commun.*, **17**, 2120–2121.
63. Zheng, Y., Long, H., Schatz, G.C., and Lewis, F.D. (2006) *Chem. Commun.*, **36**, 3830–3832.
64. Wang, W., Wan, W., Zhou, H.-H., Niu, S., and Li, A.D.Q. (2003) *J. Am. Chem. Soc.*, **125** (18), 5248–5249.
65. Balakrishnan, K., Datar, A., Naddo, T., Huang, J., Oitker, R., Yen, M., Zhao, J., and Zang, L. (2006) *J. Am. Chem. Soc.*, **128** (22), 7390–7398.
66. Chen, Z., Stepanenko, V., Dehm, V., Prins, P., Siebbeles, L.D.A., Seibt, J., Marquetand, P., Engel, V., and Würthner, F. (2007) *Chem. Eur. J.*, **13** (2), 436–449.
67. Chen, Z., Baumeister, U., Tschierske, C., and Würthner, F. (2007) *Chem. Eur. J.*, **13** (2), 450–465.

68. Neuteboom, E.E., Meskers, S.C.J., Meijer, E.W., and Janssen, R.A.J. (2004) *Macromol. Chem. Phys.*, **205** (2), 217–222.
69. Chen, Z., Lohr, A., Saha-Möller, C.R., and Würthner, F. (2009) *Chem. Soc. Rev.*, **38** (2), 564–584.
70. Abdalla, M.A., Bayer, J., Rädler, J.O., and Müllen, K. (2004) *Angew. Chem. Int. Ed.*, **43**, 3967–3970.
71. (a) Hartnagel, U., Balbinot, D., and Jux, N. (2006) *Org. Biomol. Chem.*, **4**, 1785–1795; (b) Balbinot, D. (2006) Synthesis and aggregation properties of highly charged, water soluble porphyrins, Dissertation, University of Erlangen-Nuremberg; (c) Sarova, G.H., Hartnagel, U., Balbinot, D., Sali, S., Jux, N., Hirsch, A., and Guldi, D.M. (2008) *Chem. Eur. J.*, **14** (10), 3137–3145.
72. Brettreich, M. and Hirsch, A. (1998) *Tetrahedron Lett.*, **39** (11), 8884–8891.
73. Newkome, G.R., Behera, R.K., Moorefield, C.N., and Baker, G.R. (1991) *J. Org. Chem.*, **56**, 7162–7167.
74. Newkome, G.R. and Weis, C.D. (1996) *Org. Prep. Proced. Int.*, **28** (4), 495–498.
75. Brettreich, M. and Hirsch, A. (1998) *Synlett*, **12**, 1396–1138.
76. Kellermann, M., Hirsch, A., Bauer, W., Schade, B., Ludwig, K., and Böttcher, C. (2004) *Angew. Chem. Int. Ed.*, **43** (22), 2959–2963.
77. Braun, M., Hartnagel, U., Ravelli, E., Schade, B., Böttcher, C., Vostrowsky, O., and Hirsch, A. (2004) *Eur. J. Org. Chem.*, **9**, 1983–2001.
78. Schmidt, C.D., Böttcher, C., and Hirsch, A. (2007) *Eur. J. Org. Chem.*, **33**, 5497–5505.
79. (a) Li, A.D.Q., Wang, W., and Wang, L.-Q. (2003) *Chem. Eur. J.*, **9** (19), 4594–4601; (b) Wang, W., Han, J.J., Wang, L.-Q., Li, L.-S., Shaw, W.J., and Li, A.D.Q. (2003) *Nano Lett.*, **3** (4), 455–458.
80. Ehli, C., Oelsner, C., Alonso, M., Prato, M., Schmidt, C.D., Backes, C., Hauke, F., Hirsch, A., and Guldi, D.M. (2009) *Nature Chem.*, **1** (3), 243–249.
81. Backes, C., Schmidt, C.D., Hauke, F., Böttcher, C., and Hirsch, A. (2009) *J. Am. Chem. Soc.*, **131** (6), 2172–2184.
82. Backes, C., Hauke, F., Schmidt, C.D., and Hirsch, A. (2009) *Chem. Commun.*, **19**, 2643–2645.
83. Englert, J.M., Röhrl, J., Schmidt, C.D., Graupner, R., Hundhausen, M., Hauke, F., and Hirsch, A. (2009) *Adv. Mater.*, **21** (42), 4265–4269.
84. Lynn, D.M. (2007) *Adv. Mater.*, **19** (23), 4118–4130.
85. Khopade, A.J. and Caruso, F. (2002) *Nano Lett.*, **2** (4), 415–418.
86. Chung, A.J. and Rubner, M.F. (2002) *Langmuir*, **18** (4), 1176–1183.
87. Rosenlehner, K., Schunk, T., Jux, N., Brettreich, M., and Hirsch, A. (2008) *Org. Biomol. Chem.*, **6**, 2697–2705.
88. Decher, G. (1997) *Science*, **277**, 1232–1237.
89. Ponader, S., Rosenlehner, K., Schmidt, C.D., Schunk, T., Vairaktaris, E., von Wilmowsky, C., Schlegel, K.A., Neukam, F.W., Hirsch, A., and Nkenke, E. (2009) *J. Mater. Sci.: Mater. Med.*, **20** (12), 2455–2463.
90. Würthner, F., Thalacker, C., Diele, S., and Tschierske, C. (2001) *Chem. Eur. J.*, **7** (10), 2245–2253.
91. Fink, R.F., Seibt, J., Engel, V., Renz, M., Kaupp, M., Lochbrunner, S., Zhao, H.-M., Pfister, J., Würthner, F., and Engels, B. (2008) *J. Am. Chem. Soc.*, **130** (39), 12858–12859.
92. von Herrikhuyzen, J., Syamakumari, A., Schenning, A.P.H.J., and Meijer, E.W. (2004) *J. Am. Chem. Soc.*, **126** (32), 10021–10027.
93. Dehm, V., Chen, Z., Baumeister, U., Prins, P., Siebbeles, L.D.A., and Würthner, F. (2007) *Org. Lett.*, **9** (6), 1085–1088.
94. Franke, D., Vos, M., Antonietti, M., Sommerdijk, N.A.J.M., and Faul, C.F.J. (2006) *Chem. Mater.*, **18** (7), 1839–1847.
95. Sun, R., Xue, C., Owaka, M., Peetza, R.M., and Jin, S. (2007) *Tetrahedron Lett.*, **48** (38), 6696–6699.
96. Schmidt, C.D., Böttcher, C., and Hirsch, A. (2009) *Eur. J. Org. Chem.*, **31**, 5337–5349.
97. Thalacker, C. and Würthner, F. (2002) *Adv. Funct. Mat.*, **12** (3), 209–218.

Index

a

abiotic processes 203
absorption spectrum 207
– of lake water 207, 214–216
– use of 214
acetonitrile 71
activity-based protein profiling (ABPP) 143
aldehydes 6–8
– α-selenenylation of 75–77
– tertiary benzylamines to 13–14
aldol reaction 69–73, 79–82
– prolinamide-supported polystyrenes 82–84
– proline-catalyzed 70, 72
– transition state model 82
alkanes, selective halogenation of 16–17
alkylaromatics 19–20
alkynyl cyclopropanes, cycloadditions of 34–35
AM-11 173–174
amidoalcohols 119–122
– enantioselectivity, impact of acidity on 120–122
– single-point vs. two-point activation 119–120
aminoborane 100
aminothiourea catalysis 122–125
arrhenius equation 193
artificial nucleosides 288
arynes 242
– cyclotrimerization of 242
asymmetric aldol reaction 81
asymmetric organocatalysis, double hydrogen bonding in 117–138
– amidoalcohols 119–122
– diols 119–122
– guanidiniums 134–135
– phosphoric acids 129–134

– quinolinium thioamides 136–137
– squaramides 135–136
– thioureas 122–129
– ureas 122–129
asymmetric reactions 79–82
atom-economy 189
atrazine 222
azides, [3+2] cycloaddition of 53–57

b

backbone thioester exchange (BTE) 146
– thermodynamic equilibria involved in 147
Baylis–Hillman reaction 77–79
benzophenone 227
bicyclo [4.2.1] nonanes 197
– synthesis of 197
bifunctionality 130–132
biotic processes 203
biphasic catalysts 69
bond dissociation enthalpy (BDE) 4
– O–H bonds 4
bromide 221

c

C9-*epi*-Cinchona thiourea catalyst 126–127
caproic acid 292
captodative effect 11
carbene transfer reactions 47
carbohydrates 167–169
carbon nanotubes 293–296
carboxylic acid 288
cascade reaction 190
cata-condensed polyarenes 240
catalyst discovery 152–155
chiral dyes 299
chlorobenzene 16
circumanthracene 251
Clar reaction 242

Click chemistry 53–55
colored dissolved organic matter (CDOM) 203
– absorption spectra of 215
– components of 204
– irradiation of 218–223
complex-induced proximity effect (CIPE) 95–111
– asymmetric deprotonation via precoordination 98
– reaction barriers 98
consecutive reactions 189, 190
copper-catalyzed reactions, N-heterocyclic carbenes in 43–58
copper(I) triflate 43
coupling 284
covalent bond 188
covalent capture 143
covalently supported catalysts 69
crystalline heterometallic assemblies, sequential construction of 263–278
– dipyrrin ligands 273–277
– – Ag–π interaction 274–277
– – metallatectons 274
– dithiolate ligands 268–273
– – discrete assemblies 271–273
– – infinite chains 270–271
– – metallatectons 268–270
crystalline microporous silicates 173–174
cumyl alcohol 16
cumyl hydroperoxide (CHP) 19
[3+2] Cycloaddition 53–57
– latent catalyst 55
cycloadditions 194
– gold-catalyzed intermolecular 34–40
cycloalkane 17–18
cycloaromatization
– push–pull dienyne acids
– – 2,3-disubstituted phenols, synthesis of 29–34
cycloheptatriene 195
– complexation of 195
cyclohexane 14
cyclohexyl acetate 14
cycloisomerizations 26
cyclopropanation reactions 47
cyclotrimerization 239

d

decahelicene, synthesis of 241
decastarphene, synthesis of 241
dehydration process 170
– kinetics of 170
– mechanism of 171

deprotection 152
deprotonation reactions
– reaction barrier 102
– of tertiary amines 101–104
Diacids 17–18
dibenzooctacene, retrosynthesis of 242
dibenzoterrylene, synthesis of 243
4,5-diazafluorene 269
diboration reactions 48
dicyanobenzoquinone (DDQ) 252
dicyclohexylcarbodiimide (DCC) 289
1,2-dichloroethane (DCE) 36
Diels–Alder cycloaddition 241, 254
2,4-dien-6-yne carboxylic acid 30
dimethyl sulfoxide (DMSO) 174
dinaphthocoronene 247
diols 119–122
– enantioselectivity, impact of acidity on 120–122
– single-point vs. two-point activation 119–120
dioxiranes 18
diphenols, aerobic synthesis of 16
diphosphines 53
dipyrrin ligands 273–277
– Ag–π interaction 274–277
– metallatectons 274
dissolved organic carbon (DOC) 214
dissolved organic matter (DOM) 203
– fraction of 203
2,3-disubstituted phenols 29–34
dithiolate ligands 268–273
– discrete assemblies 271–273
– infinite chains 270–271
– metallatectons 268–270
diuron 222
domino reactions 190–191
double hydrogen bonding 117–138
D-xylose 178
dynamic covalent capture 143–162
– catalyst discovery 152–155
– complex chemical systems, analysis of 156–160
– – ^{1}H-^{13}C HSQC NMR spectroscopy 156–158
– – UV/Vis spectroscopy 158–160
– drug discovery 150–152
– hydrogen bond–driven self-assembly
– – in aqueous solution 144–146
– perspective 160–162
– stability and order measurement 146–150
– – bilayer membranes 149–150
– – peptides 146–148

e

eco-compatible chemistry   189
Elbs pyrolysis   241
enantioselective Henry reaction   124
enantioselective organocatalytic reactions   67
– acidity impact   120–122
endo-peroxide derivative   247
enyne cycloisomerizations, gold-catalyzed   26–35
– alkyne substituent, influence of   26–29
– ene-ynamide   8, 29
– enyne   10, 29
epoxidation   18–19
EPR radical equilibration technique   4
ethyl diazoacetate (EDA)   47

f

Fenton process   208
fenuron   222
ferromagnetic chain   267
first-order kinetics   225
fluorescence dyes   286
formic acid   289
fossil oil   167
Friedel–Crafts acylation   239
fulvic acids   203
furfural   167–182
– advantages of   170
– applications of   169–170
– based industrial chemicals   167–182
– carbohydrates for life   167–169
– evolution over nearly two centuries   169
– pentosans, conversion of   170–172
– production of   172–180
– – crystalline microporous silicates   173–174
– – functionalized mesoporous silicas   174–180
– – industrial processes of   173
– – transition metal oxide nanosheets   180
furfuryl alcohol   170
fused benzene rings   239–256

g

Gibbs free energy   288
gold-catalyzed isomerization   28
graphene   237–238
greenhouse effect   167
Grignard reagent   245
guanidiniums   134–135

h

Halex reaction   286
half-life time   226
– assessment of   228
Hantzsch esters   131
Henry reaction   124, 125
hetero domino reaction   190
hetero-dehydro-Diels–Alder reaction (HDDAR)   35–40
heteroleptic complexes   271
heteropolyacids   174
– advantages of   174
– concentration of   175
hexabenzotriphenylene   239
– synthesis of   242
hexafluoroacetylacetonate (HFAC)   275
homo domino reaction   190
^{1}H-^{13}C HSQC NMR spectroscopy   156–158
H-Pro-Pro-Asp-NH2   69–73
humic acids   203
hydrazones   153
– detection protocol   159
hydrolysis process   204
hydrolytic reaction   204
hydroperoxides   19
hydrophobic effects   288
hydrosilylation
– of carbonyl compounds   49–51
– – catalysts comparison   51
hydroxymethylfurfural (HMF)   168

i

imidazole   77
imidazolidinones   68
indirect phototransformation reactions, modeling of   203–231
– indirect photolysis processes   205–213
– photochemistry of surface waters, modelling   214–229
– – formation of ^{3}CDOM* in   226–229
– – formation of $CO_3^{-\bullet}$ 223–226
– – generation of $^\bullet$OH   218–223
– – lake water, absorption spectrum of   214–216
– – mass vs. concentration   216–217
– – oxidation of CO_2^{-3} by ^{3}CDOM*   224
– – oxidation of CO_3^{2-} by $^\bullet$OH   224
– – oxidation of $HCO_3^{-\bullet}$ by $^\bullet$OH   224
– – photoactive water components, radiation by   217–218
– – reactivity of $^\bullet$OH   218–223
– – reactivity of ^{3}CDOM* in   226–229
– – reactivity of $CO_3^{-\bullet}$ 225–226
– surface waters, in   205–213

interleukin-2 (IL-2)   150
intermolecular cycloadditions, gold-catalyzed   34–40
– alkynyl cyclopropanes   34–35
– enynes   34–35
– hetero-dehydro-Diels–Alder reaction (HDDAR)   35–40
– propargyl acetylenes   34–35
intersystem crossing (ISC)   206
iodocyclohexyl acetate   14
ionic liquid–modified silica gels   69–73
– advantages   70

k

Keggin acids   175
ketones   6–8

l

lactams   12
lactic acid   172
Lambert–Beer law   217
laser dyes   286
4-lauryloxycarbonyl-N-hydroxyphthalimide   17
L-proline   69–73
Lewis base   97
– in organolithium chemistry   99
Lewis base–Brønsted acid (LBBA)   130
lipid hydroperoxides   9
lipid rafts   149
α-lithiated tertiary amines   99–101
– methyl amines preparation   99
α-lithiation
– vs. β-lithiation   108–110
– regioselective   104–108
β-lithiation, vs. α-lithiation   108–110

m

mesoporous silica   174–180
metallatectons   268–270
metal-organic frameworks (MOFs)   263
metathesis reaction   188
methylenation reactions   48
methyl isobutyl ketone (MIBK)   174
3′-methoxyacetophenone (MAP)   227
microwave-assisted organic chemistry (MAOS)   193
mixed domino reaction   190
mixed-ketone Clar reaction   244
m-terphenyl   33
multicomponent reactions   191
multiple bond-forming transformations (MBFTs)   187–200
– anions, involving   196–200
– consecutive reactions   190
– domino reactions   190–191
– metals, involving   194–196
– multicomponent reactions   191
– one-pot reactions   191–192
– synthesis, science of   187–189
– Wolff rearrangement, involving a   192–194
multiple-component condensations   191

n

N-(phenylseleno)phthalimide (NPSP)   76
N,N,N′,N″,N‴-pentamethyldiethylenetriamine (PMDTA)   100
N,N′,N″-tetraethylethylenediamine (TEEDA)   109
N,N,N′,N′-tetramethylethylenediamine (TMEDA)   100
N,N,N′,N′-tetramethylmethylenediamine (TMMDA)   100
N-alkylamides   12–13
nanographene
– alternant polyarenes with
– – 10 fused benzene rings   239–244
– – 11 fused benzene rings   244–248
– – 12 fused benzene rings   248–249
– – 13 fused benzene rings   249–252
– – 14 fused benzene rings   252
– – 15 fused benzene rings   252–256
– bottom-up approaches through organic synthesis   237–257
nanoribbons   239
naphthalene units   283
naphthopyrenopyrene, synthesis of   243
natural fibers   167
nearest-neighbor recognition (NNR)   149
Newkome dendrimers   289
N-aryl ketimine   132
N-heteroaromatic base   15–16
N-heterocyclic carbenes
– in copper-catalyzed reactions   43–58
– copper center reactivity, modulation of   54
– copper complexes preparation   43–46
– – derivatization of   45
– – main applications in catalysis   46–49
– copper hydride-mediated reactions   49–53
– – hydrosilylation of carbonyl compounds   49–51
– – mechanistic considerations   51–52
– – related transformations   52–53
– [3+2] cycloaddition of azides and alkynes   53–57
– – click chemistry   53–55
– – internal alkynes, use of   55–57

N-hydroxy derivatives (NHD) 3–20
– aerobic N-hydroxy amines, oxidation catalyzed by 6–8
– – alcohols to aldehydes and ketones 6–8
– general reactivity of 3–6
– N-hydroxy amides, aerobic oxidation catalyzed by 9–10
– – polyunsaturated fatty acids, peroxidation of 9–10
– N-hydroxy imides, aerobic oxidation catalyzed by 10–20
– – alkanes, selective halogenation of 16–17
– – alkylaromatics 19–20
– – benzylalcohols to aldehydes 10–12
– – cycloalkanes to diacids 17–18
– – diphenols, aerobic synthesis of 16
– – epoxidation of olefins 18–19
– – N-heteroaromatic bases, oxidative acylation of 15–16
– – oxidative functionalization of alkylaromatics 14–15
– – p-hydroxybenzoic acids, aerobic synthesis of 16
– – silanes, 12 N-alkylamides 12–13
– – tertiary benzylamines to aldehydes 13–14
N-hydroxyphthalimide (NHPI) 3
– aerobic oxidation of primary benzylic alcohols 10–12
– substituent effect 11
N-hydroxysaccharin (NHS) 17
N-hydroxysuccinimide (NHSI) 13
N-methylbenzohydroxamic acid 9
N-methylpiperidine 100
nitrate, irradiation of 218–223
nitrite, irradiation of 218–223
nonalternant polyarenes 238
noncovalently supported catalysts 69
nonpurgeable organic carbon (NPOC) 214

o

octanuclear cubic complex 269
Olefins 18–19
– hydration of 174
one-pot reactions 191–192
– definition of 192
o-phenanthroline 18
orange tetrabenzoperopyrene 247
organic light emitting diodes (OLEDs) 286
organocatalysis 67
– asymmetric 117–13
organolithium chemistry
– applications 95
– complex-induced proximity effect in 95–111
– latest developments 101–110
– – direct deprotonation of tertiary amines 101–104
– – α-lithiation versus β-lithiation 108–110
– – regioselective α-lithiation 104–108
– state of the art 96–101
– – α-lithiated tertiary amines 99–101
– – structure formation patterns 96–97
organolithium compounds, structure formation patterns 96–97
osteoblastic cells 296
oxalic acid 204
oxidative process 3
oxyanion hole 118

p

palladium-catalyst 239
Pauson–Khand cycloaddition 194–195
pedogenic 203
pentosans 170–172
– conversion into furfural 170–172
– – mechanism of the dehydration 171
– – net reaction of 170
peptides 146–148
peri-fused polyarenes 239
perylene 246
– scholl reaction of 246
perylene diimides (PDIs) 284
– chiral water-soluble 297–300
– M-helical arrangement 299
– P-helical arrangement 299
– as reporter electrolytes for implantation technology 296–297
– symmetric water-soluble 289–292
– unsymmetric 292–297
perylene skeletal structure 283
perylene-3,4,9,10-tetracarboxylic acid dianhydride (PTCDA) 283
– possible reactions of 285
– properties of 286
– transformations of 283
pH, function of 209
phenalenones 246
phenylacetates 155
phenyllithium 104
phenylurea pesticides 212, 227
phosphoric acids 129–134
– bifunctionality 130–132
– competent substrates 133–134
photochemical significance, parameters of 219
photochemistry 214–229

photo-Fenton reactions 207
photolysis process, indirect 205–213
– reactions induced by
– – $CO_3^{-\bullet}$ 209–210
– – CDOM* 211–212
– – $^\bullet OH$ 206–209
– – other reactions 212–213
photolysis processes 205
photon flux density 217
photosensitizers 205
phthalimide-N-oxyl (PINO) 3
p-hydroxybenzoic acids, aerobic synthesis of 16
Pictet–Spengler-type cyclization 128–129
platinum-catalyzed cycloisomerization 28
polyarenes 238
– cata-condensed 245
– 10 fused benzene rings 239–244
– 11 fused benzene rings 244–248
– 12 fused benzene rings 248–249
– 13 fused benzene rings 249–252
– 14 fused benzene rings 252
– 15 fused benzene rings 252–256
– retrosyntheses of 248
– synthesis of 239
polycyclic aromatic hydrocarbons (PAHs) 237, 245
polyethylene oxide (PEO) 288
polypyrrolic derivative 277
polystyrene-supported proline 73–82
– asymmetric reactions 79–82
– – aldol reaction 79–82
– Baylis–Hillman reaction 77–79
– α-selenenylation of aldehydes 75–77, 79
polyunsaturated fatty acids (PUFAs) 9–10
porphyrin 266, 267
prolinamide-supported polystyrenes, aldol reaction 82–84
proline 68
propargyl acetylenes, cycloadditions of 34–35
Prussian blue analogs 263
push–pull dienynes
– gold-catalyzed intra- and intermolecular cycloadditions 25–40
– – 2,3-disubstituted phenols, synthesis of 29–34
– – enyne cycloisomerizations 26–34
– – intermolecular cycloadditions 34–40
– pyridines 35–40
pyrrole ring 274

q
quinazoline 15
quinolinium thioamides 136–137

r
redox cycle 207
regioselective synthesis, of pyridines 35–40
regioselective α-lithiation 104–108
resins
– inactivation of 83
– reactivation of 83
rylene dyes 285
– electron affinity of 286
– general structure of 284

s
saccharides 173
– conversion of 173
salicylic acid 204
scanning tunneling microscopy (STM) 250
scattering process 214
Schmidt reaction 244
Scholl reaction 246, 283
α-selenenylation 75–77
– proposed mechanism of aldehydes 78
silanes 12
silica-supported ionic liquids 73
silver–π interaction 276
single-point activation, vs. two-point activation 119–120
single-wall carbon nanotubes (SWNTs) 293
– applications of 293
– properties of 294
– types of 293
solar cells 286
solvatochromic behavior 288
squaramides 135–136
square-planar complexes 267
staining cells 288
stannyldiazoacetate esters 48
s-trans-s-cis isomerization 32
Stryker's reagent 53
sulfated zirconia 177
sulfonation reactions 176
2-sulfonymidoylmethylenetetrahydrofuranes 197
sulfonic acid 287
summer sunny day (SSD) 217
supertriphenylene 249
supported organocatalysts
– future perspectives 84–85
– immobilization of 69
– – biphasic catalysts 69
– – covalently supported catalysts 69

– – noncovalently supported catalysts 69
– ionic liquid–modified silica gels, supported on
– – H-Pro-Pro-Asp-NH2 69–73
– – L-proline 69–73
– organic synthesis, as a powerful tool in 67–85
– polystyrene-supported proline 73–82
– prolinamide-supported polystyrenes 82–84
– selection of 68

t
tandem reaction 190
terrylene dyes 287
2,2,6,6-tetramethylpiperidine-1-oxyl (TEMPO) 3
tert-butyllithium 96–97
tertiary benzylamines 13–14
tethering strategy 151–152
tetraanionic octanuclear complex 269
tetrabenzocoronene 246
tetrabenzoperopyrene 251
thioureas 122–129
– aminothiourea catalysis, mechanistic duality in 122–125
– catalyst self-association 126–127
– halide binding 127–129
– mono- *vs.* bidentate coordination 125–126
thymidylate synthase 150
titania acid 177
tosyl cyanide 36
transition metal oxide nanosheets 180
transmission electron microscopy (TEM) 292
trichloroacetonitrile 36
trifluoroacetic acid 289
trinuclear heterobimetallic discrete complex 272
tungstophosphoric acid 176

two-point activation, *vs.* single-point activation 119–120

u
ureas 122–129
– catalyst self-association 126–127
– halide binding 127–129
– mono- *vs.* bidentate coordination 125–126
UV/Vis spectroscopy 158–160

v
van der Waals radii 272

w
water-soluble perylene dyes 283–301
– advantages of 286–289
– applications of 286–289
– arrangements of 299
– chiral non-water soluble 298
– – examples of 298
– chiral 297–300
– derivatization of 284–286
– functionalization of 283–284
– history of 283–284
– nomenclature of 284
– properties of 286
– symmetric perylene diimides 289–292
– – synthesis of 290
– unsymmetric perylene diimides 292–297
– – carbon nanotubes, noncovalent functionalization of 293–296
– – implantation technology, reporter electrolytes for 296–297
Wittig reaction 48
Wolff rearrangement 192–194

x
xenobiotics 203
xylose 170
– conversion 176